2nd Scientific Conference on Fundamental Information Security Problems in Terms of the Digital Transformation (FISP 2020)

IOP Conference Series: Materials Science and Engineering
Volume 1069

Stavropol, Russia
30 November 2020

ISBN: 978-1-7138-3344-4
ISSN: 1757-8981

Printed from e-media with permission by:

Curran Associates, Inc.
57 Morehouse Lane
Red Hook, NY 12571

proceedings.com

Some format issues inherent in the e-media version may also appear in this print version.

Institute of Physics
Dirac House, Temple Back
Bristol BS1 6BE UK

Phone: 44 1 17 929 7481
Fax: 44 1 17 920 0979

techtracking@iop.org

Additional copies of this publication are available from:

Curran Associates, Inc.
57 Morehouse Lane
Red Hook, NY 12571 USA
Phone: 845-758-0400
Fax: 845-758-2633
Email: curran@proceedings.com
Web: www.proceedings.com

TABLE OF CONTENTS

NCFU

North-Caucasus Federal University

The 2-nd Scientific Conference on Fundamental Information Security Problems in terms of the Digital Transformation FISP-2020

Stavropol, Russia

30 November, 2020

On the basis of submitted works at the Conference, 48 selected peer reviewed papers have been included in this volume of IOP Conference Series: Materials Science and Engineering.

CONFERENCE COORDINATORS

Vyacheslav Petrenko, North-Caucasus Federal University
Fariza Tebueva, North-Caucasus Federal University

IOP Conf. Series: Materials Science and Engineering 1069 (2021) 011001 doi:10.1088/1757-899X/1069/1/011001

The aim of The 2-nd Scientific Conference on Fundamental Information Security Problems in terms of the Digital Transformation (FISP-2020) was discussing a wide range of issues related to the latest achievements of science and technology in information security. It was held on November 30, 2020 in Stavropol, Russia by North-Caucasus Federal University. Due to nowadays difficult situation imposed by SARS-CoV-2 pandemic, FISP-2020 Conference had to be organized in online mode in BigBlueButton platform.

The authors of the works highlighted the problems of digital data security, threats to individual, business and government, problems of computer crime, effectiveness of scientific research in the field of information security, issues of education in the field of data safety. The sections of the Conference were:

1. Humanitarian problems of ensuring information security of the Russian Federation
2. Artificial intelligence and cybersecurity issues
3. Security of information systems in the context of digital transformation
4. Cryptographic information security methods, Internet broadcasting and blockchain technologies
5. Situational management of information security incidents in organization
6. Investigation methods of cyber crime
7. Competition of scientific reports of students and graduate students.

The papers submitted to FISP-2020 were selected for the Conference by the Scientific Committee and were double-blind peer-reviewed. The conference was attended by more than 150 authors from the leading universities, such as International Institute for Applied Systems Analysis (Laxenburg, Austria), La Sapienza University of Rome (Rome, Italy), Technical University of Dresden (Dresden, Germany), University of Stuttgart (Stuttgart, Germany), L. N. Gumilyov Eurasian National University (Nur-Sultan, Kazakhstan), Lomonosov Moscow State University (Moscow, Russia), Moscow Institute of Physics and Technology (Moscow, Russia), National Research University "Higher School of Economics" (Moscow, Russia), National Research University "Moscow Institute of Electronic Technology" (Moscow, Russia), National Research Nuclear University "MEPhI" (Moscow, Russia), Military Academy of Communications named after Marshal of the Soviet Union S. M. Budyonny (St. Petersburg, Russia), Far Eastern Federal University (Vladivostok, Russia), Leningrad State University named after A. S. Pushkin (St. Petersburg, Russia), Ural Federal University (Yekaterinburg, Russia), Ufa State Aviation Technical University (Ufa, Russia), Southern Federal University (Taganrog, Russia) and others.

Venue:

Institute of Mathematics and Information Technologies named after Professor N.I. Chervyakov, North Caucasus Federal University, Russia, Stavropol, Pushkina str., 1

Committee:

Chairman of the Organizing Committee: Mezentseva Oksana Stanislavovna, Vice-rector at the North Caucasus Federal University, Professor.

Deputy Chairman of the Organizing Committee: Petrenko Vyacheslav Ivanovich, Director of the Institute of Mathematics and Information Technologies named after Professor N.I. Chervyakov at the North Caucasus Federal University, Associate Professor.

Chairman of the Program Committee: Bespalov Dmitry Nikolaevich, Rector of the North Caucasus Federal University

Vice-chair of the program Committee:

Tebueva Fariza Bilyalovna, Head of the Department of Applied Mathematics and Computer Security at the North Caucasus Federal University.

Keynote speakers:

Malkevich Alexander Alexandrovich, Associate Professor, First Deputy Chairman of the Commission for the Development of the Information Community, Mass Media and Mass Communications of the Public Chamber of the Russian Federation; President of the Foundation for the Protection of National Values.

Baranov Alexander Pavlovich, Professor, Head of the Department of Integrated Safety of Critical Facilities, Russian State University of Oil and Gas (National Research University) named after I.M. Gubkin, Deputy General Director of JSC "Main Scientific Innovation Implementation Center".

Streltsov Anatoly Alexandrovich, Professor, Honored Scientist of the Russian Federation, Acting State Councilor of the Russian Federation of the 3rd class, Member of the Presidium National Association for International Information Security, Head of the Department of Information Security of Computer Systems, Institute for Information Security Problems, Lomonosov Moscow State University.

Yusupova Nafisa Islamovna, Professor, Dean of the Faculty of Informatics and Robotics at the Ufa State Aviation Technical University.

Elin Vladimir Mikhailovich, Associate Professor, Associate Professor of the Department of Information Security, Russian State University of Oil and Gas (National Research University) named after I.M. Gubkina.

Petrenko Vyacheslav Ivanovich, Director of the Institute of Information Technologies and Telecommunications at the North Caucasus Federal University.

Peer review declaration

All papers published in this volume of *IOP Conference Series: Materials Science and Engineering* have been peer reviewed through processes administered by the Editors. Reviews were conducted by expert referees to the professional and scientific standards expected of a proceedings journal published by IOP Publishing.

- **Type of peer review: Double-blind**
- **Conference submission management system: by e-mail *fisp.ncfu@gmail.com***
- **Number of submissions received: 66**
- **Number of submissions sent for review: 60**
- **Number of submissions accepted: 52**
- **Acceptance Rate (Number of Submissions Accepted / Number of Submissions Received X 100): 78**
- **Average number of reviews per paper: 1**
- **Total number of reviewers involved: 8**
- **Any additional info on review process:** The review process was the following:
  The editors of the Conference received papers by e-mail fisp.ncfu@gmail.com. The papers were preliminary checked by formal criteria and in case of conformity they were sent for double-blind review (by e-mail). The reviewers evaluated articles according to the criteria (Review form is in an attached file). If the article was recommended by the reviewer, but he left comments to change anything in material, the authors could correct it according to recommendations of reviewers. Afterwards, the articles were accepted.
  Generally, there was 1 review per paper.
- **Contact person for queries: Irina Samoylenko ivsamoilenko@stgau.ru**

Model of a domain-specific profiling system based on Explainable AI technologies

V.V. Antonov[1], Z.I. Kharisova[2], L.E. Rodionova[1], G.G. Kulikov[1,3]

[1]Ufa State Aviation Technical University, Ufa, Karl Marx. str., 12, 450008, Russia

[2]Ufa Law Institute of the Ministry of Internal Affairs, Ufa, Muksinova str., 2, 450103, Russia

[3] CJSC «Ufa Scientific and Production Enterprise "Molniya", Ufa, Zentsova. str., 70, 450052, Russia

E-mail: antonov.v@bashkortostan.ru, zarinaid@mail.ru, lurik@mail.ru, grisha@molniya-ufa.ru

Abstract. The article offers a system model of intelligent profiling system that opens up opportunities for both staff selection and effective training and rotation of existing employees of the organization. To implement this paradigm, state registers of open information resources are defined and created. From the point of view of science, the system model corresponds to the structure of the expert system of explainable artificial intelligence (XAI), which has the task of forming many solutions. The task is set to provide validation and verification of the models taking into account the uncertainty. The tool to improve this approach is a software analytical complex, the development and implementation of which will allow to search and effective analysis of data, forming indicators for their subsequent use in decision-making in the management of organization processes. Presentation of knowledge about the subject area under study in the form of system model allows applying formal logical rules, graphoanalytic metalanguage and rules for development and design of program analytical complexes.

Keywords: artificial intelligence; explainable artificial intelligence; ontological model; neural network technologies, profiling; software analytical complex

1. Introduction

It should be noted that the problem of training and effective use of highly qualified specialists is a strategic state task regularly outlined in the decrees of the President of the Russian Federation. The requirements currently imposed on employees of most enterprises and organizations are based on the need to possess a complete system of knowledge, skills and competencies that are formed in the course of training and its practical activities, as well as the need for professionally significant psychological qualities of the individual, psychological stability and readiness to effectively perform production and service tasks.

Increased requirements for employees pose complex tasks for HR employees, who must not only find candidates who meet their professional qualities, but also predict their behavior for compliance with the law, moral, ethical and other necessary requirements.

The proposed concept of a system for profiling based on the structure of hybrid expert systems will effectively solve the problems of personnel selection of employees, the formation of a personnel reserve system, as well as high-quality training and retraining of employees.

Under profiling is commonly understood as the direction in psychological science to reveal the components of verbal and nonverbal behavior of the character used in interpersonal communication, including in the form of a be-contact detection of deception and psychological observations of subjects to identify non-standard reactions [1]. Profiling methods are based on the analysis of characteristic informative features of appearance, behavior, and the use of observation and survey technologies in order to identify individuals with tendencies to deviant behavior. Predictive identification of such individuals for proactive actions is a priority for most organizations according to Russian Federal Law of 30.11.2011 № 342 "On Service in the Internal Affairs Bodies of the Russian Federation and Amending Certain Legislative Acts of the Russian Federation".

2. Profiling and personnel selection

In accordance with the existing requirements, candidates accepted for certain positions, specialties and a number of departments undergo special psychophysiological studies and testing, the results of which allow us to conclude that it is appropriate to accept the candidate in question. However, along with the possible error of the interpreter, there is a problem of well-known tests and research methods, which significantly reduces the effectiveness of personnel selection.

Currently, a number of commercial structures are widely implemented so-called Data Leak Prevention (DLP) systems as technologies to prevent leaks of confidential data from information systems [2]. The use of DLP systems is aimed at: identifying the destructive behavior of employees; minimizing the risks associated with the disclosure of official information; improving the efficiency of work related to the processing of information in automated systems; conducting data analysis and interpretation [3]. The presented technology is very effective in analyzing information about a potential candidate. This is about the candidate's behavior model. In this case, there is a restriction on the reliability of the result, since the model is an abstraction, where not all parameters-properties of the candidate are taken into account.

The concept of the proposed system lies in the availability of semi-structured data in open information sources hypertext digital environment (such as Internet, Intranet, etc.), synthesis technology profiling systems artificial intelligence(SAI) and presents a profile system based on technologies XAI, designed to detect destructive behavior, and the neural subsystem that is used to test the reliability of the received data. Currently used neural network technologies SAI [4-6] have proven themselves in classification problems when working with big data and for this reason are used in the system under consideration for data verification.

Using the data available in memory, SAI are able to learn and present a forecast based on the curves of probable solutions. Accordingly, it is not difficult to imagine a system that takes into account the characteristics of a potential candidate for the service, work experience and interview questions with candidates, as well as predicts how likely it is to match the position being filled [7].

When selecting candidates for positions, the number of positions and their requirements (full name, position, education, age, and work experience) are first determined. After that, an order is drawn up, which specifies the terms for providing a list of employees formed for a particular position. Then the employee's personal qualities and professional qualifications are evaluated using a personal card, autobiography, characteristics, and test results. Also, an interview is organized to identify the employee's motivation, aspirations, goals, and distinctive individual competencies. The employee's labor productivity over a certain period of time is considered. Then, applicants for certain positions of the personnel reserve are compared by evaluating professional knowledge, experience and skills in the form of testing (evaluation forms), the results obtained are processed, and the competition Commission forms a set of recommendations for each applicant. After the list of candidates for the position is formed, it is adjusted. Then the lists of applicants are approved by the heads of structural divisions and heads of enterprises. Next, an individual development plan of the employee, an individual internship plan is drawn up. All data is stored in heterogeneous storage, ensuring integrity, order, and consistency. Analytical data processing will allow you to extract the necessary information for making a decision.

In addition to the exhaustive number of required skills for a specific position, it is necessary to take into account emotional and psychological characteristics, such as: ambition, learning ability, purposefulness, etc., determined by the profile subsystem, which is collected through testing. Through this SAI, the applied software and analytical complex (SAC) will be able to identify suitable candidates and highly effective employees, as well as provide a transcript of video recordings of interviews. In addition, the SAI is able to identify stress zones, possible violations of rules, norms and other forms of risk, and non-compliance with established requirements in the service. Separately, you can emphasize the possibility of analyzing the experience, skills and results of certification of each candidate, and then determine the skills that need to be improved in order to increase the degree of compliance with the position.

3. System models and profiling directions

The concept of the proposed system allows you to simultaneously view the results for various categories of employees and employees, and form expert opinions both individually and by Department. When creating a conclusion, the expert can use automatically generated data: estimates received by the employee, automatically created text interpretations of the results.

The intelligent system allows you to prepare a set of related data, based on which it is possible to analyze the communication characteristics of an employee and forecast the assessment of changes in working behavior (slow response to tasks, failure to meet deadlines, etc.), comparing this data with typical characteristics stored in the system memory.

We will conduct an ontological analysis of modern profiling trends. Criminalistic direction, in a broad sense, is defined as a set of psychological methods and methods that determine the character, temperament, intelligence of the person who committed an offense or crime [8-10]. Forensic profiling has several main levels: lie detection (for verbal and non-verbal communications); diagnostics of lies based on suspicious behavioral signs, such as physiological symptoms of lying (changes in the autonomic nervous system – sighs, redness of the skin, excessive sweating of the forehead and palms, apnea, obviously shaking hands, pulsation of the carotid artery, sitting on the edge of a chair, trembling in the knees, body turned away from the employee), facial expressions and gestures during deception (scratching different parts of the body), as well as speech structures that give out lies, and eye signals), identifying the congruence of linguistic information, non-verbal communications [11]. In his works, Paul Ekman points out the signals of emotions that need to be paid attention to when determining the truthfulness of testimony – facial expression, voice, gestures, manifestations of the autonomic nervous system, since there is a possibility of verifier errors [12].

Forensic profiling is mainly used in the detection and investigation of offenses and crimes. One of the directions of this profiling is the preparation of a psychological and forensic portrait [13]. Thus, the Russian scientist A. I. Anfinogenov States that " The psychological portrait of a criminal is a psychological and forensic method and the result of knowledge of a criminal event, focused on identifying a set of information about the individual characteristics and personality features of the subject of the crime, manifested in the totality of circumstances and traces of criminal activity." The main characteristics of profiling are: gestures, emotions, psychology, facial expressions. There are personnel, forensic and commercial profiling, the ontological structure is shown in the figure 1.

This ontological metadata structure is used in the PAK structure for staff selection and it takes the following form:

$$M^{onto} = \left\{C^M, P^M, K^M\right\},$$

where M^{onto} – metadata ontology;

$C^M = \left\{C_1^M, ..., C_l^M\right\}$ – many metadata concepts;

$P^M = \left\{P_1^M, ..., P_j^M\right\}$ – multiple relationships between metadata concepts;

$K^M = \{K_1^M, ..., K_h^M\}$ – many properties of metadata classes.

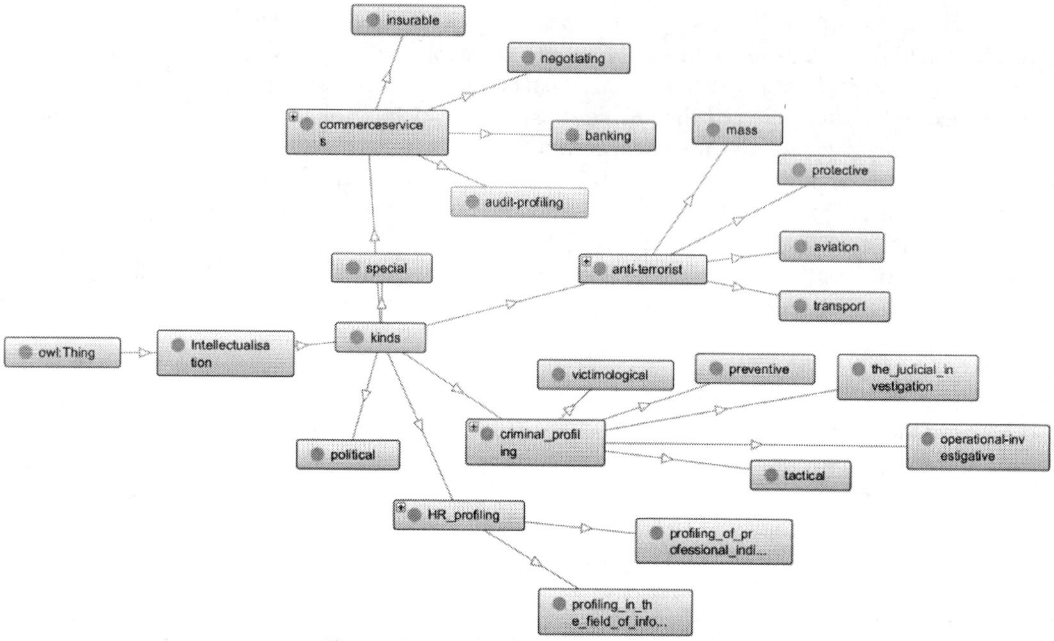

Figure 1. Ontological structure of profiling

"Metadata Ontology is a top-level ontology that contains the basic concepts and relationships between them, which are then used to build spatial and attributive ontologies", as defined by the definition borrowed from the [14].

The Ontology of Profiling describes the main concepts and types of profiling (human resources, forensic, political, etc.) as well as the objects that serve as sources of knowledge for the organization (figure 2). The Ontology of Profiling Structure contains the concepts and relationships required to form a hierarchy of knowledge areas and then use this hierarchy of SAC. This hierarchy reflects the subject areas related to different activities of the organization, i.e. those areas of knowledge in which the company is interested.

With the help of the Protégé ontology engineering tool in the OWL language, a model system was created for ontology visualization. As a result, classes of diagrams of the developed ontologies have been obtained which can be used to demonstrate the results of the simulation.

Some authors highlight the research direction - this understanding corresponds to the method of systematic classification of facial expressions. Attention should also be paid to the medical and psychological area - in the scientific world this area is based on the work of Russian psychiatrist Peter Gannushkin and German clinician Karl Leongard. In addition, there is a special area related to the work of special services around the world and the existence of specific files of persons in databases. The psycho-technological direction is actively used - based on the use of neurolinguistic programming for the subsequent development of a program profile of a person, his or her system of values and beliefs [15]. The process of communication in research consists of such components as subjects, means, needs, motivations and goals, ways of interaction and mutual influences, results. In communication, a person acts as a "subject", using means of communication such as speech, mimicry, pantomime (posture, movement, gait, gestures) and manner of holding on. In order for professional communication, understanding and perception to take place, it is necessary that non-verbal and verbal factors have a complex effect.

A number of features of professional communication can be highlighted, such as:

- communication orientation (personal or social orientation);

- quantitative parameters of communication (degree of mediocrity) - direct face-to-face communication or indirect communication of documents;
- regulation of communication (legal deontology norms);
- dynamics of communication (change in psychological content from contact to contact).

Thus, it is possible to talk about relations between objects, where both the objects themselves and the rules of relations can be presented as a semantic model of interactions in the information environment.

The task of recognition through non-verbal channels arises, such as seeing that the person speaking is lying or telling the truth. Non-verbal signals from the object of analysis are an important source of data, for example when assessing the validity of the information provided to them. For example, in the case of conscious lies, certain stress markers appear, e.g. the voice sounds higher, pauses become more frequent in order to speak and invent at the same time. Using artistic techniques, you can demonstrate unnatural attacks of coughing, yawning, crying or laughing. The crossed arms, whose hands lie on the shoulder, speak for the nervousness of the person. Gestures such as "crossed arms" and "crossed legs" testify to the closedness of a citizen, which does not always have a positive effect on the outcome of an important conversation. A person has much less control over his or her gestures and postures than over his or her words. To form a primary opinion, it takes 2-4 minutes to evaluate not only what a person is wearing, but also their manner of holding on, their gaze, smile and facial expressions, which often reveal hidden intentions and thoughts [16]. In the profiling, elements of visual psychodiagnostics are used, which highlight the pattern of human perception. When a person meets, the first thing they consider instantly and unconsciously is gender, age and build. The human perception scheme is based on the principle of the general to the particular, evaluating external features, facial features (proportions, features, smile, look), gait, posture, gestures (features of hands, fingers), clothing (colour preference), makeup, etc. Visual psychodiagnostics uses various visual aids such as observation, video recording and document examination [17,18].

A result of SAC's work, it is possible to obtain an effective tool for selecting and analysing candidates for the service (work), presented in tabular or graphical form. This SAC primarily solves the problem of ensuring the storage and collection of data, processing and structuring of data clusters, integration into other systems. The main principles of subsystems implemented in the SAC are the principle of developing structure, adaptability and independence of the operated subsystems and databases.

SAC, according to ISO/IEC/IEEE 15288 [19] system engineering standard, includes the following stages:
- conception (M_1);
- developments (M_2);
- production (M_3);
- applications (M_4);
- application support (M_5);
- termination of application and write-off (M_6).

The application of virtual models and objects describing the possible behaviour of a person or his or her condition will make it possible to monitor and take preventive measures in real time and forecast possible negative actions and phenomena. Considering that the information system during the life cycle (LC) consistently passes through stages that are milestones that record more significant changes in the state of the system. The LC model can be represented as a sequence of stages [20], and the structure of the system within each stage remains constant.

The scheme of relationships between subprocesses in this case is fully consistent with the provisions of ISO/IEC/IEEE 15288 international standard for system engineering [21]. In other words, there is a category set mapping that preserves the structure of these category sets, and the SAC system model can also be fully formed according to the requirements of this system engineering standard.

ISO/IEC/IEEE 15288 system engineering standard is applied to describe the LC in a BPMN notation. According to the designations entered, which are presented in Figure 2.

Let us look at the application stage in more detail [20].

The application phase M_4 begins after the software system has been installed and handed over for use by the customer [20].

Planning begins at previous stages.

The purpose of this stage is to ensure that the system is functioning with guaranteed performance.

The parameters under study:

1. Production stage result M_3.

2. Performance monitoring, identification and classification indicators $MM_2 = \{mm_2^1, ..., mm_2^{i_2}\}$:

3. Reports on deviations, disadvantages and failures $MM_3 = \{mm_3^1, ..., mm_3^{i_3}\}$.

4. Information on necessary changes to the product configuration (version) $MM_4 = \{mm_4^1, ..., mm_4^{i_4}\}$.

5. Stage results used for recursive analysis $MM_5 = \{mm_5^1, ..., mm_5^5\}$.

– experienced personnel with the level of competence required to perform the functions of operators in the software system in question and to provide relevant services are identified;

– an established software system capable of providing functional services is applied;

– compliance with the system's objectives is monitored;

– problems are identified, with users, developers and manufacturers getting to know each other and making adjustments;

– plans are made and criteria for moving to the withdrawal and write-off phase are defined.

6. Stage results to be applied in subsequent stages (problems detected are resolved):

– inactivity;

– software system maintenance;

– slight reengineering;

– determining the transition to the next stage of withdrawal and write-offs.

7. Results applied by the development M_2 and M_3 production phases (actions taken when errors are detected) $MX_6 = \{mx_6^1, mx_6^2\}$:

– reconfiguration of the system in order to extend its LC;

– Identification of new opportunities for upgrading the software system by applying feedback from the rights holder.

Considering this process of the application stage in Figure 3 as relations, we can show that relations and conclusions have an iterative component (recursion). When designing a software analytical complex, this process allows us to build an iterative model. At the same time, the time it takes to design and develop a SAC is reduced by introducing feedback. ISO/IEC/IEEE 15288 identifies iterative relationships as functions (i.e. how inputs/outputs are described), and input/output relationships are not formalized in this standard. The relationship can be considered as an analogue of the Zahman matrix, where columns are categories and rows are categories, when columns and rows are crossed the cells that form new categories. Each stage in ISO/IEC/IEEE 15288 is treated as a separate category. The methodology of this standard makes it possible to link each stage by introducing a recurrence relationship [24].

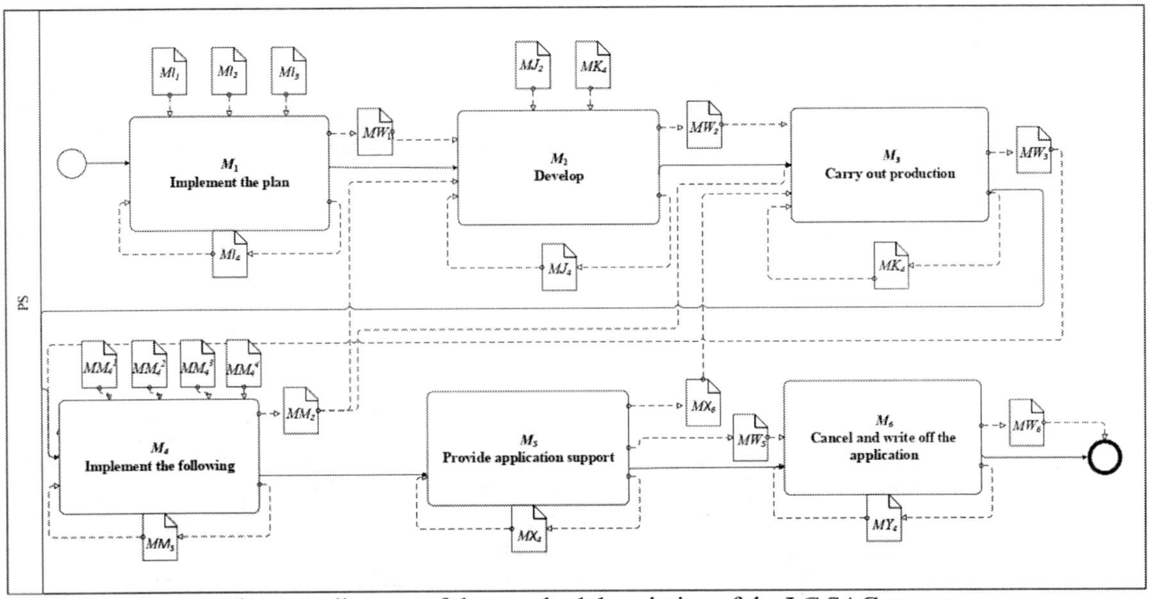

Figure 2. diagram of the standard description of the LC SAC stages

SAC modules are used as information objects, which allows for the construction of an adaptive system that is universal for changing external data. Such a mechanism helps to examine the subject area from different angles and develop a compact and universal software system, as the programme works according to a given algorithm and can be changed by adding only other parameters to the knowledge base (KB).

As a result, it is possible to create a virtual physical and digital environment in which people and information systems interact and evolve from isolated solutions to the formation of "smart information spaces".

Smart Information Space refers to a set of open Internet resources that use common languages and development methodologies for joint decision making.

To implement this model, we will define the following register of open information resources included in the SAC structure, including Google, Yandex search engines, personal Windows, Mac OS and Linux operating systems. Various cross-platform technologies that allow recruitment of Potok, Amazing Hiring, HireVue, Experium and others. Data from open Internet resources are included in the PAC system model, which corresponds to the structure of the expert system based on Explainable AI technologies.

4. Conclusion

The experience of foreign countries in introducing SAI in activities related to personnel management shows its effectiveness [23-27]. The use of SAI for various types of law enforcement activities, including forecasting, identification of potentially dangerous situations, etc., is promising [25-27]. Synthesis of such systems leads to the formation of "smart information spaces" that allow to solve a variety of tasks to achieve the goals of effective activity of organizations both in terms of formation of personnel policy and implementation of production activities. At the heart of such "smart information space" is a subsystem of profiling, which has the ability to predict the behavior of the evaluated individual, his potential, weaknesses and propensities based on private features such as appearance, verbal and non-verbal behavior, etc.

The question of rules for the use of such systems remains unresolved, even if a high degree of compliance of the forecast with the real consequences, a person's choice as an integral property of the personality should always be taken into account when making final decisions based on this forecast. Therefore, the use of XAI technologies to provide explanations in the form of "evidence or justification for each result", with the possibility of detailed explanation of how the PAK generated the final result significantly expands the scope of use.

The proposed systemic approach allows, among other things, taking into account and formalizing many behavioral and cognitive aspects of human activity models. Combining groups of objects (people and information systems) into separate new virtual objects - categories, description of their properties and relations through categorical relations taking into account ontological (causal - investigative) aspect [20, 21]. It makes it possible to form new synthetic models of so-called small groups - of production, service or scientific nature.

5. Acknowledgment

The research is being carried out with the financial support of the Ministry of Education and Science of the Russian Federation as part of the main government task for higher education institutions № FEUE-2020-0007.

References

[1] Perry, Walter L., Brian McInnis, Carter C. Price, Susan Smith, and John S. 2013 *Hollywood, Predictive Policing: The Role of Crime Forecasting in Law Enforcement Operations*, Santa Monica, Calif.: RAND Corporation, RR-23 NIJ, As of August 25, 2014: http://www.rand.org/pubs/research_reports/RR233.html.

[2] Harisova Z 2020 Certificate of state. registration of computer program No. 2020610510. BASIS - Basic analysis of the security of information systems, / No. 2020610510; declared 12/17/2019; publ. 01/15/2020.

[3] Antonov V, Lonshchakova A, Kharisova Z 2019 Ensuring the reliability and information security of psychophysiological research in criminal proceedings in the Russian Federation and abroad / *Eurasian Law Journal.* [In Russian] 9 (136) 240-242.

[4] Analytical Center Anti-Malware.ru. *Official portal of the Anti-Malware.ru Analytical Center*, 2020. https://www.anti-malware.ru/team

[5] Fetisov V, Kharisova Z, Dmitriyev O, Melnichuk O 2017 Rapid particle size analysis of suspensions based on video technology and artificial neural network with additional training during operation *International Journal of Applied Engineering Research*; 12(7) 1271-1278.

[6] Hagan M T, Demuth H B, Beale M H and De Jesús O., "Neural Network design,", 2nd ed., 11-7.

[7] Ponkin I, Redkina A 2018Artificial intelligence from the point of view of law. *PFUR Newsletter. Series: Legal Sciences.* №1 doi: 10.22363/2313-2337-2018-22-1-91-109

[8] Ilnitsky S 2018 Staff training as a strategic factor in the organization's personnel management system. *Young scientist.* 3 418-423.

[9] Amirkhanova L, Bikmetov E, Zinnurov U, Kharisova A 2017 Management of university competitiveness based on image creating research *Banach Journal of Mathematical Analysis* 11(2) 40-48.

[10] Lonshchakova A 2016 Features of the use of modern scientific and technical means in the production of interrogation *Actual problems of law and state in the XXI century* 8 (4) 76-81.

[11] Kiryukhin D 2015 Geographic profiling - help in compiling the psychological profile of the offender and finding places to hide the corpses *Forensic expert* 4 6-8.

[12] Ekman P 2013 Psychology of emotions: book. - St. Petersburg 240.

[13] Anfinogenov AI. Psychological portrait of the criminal, his development in the process of investigating crimes: autoref. dis. psychol. sciences. - M., 1997: 26.

[14] Pavlov S, Efremova O 2017 Ontological model for integration of structurally heterogeneous spatial databases of various subject areas into a uniform regional database *Ontology of designing* 7(3) 323-333. DOI: 10.18287/2223-9537-2017-7-3-323-333.

[15] Ganushkin P, Ovcharenko V, Leibin V 1999 Anthology of Russian Psychoanalysis *Moscow Psy-Cold-Social Institute* 2 516-517. DOI: 10.24411/2541-7673-2018-10533

[16] Leogard K 2001 Accentuated personalities. - M.: *Eksmo-Press* 358.

[17] Volynsky-Basmanov Yu, Volynsky V, Kameneva M [and others] 2017 Profiling. Technologies for preventing illegal actions: a textbook for university students enrolled in the specialties of "Law" and "Law Enforcement" *edited by Yu. M. Volynsky-Basmanov, N. D. Eriashvili*. - 3rd ed. – Moscow
[18] Corporate data warehouses. System integration. Project documentation Electronic resource Official site of Project Experience. Corporate data warehouses. System integration. Project documentation., 2020. [In Russian]. https://www.prj-exp.ru/patterns/pattern_draft_project.php.
[19] ISO/IEC/IEEE 15288 System and software engineering – System life cycle processes
[20] Antonov V, Kulikov G, Antonov D 2011 Theoretical and applied aspects of building models of information systems *LAP LAMBERT Academic Publishing GmbH & Co.KG*, Germany 134.
[21] Kulikov G, Antonov V, Kharisova Z, Rodionova L 2020The set-theoretic approach to the construction of the dual system model of the pack for the studied area of activity with mixed real and virtual objects *Vestnik SUSU. Series: "Computer Technologies, Management, Radio Electronics"* 20 (1) 5 –15. DOI: 10.14529 / ctcr200101
[22] Rodionova L, Antonov V, Kulikov G [and others] 2019 Software analytical complex architecture for December closed logics // Certificate on state registration of computer programs No. 2019611359 / copyright holders; M: Rospatent,
[23] Milliez G, Lallement R, Fiore M, Alami R. 2016 Using human knowledge awareness to adapt collaborative plan generation, explanation and monitoring *The Eleventh ACM/IEEE International Conference on Human Robot Interaction, IEEE Press* 43-50
[24] Trafton G, Hiatt L, Harrison A, Tamborello F, Khemlani S, Schultz A 2013 ACT-R/E: an embodied cognitive architecture for human–robot interaction *J. Hum-Robot Interact.* 2 (1) 30-55
[25] P. Matsa, K. Gullamajji 2019 To Study Impact of Artificial Intelligence on Human Resource Management // International Research Journal of Engineering and Technology 08 1229 – 1238
[26] Aral, Sinan, Brynjolfsson E, Alstyne V 2012 "Information, Technology and Information Worker Productivity," *Information Systems Research*, March
[27] Fominykh I 2018 Mental image engineering, creative problems, emotional evaluations *Ontology of designing.*; 8(2) 175-189. - DOI: 10.18287/2223-9537-2018-8-2-175-189.

Modeling problems legal regulation of the field of artificial intelligence

V.V. Antonov [1], Z.I. Kharisova [2], N.R. Kalimullin [2], A.I. Abdulnagimov [1]

[1] Ufa State Aviation Technical University, Ufa, K. Marks str., 12, 450008, Russia

[2] Ufa Law Institute of the Ministry of Internal Affairs of Russia, Ufa, Muksinov str., 2, 450103, Russia

E-mail Zarinaid@mail.ru

Abstract. The analysis of the current state of regulation issues at various levels of the development and application of artificial intelligence systems is presented. Key issues in the field of regulation of robotic systems are considered. Possible threats are described, a model for the formation of artificial intelligence systems, including the ones for high-risk facilities is proposed.

Keywords: Artificial intelligence; Law; Neural networks; Robot ethics

1. Introduction

The active development of artificial intelligence (AI)-based systems creates new challenges for society, the discussion of which is necessary, moreover, with the participation of representatives of completely different spheres of public life: from developers and technical specialists to legislators, law enforcement bodies [1,2] and art workers. The issues of regulation of both the activities of developers of AI systems and the results of AI activities require vigorous discussion. Today, there is no clear understanding of the legal status of AI and the results of its activities, i.e. it is unclear who will own this result, who will be responsible for socially dangerous consequences and violations of the law. The introduction of self-learning systems into various systems [3,4] creates new sources of increased danger, for example, transport with an AI-based control system. The issue of legal regulation of such systems is relevant primarily in connection with industry-specific legal problems that cannot be resolved within the framework of current legislation (which hinders the development of technologies). In addition, the introduction of new technologies is intensifying within the digital economy, which requires appropriate regulatory mechanisms.

The regulation of AI should not hinder the development of technologies both in our country and around the world. The importance of AI development has been repeatedly emphasized at the political level as well.

Today, there are elaborations of national strategies for the development of artificial intelligence in many countries [5,6], the goals of which are to maintain leadership positions in this field at the international level, to develop safe and comfortable cities, to provide affordable and high-quality medical, logistics and educational services, to create a developed transport system in the country, as well as the exploration of outer space and the oceans.

Due to the emerging opportunities for solving many problems in the economic and social spheres, it becomes necessary to determine the legal status of artificial intelligence, taking into account the general trends in the regulation of robotics in the world.

2. AI terminology

Participants in different discussions do not always mean the same thing when discussing AI. Thus, an important step is the development of a generally accepted conceptual apparatus. This article discusses the concept of "artificial intelligence" as a complex of technological solutions that allows you to simulate human cognitive functions (including self-learning and search for solutions without a predetermined algorithm) and obtain specific tasks results comparable at least with the results of human intellectual activity. The complex of technological solutions includes information and communication infrastructure, software (including those employing machine learning methods), processes and services for data processing and finding solutions. Thus, AI is a system, for the creation of which a complex technological chain is required: from the creation of specific processors and computing centers (if we talk about AI for mass or industrial use), other hardware, algorithms to databases for training, etc. So, it becomes apparent that many parties are involved in the creation of AI.

Promising methods of artificial intelligence are the methods aimed at creating fundamentally new scientific and technical products, including those for the development of universal (strong) artificial intelligence (autonomous solution of various problems, automatic design of physical objects, automatic machine learning, algorithms for solving problems based on data with partial markup and (or) insignificant amounts of data, information processing based on new types of computing systems, interpreted data processing and other methods) [7,8]. The AI development strategy assumes the emergence of the so-called strong artificial intelligence, which becomes more likely with the development of quantum computers, removing the limitations of the architectures with binary logic used today.

Thus, the term "artificial intelligence" is suitable for two types of intelligent systems, which have one common property - learning and self-learning, but which have an incomparably different level of capabilities and consequences, including legal ones.

3. Ethical aspects of artificial intelligence regulation

The formation of man and society took place over millennia, while the human psyche was formed with the ability to distinguish between good and evil, to sympathize, to predict implicit consequences, etc. Artificial intelligence (if we talk about "strong" AI) without going through such a long evolutionary path will possess the properties obtained during training and self-learning. Thus, those behavioral models, which seem to be "disadvantageous" at first glance, should be considered as a mandatory element of learning, otherwise AI behavior will be sociopathic in nature.

Despite the lack of the ability to create "strong" AI in the near future, the existence of existential risks of such a rapid development of technologies is quite obvious [9]. In these conditions, the adoption of the Asilomar AI Principles in January 2017 at the conference in Asilomar, USA (2017 Asilomar conference) was the first and very significant step to the responsible approach of humanity to the development of AI and robotics. The Asilomar AI Principles were developed and adopted following the conference of AI developers and researchers. To date, over 4,000 scientists, developers, entrepreneurs and experts have signed them. Among them are Elon Musk, Stephen Hawking and Ray Kurzweil, representatives from Google, Apple, Facebook, IBM, Microsoft, etc. The first paragraph of these principles proclaims the goal of AI research: the goal should be to create useful intelligence rather than uncontrollable intelligence. In total, 23 principles are presented, each of which emphasizes the responsibility of developers, politicians, businessmen towards humanity in the field of AI development.

Among the aspects of this document that deserve special attention are:

- the provision on "useful financing", which implies additional investments for research in the field of AI in order to increase the effectiveness of implementing AI in the field of the most pressing problems;
- the provision on "scientific and political communication", which implies close cooperation between developers of AI systems and persons responsible for the legal regulation of AI;
- the provision on "security and transparency of errors", as well as the application of restrictions in software implementations of AI related to the principles of universal morality.

Software developers based on AI principles are one of the key "players" and are responsible for erroneous decisions based on AI, as well as the formation of moral costs from the use of AI software systems [10, 11]; moreover, these developers can influence such consequences. At the same time, the Asilomar principles of research in the field of AI declare the so-called anthropocentric path, which says that everyone should be able to choose in what capacity to use AI-based software systems for what purpose. It is possible for a number of purposes – do not apply at all. At the same time, the emerging problem of using personal data in AI-based solutions is, also given special attention, since this is, directly related to ensuring basic human rights and society as a whole. In the long term, the development of the theory of strong AI, when software systems in the field of AI begin to acquire the ability to recognize "themselves "as" individuals", is declared with significant restrictions - supervision, compliance with ethical standards and minimizing the above risks, and even at the stages of forecasting[12].

The proposals put forward by the experts are essentially the example of the first attempts at self-regulation, when organizations or individuals are developing rules before the state has developed binding laws. One of the characteristic attributes of self-regulation is an association. Its rules become binding on all its members. An example of such association is the Partnership on AI, the partnership that includes Amazon, Apple, Google, IBM, Facebook, Microsoft and others. It was created in 2017 and it is too early to talk about specific rules, but we see one of the unique examples of association of the largest companies in the world (by the way, competitors in a number of areas) around the problem of artificial intelligence development.

4. Possible threats and legal regulation of systems based on artificial intelligence

The civil liability system, especially in European countries, covers most of the future scenarios for the development of new technologies, including those based on artificial intelligence, however, in some countries, especially in developing ones, there are a number of unsolved problems. In general, the problems are associated not with the proliferation of systems based on AI, but with the problem of identifying the actions of such systems and correlating them with deliberately inherent errors or, in the worst case, the development and additional training of such systems, including, for unknown reasons, according to an unlawful scenario.

At the same time, it is important to regulate this industry not in separate countries or exclusively on the territory of the European Union [13], but at the international level.

Since artificial intelligence algorithms can process large amounts of data with great precision and speed, often significantly exceeding human potential, AI applications are becoming more common today. However, the dependence on AI carries risks, especially when a decision-making process is totally automated. Problems arise when the available algorithms reflect the prejudgement, for example, of the developer company in relation to any category of people, race or even gender, as a result of which, just as in the case of "mechanical" solution of problems by the person himself, one may encounter "human" as well as a "robotic" factor. For this reason, the openness of a particular development is important, as well as its evaluation and recognition as a safe, impartial and effective system by an independent group of experts in this industry. Thus, the regulations are needed to recognize the development as objective, effective and qualitative with the possibility of continuous training, adaptation and data protection with encryption [14].

IOP Conf. Series: Materials Science and Engineering 1069 (2021) 012002 doi:10.1088/1757-899X/1069/1/012002

The accuracy and reliability of AI algorithms and automated decision-making primarily depends on the operator (trainer), at the same time, the level of responsibility is often assigned to him in proportion to the contribution to the system, since it is too early to talk about establishing responsibility for a robot. In this case, there is an obvious need for insurance of AI systems as a guarantor of compensation for damages.

However, in the case of full assignment of responsibility for the actions of the robot to the developer or operator, the factor of stimulating the development of modern systems based on AI is lost. An alternative solution is to establish a separate category of law for robotic systems with the status of an electronic person, making him responsible for all actions [15].

Thus, it can be noted the need to consider the following issues in the field of legal regulation of robotic systems:

- formation of an insurance fund to compensate for damage caused by AI systems;
- introduction of a licensing system for AI developments;
- establishing legal entity status for AI-based systems;
- establishing uniform international legal norms for neural network developments;
- imposition of responsibility for damages (violation of production technology, operation and information security) on the operator, developer or owner.

5. AI information security threat model

The widespread use of AI generates many potential threats. However, already today information systems used in certain areas are regulated by law, for example, in Russia, the "Requirements for ensuring the protection of information in automated control systems for production and technological processes at critical facilities, potentially hazardous facilities, as well as the facilities posing an increased danger to life and health of people and to the environment" and the "Requirements for ensuring the safety of significant objects of the critical information infrastructure of the Russian Federation" have been approved. AI systems introduced in these areas, or forming high-risk facilities, naturally fall under such requirements.

One of the requirements for such information systems is the availability of the Information Security Threat Model. A threat model is a fundamental document in building an information system. It takes into account the peculiarities of the information system: software, hardware, software and hardware tools and information processing [16, 17]. A threat model is a document that regulates threats to an information system and the security of information in it. The formation of threats by the information system itself is not reflected in the threat model. However, an established and proven control system can be a reliable basis for the formation of a basic methodology and standards in designing AI systems from the standpoint of safety.

When building a threat model of information systems, a certain methodology is used, which includes the following stages:

- description of the information system;
- structural and functional characteristics;
- description of security threats;
- violator model;
- possible vulnerabilities;
- ways of implementing threats;
- consequences of violation of information security properties.

To adapt the methodology for describing AI systems, the description of security threats and the threat bank, presented on the relevant information Internet resources of the regulators, should be supplemented accordingly.

The compliance of the information system with the developed threat model will require constant or periodic monitoring. For this, it is necessary to develop a verification concept, including the following methods:

- expertise;
- statistical analysis;
- formal verification methods;
- dynamic verification methods;
- synthetic verification methods.

6. Conclusion

Thus, it can be noted that it is necessary to consider the following issues in the field of regulation of robotic systems:

- development of the concept of ethical and legal regulation of AI-based systems;
- establishing the status of a legal entity for AI-based systems and uniform international legal norms for neural network developments;
- introduction of a licensing system for AI developments;
- imposition of responsibility for damages (violation of production technology, operation and information security) on the operator, developer or owner;
- development of a security threat model of AI-based systems in accordance with the proposed methodology;
- formation of a specialized bank of threats to AI-based security systems, as well as the descriptions of typical security threats;
- development of a threat model verification concept.

The proposed measures will make it possible to establish the fundamental basis of legal regulation of AI-based systems, as well as to ensure compliance with the safety requirements of functioning based on the proposed methodology.

7. Acknowledgment

The study is carried out with the financial support of the Ministry of Education and Science of the Russian Federation within the framework of the main part of the state assignment to higher educational institutions No. FEUE-2020-0007.

References

[1] Cerka P, Grigiene J and Sirbikyte G 2017 Is it possible to grant legal personality to artificial intelligence software systems. *Computer Law & Security Review* 33 685.- DOI: 10.1016/ j.clsr.2017.03.022

[2] Carrillo M 2020 Artificial intelligence: From ethics to law *Telecommunications Policy* 44.- DOI: 10.1016/ j.telpol.2020.101937

[3] Fetisov V, Kharisova Z, Dmitriyev O and Melnichuk O 2017 Rapid particle size analysis of suspensions based on video technology and artificial neural network with additional training during operation *International Journal of Applied Engineering Research* 12(7) 1271.

[4] Peng Y, Li S and Hu Z 2019 A self-learning dynamic path planning method for evacuation in large public buildings based on neural networks *Neurocomputing* 365 71.- DOI: 10.1016/ j.neucom.2019.06.099

[5] Borges F, Laurindo J, Spínola M, Gonçalves R and Mattos C 2020 The strategic use of artificial intelligence in the digital era: Systematic literature review and future research directions *International Journal of Information Management*.- DOI: 10.1016/ j.ijinfomgt.2020.102225
[6] Trappey A, Lupu M and Stjepandic J Embrace artificial intelligence technologies for advanced analytics and management of intellectual properties *World Patent Information* 61.- DOI: 10.1016/ j.wpi.2020.101970
[7] Noriega M. 2020 The application of artificial intelligence in police interrogations: An analysis addressing the proposed effect AI has on racial and gender bias, cooperation, and false confessions *Futures* 117.- DOI: 10.1016/ j.futures.2019.102510
[8] Gasimova R and Abbasli R 2020 Advancement of the search process for digital heritage by utilizing artificial intelligence algorithms *Expert Systems with Applications* 158.- DOI: 10.1016/ j.eswa.2020.113559
[9] Sergievskii V 2020 Super Strong Artificial Intelligence and Human Mind *Procedia Computer Science* 169 458.- DOI: 10.1016/ j.procs.2020.02.225
[10] Samsonovich A 2020 Socially emotional brain-inspired cognitive architecture framework for artificial intelligence *Cognitive Systems Research* 60 57.- DOI: 10.1016/ j.cogsys.2019.12.002
[11] Shank D and Desanti A 2018 Attributions of morality and mind to artificial intelligence after real-world moral violations *Computers in Human Behavior* 86 401.- DOI: 10.1016/ j.chb.2018.05.014
[12] Kouziokas G 2017 The application of artificial intelligence in public administration for forecasting high crime risk transportation areas in urban environment *Transportation Research Procedia* 24 467.- DOI: 10.1016/ j.trpro.2017.05.083
[13] Cohen G Evgeniou T, Gerke S and Minssen T 2020 The European artificial intelligence strategy: implications and challenges for digital health *The Lancet Digital Health* 2 376.- DOI:10.1016/ S2589-7500(20)30112-6
[14] Rahman S, Khalil I, Atiquzzaman M and Yi X 2020 Towards privacy preserving AI based composition framework in edge networks using fully homomorphic encryption *Engineering Applications of Artificial Intelligence* 94.- DOI: 10.1016/ j.engappai.2020.103737
[15] Villaronga E and Golia 2019 A Robots, standards and the law: Rivalries between private standards and public policymaking for robot governance *Computer Law & Security Review* 35 129.- DOI: 10.1016/ j.clsr.2018.12.009
[16] Figueira P, Bravo C and Lopez J 2020 Improving information security risk analysis by including threat-occurrence predictive models *Computers & Security* 88.- DOI: 10.1016/ j.cose.2019.101609
[17] Kulikov G, Antonov V, Kharisova Z and Rodionova L 2020 The set-theoretic approach to the construction of the dual system model of the pack for the studied area of activity with mixed real and virtual objects *Vestnik SUSU. Computer Technologies, Management, Radio Electronics* 20(1) 5–15.- DOI: 10.14529/ ctcr200101

Telemetry data integrity monitoring system

V V Berkholts[1], A M Vulfin[1], A I Frid[1]

[1] Ufa State Aviation Technical University, Ufa, Karla Marxa str., 12, 450000, Russia

E-mail torina4@yandex.ru

Abstract. The article presents a way to improve the system for monitoring the integrity of telemetric information received from a mobile object. The method is based on comparing telemetry information received from a mobile object and information generated by a model. The comparison is carried out on the basis of proximity metrics of technological time series using deep learning neural networks

Keywords: Data integrity; Information Security; time series, neural networks

1. Introduction

Safety requirements for mobile objects oblige to register and store data about their state using on-board control devices and recorders. Similar requirements apply to both ground transport [1] and air [2]. Ensuring the reliability of data transmission systems about an object - telemetric information (TMI), is one of the important tasks of ensuring the effective functioning of a mobile object. The possibility of transmitting TMI about the actual state of individual modules during operation and the entire complex of equipment to the manufacturer makes it possible to increase the efficiency of operating a mobile object in a normal state, in the event of failures or attacks by intruders, during incident investigation.

The need to transfer recorded data on the state of a mobile object (MO) to a manufacturer for the purpose of subsequent analysis is one of the possible ways to improve promising on-board systems [3, 4, 5] and allows the MO to gain a competitive advantage. The problem of building secure systems for collecting, storing and processing data on the state of a mobile object in such systems is urgent.

Monitoring the integrity of data received from a mobile object to a manufacturer was previously considered by the authors in [6]. The method presented there was based on a comparison of technological time series (TTS) [7], characterizing the behavior of the parameters of a mobile object and the model of the same object installed at the manufacturer. Comparison of two TBRs was carried out using a set of fuzzy logic rules proposed by the authors.

The disadvantage of this approach is the low probability of detecting an intruder, due to the fact that the proposed rules for making decisions for a large number of MO modes of operation are not a sufficiently flexible mechanism for assessing the proximity of TBP, and therefore for establishing the fact of intruder.

An additional factor that reduces the likelihood of detecting intruders is that the correlation and determination coefficients used in [6] illustrate the proportion of variance, and the consistency of TBR is actually determined by only two parameters: the proportion of variance and the average percentage of deviation, which leads to a decrease in the accuracy of the estimation of consistency. TVR.

The purpose of this article is to increase the likelihood of making the right decision about the fact of an attack on data integrity.

To achieve this goal, the following tasks were solved:

1. a classifier of operating modes of a mobile object has been developed;
3. mathematical models of possible attacks of an intruder have been determined;
4. A classifier of attacks by an intruder has been developed;
5. a block for making a decision on the presence or absence of an attacker in the current time window has been developed.

2. Classifier of operating modes of a mobile object

In the previous work of the authors [6], the main factor that reduces the likelihood of detecting an attacker was the partitioning of data into multiple operating modes. This partitioning is redundant and complicates the set of rules for deciding whether an attack has occurred. Thus, it is necessary to determine the minimum required number of operating modes of a mobile object.

As a mobile object, we will consider a gas turbine engine (GTE) with an automatic control system (ACS) installed on an aircraft (AC).

To solve the problem of clustering the GTE parameters by operating modes, the data recorded on board the aircraft were taken. A fragment of the flight data recording 60 minutes long was taken. These data are recorded with reference to standard atmospheric conditions. The sampling period is 0.25 seconds.

For the task of selecting the number of modes, 2 parameters were identified that are most sensitive to changes in the ACS GTE operating modes: specific fuel consumption G and high-pressure rotor speed n2, and it is more informative to analyze the increment of these parameters.

To solve the problem of classifying the operating modes of the ACS GTE, the clustering process is carried out in a sliding time window. For a qualitative classification, the width of the time window must be at least five samples in order to recognize the classes of states of the ACS GTE.

Clustering was carried out on a sliding time window of 15 counts. Table 1 provides an example of such time windows.

Table 1. Example of time windows

№ window	ΔG1	…	ΔG14	ΔN1	…	ΔN14
1	IG_2-G_1I		IG_{15}-$G_{14}I$	IN_2-N_1I		IN_{15}-$N1_4I$
2	IG_3-G_2I		IG_{16}-$G_{15}I$	IN_3-N_2I		IN_{16}-$N_{15}I$
3	IG_4-G_3I		IG_{17}-$G_{16}I$	IN_4-N_3I		IN_{17}-$N_{16}I$
…						
140000	IG_{14002}-G_{14001}		IG_{14015}-G_{14014}	IN_{14002}-$N_{14001}I$		IN_{14015}-$N_{14014}I$

The selection of the number of operating modes (clusters) was carried out using the "Elbow method" [8]. The essence of the method is that it is necessary to select such a number of clusters so that adding another one does not improve the quality of clustering.

Figure 1 illustrates the application of the Elbow method to the selected parameters. The criterion by which the quality of clustering was assessed with a different number of clusters, inertia (the sum of squared errors in the context of clustering). In Figure 1, inertia is located on the ordinate, and the abscissa is the number of clusters k. The "fracture" characteristic of this method occurs on 2 clusters. Thus, 2 classes of states of the ACS GTE are distinguished. In a similar way, the number of state classes for other mobile objects can be selected.

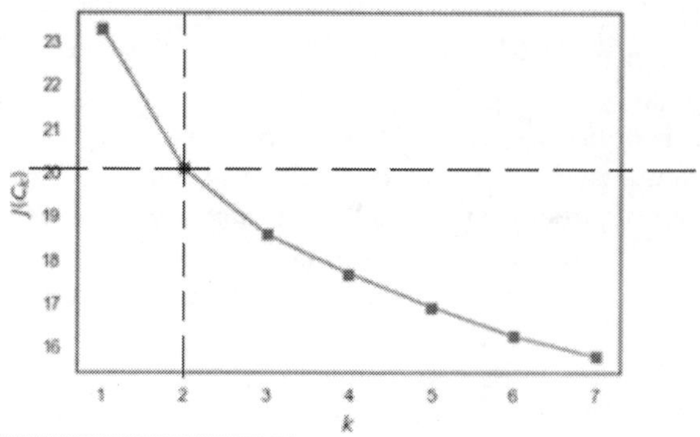

Figure 1. Selection of the number of state classes. Elbow method

To make sure the data breaks down really well into the two state classes of a mobile object, visualization of this breakdown is needed. Visualization was carried out by two methods: principal component analysis (PCA) [9] and t-distributed Stochastic Neighbor Embedding (t-SNE) [10].

PCA visualization is shown in Figure 2a, t-SNE visualization is shown in Figure 2b.

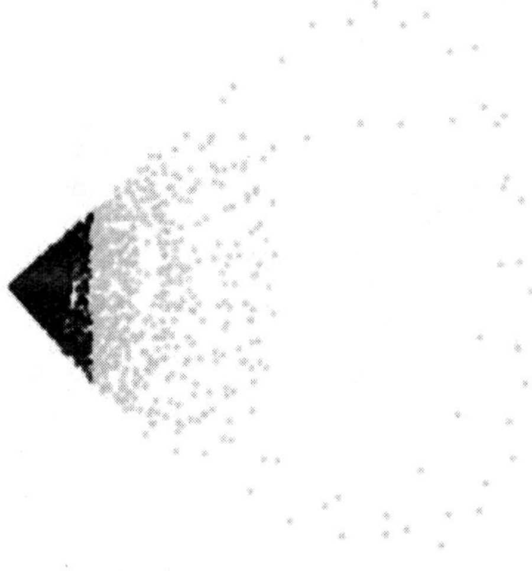

Figure 2a PCA

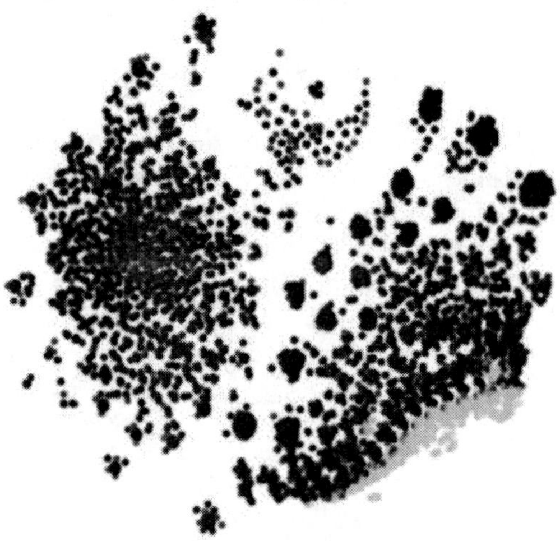

Figure 2b t-SNE

Figure 2. Visualization of clustered data

In Figure 2a clearly shows a large cluster of points in the vicinity of the coordinate (0,0). Since not the values of the parameters themselves at the moment of time t were clustered, but only their increments, it is possible to determine the physical meaning of such a distribution of points. The increments of parameters close to zero indicate the static operating mode of the ACS GTE, the increments located far from the center of coordinates indicate the dynamic operating mode of the object.

After determining the operating modes of a mobile object, it is necessary to select a method for classifying TMI, which will be used in the future to determine the fact of an attacker. The k-nearest neighbors (KNN) algorithm, one of the most popular machine learning algorithms, was chosen as the classification algorithm [11]. After testing the classification algorithm on a test sample, the classification accuracy was 0.991.

3. Mathematical models of attacks

In order to recognize attacks based on the attacker model described in [12], mathematical models of four types of attacks on TMI were developed.

1.Attack. Overlay multiplicative noise

1A. Attack in the entire time window. The monitored parameter after manipulations by the attacker takes the following value:

$$x_i^z = a_i \cdot x_r^i, \quad (1)$$

$i, i \in [1, n]$ - iteration number in a time window of length n,, x_r^i - the parameter value in the i-th iteration,

x_i^z- the value of the parameter, after manipulation by the attacker,

$a_{i,}$- coefficient, a random number, and$a_{i,} \in (-\infty; -1{,}1) \cup (1{,}1; +\infty)$

1B. Attack in part of the time window. The parameter after manipulation by the attacker takes the following value:

$$x_i^z = \begin{cases} k_i \cdot x_r^i, i \le \frac{n}{2}, k_i \epsilon (-1.1; 1.1) \\ a_i \cdot x_r^i, i > \frac{n}{2}, a_{i,} \in (-\infty; -1{,}1) \cup (1{,}1; +\infty) \end{cases} \quad (2)$$

$k_{i,}$- natural noise ratio arising from data transmission, which is a random number in a small range

1C. Attack in a small part of the time window. The parameter after manipulation by the attacker takes the following value:

$$x_i^z = \begin{cases} k_i \cdot x_r^i, i \le \frac{2n}{3}, k_i \epsilon (-1.1; 1.1) \\ a_i \cdot x_r^i, i > \frac{2n}{3}, a_{i,} \in (-\infty; -1{,}1) \cup (1{,}1; +\infty) \end{cases} \quad (3)$$

2. Substitution of data similar to real ones.

2A. Attack in the entire time window. The parameter after manipulation by the attacker takes the following value:

$$x_i^z = a_i \cdot x_r^i \pm b_i, \quad (4)$$

$a_{i,}$- coefficient, a random number, and $a_{i,} \in (0.5; 0.9) \cup (1{,}1; 1.5)$

$b_{i,}$- coefficient, a random number, and $|b_i| \ge 0.2 \cdot x_r^i$.

We also prohibit the mutually compensating values $a_{i,}$and b_i. For example, the following event cannot be perceived as an attacked signal:

$$a_{i,} = 0{,}8, x_r^i = 5, b_i = 1,$$

$$\text{so } x_i^z = 5, \text{which is equal to the original}, x_r^i.$$

2B. Attack in part of the time window. The parameter after the manipulation of the attacker takes the following value:

$$x_i^z = \begin{cases} k_i \cdot x_r^i, i \le \frac{n}{2}, k_i \epsilon (-1.1; 1.1) \\ a_i \cdot x_r^i \pm b_i, i > \frac{n}{2}, a_{i,} \in (0.5; 0.9) \cup (1{,}1; 1.5), |b_i| \ge 0.2 \cdot x_r^i \end{cases} \quad (5)$$

2C. Attack in a small part of the time window. The parameter after manipulation by the attacker takes the following value:

$$x_i^z = \begin{cases} k_i \cdot x_r^i, i \le \frac{2n}{3}, k_i \epsilon (-1.1; 1.1) \\ a_i \cdot x_r^i \pm b_i, i > \frac{2n}{3}, a_{i,} \in (0.5; 0.9) \cup (1{,}1; 1.5), |b_i| \ge 0.2 \cdot x_r^i, \end{cases} \quad (6)$$

3. Increase or decrease the signal while maintaining the behavior of the signal (Additive Noise Overlay)

3A Attack in the entire time window. The parameter after manipulation by the attacker takes the following value:

$$x_i^z = x_r^i \pm b_i, \quad (7)$$

$b_{i,}$- - coefficient, a random number, and $|b_i| \ge 0.2 \cdot x_r^i$.

3B. Attack in the entire time window. The parameter after manipulation by the attacker takes the following value:

$$x_i^z = \begin{cases} k_i \cdot x_r^i, i \le \frac{n}{2}, k_i \epsilon (-1.1; 1.1) \\ x_r^i \pm b_i, i > \frac{n}{2}, |b_i| \ge 0.2 \cdot x_r^i. \end{cases} \quad (8)$$

3C. Attack in a small part of the time window. The parameter after manipulation by the attacker takes the following value:

$$x_i^z = \begin{cases} k_i \cdot x_r^i, i \le \frac{2n}{3}, k_i \epsilon (-1.1; 1.1) \\ a_i x_r^i \pm b_i, \\ i > \frac{2n}{3}, a_{i,} \in (0.5; 0.9) \cup (1{,}1; 1.5), \\ |b_i| \ge 0.2 \cdot x_r^i \end{cases} \quad (9)$$

4. Forgery of controllers and external influences. When falsifying control and external influences, the model will generate signals from the automatic control system of the gas turbine engine that do not correspond to the actual values of the parameters. Thus, the data generated by the model can be considered attacked.

4A. Impact across the entire time window. To simulate this kind of impact, you can substitute a real signal into any function

$$.y_i^z = f(x_r^i), \quad (10)$$

y_i^z- parameter value generated by the model for forged control actions, $f(x_r^i)$- any function.

4BAttack in part of the time window. The parameter after manipulation by the attacker takes the following value:

$$y_i^z = \begin{cases} k_i \cdot x_r^i, i \le \frac{n}{2}, k_i \epsilon (-1.1; 1.1) \\ f(x_r^i), i > \frac{n}{2}, \end{cases} \quad (11)$$

4C. Attack in a small part of the time window. The parameter after manipulation by the attacker takes the following value:

$$y_i^z = \begin{cases} k_i \cdot x_r^i, i \leq \frac{2n}{3}, k_i \epsilon(-1.1; 1.1) \\ f(x_r^i), i > \frac{2n}{3}, \end{cases} \quad (12)$$

4. Construction of classifiers for the presence and type of attack on TMI

To build a classifier of the type of attack on TMI, it is necessary to form a training and test sample. To create such samples, 14000 time windows were formed, 15 samples each. For the experiments, the specific fuel consumption G was taken as the parameter modified by the attacker.

An example of some time windows for building an attack type classifier is presented in Table 2.

Table 2. Time windows

Model data		Attacked/not attacked data		State class	Attack type				
G_0	G_{14}	G_{a0}	G_{a14}	S/D	0	1	2	3	4
2,8	4,0	2,8	3,8	D	0	1	0	0	0
1,3	1,3	2,2	44.	S	0	0	0	0,4	0,6
1,3	1,	1,4	1,4	S	1	0	0	0	0

If data with different types of modifications fell into the same time window, the shares of the attacks were entered in the "Attack type" columns, as shown in line 2.

Attack type classifier is a multi-input and multi-output classifier, so conventional classification algorithms (for example, the usual kNN and Random Forest) are not suitable for this task. An algorithm is needed to build multiple regression. To solve this problem, Multioutput random forest regressor is used.

The quality of classification of attack types by this method was 0.65.

The low quality of classification by multiple regression methods forces us to switch to more complex classification algorithms. It is necessary to apply algorithms built on neural networks to this problem.

Neural networks are a universal tool for approximating functions, which allows them to be used in solving classification problems [13]. A typical task when using neural networks is choosing the neural network architecture (number of layers, activation functions, etc.) for each task. Table 3 shows the iterations done to find the attack type classifier.

Activation functions listed in the table:

1,Tanh-hyperbolic tangent

$$\sigma(x) = \tanh(x) \quad (13)$$

2. ReLu is a rectified linear unit. Less computationally intensive than hyperbolic tangent or sigmoid because it performs simpler mathematical operations. Therefore, it makes sense to use ReLu when creating deep neural networks.

$$\sigma(x) = \max(0, x) \quad (14)$$

3.Sigmoid-.sigmoid function

$$\sigma(x) = \frac{1}{1+e^{-x}} \quad (15)$$

Table 3 Finding the best attack type classifier

№	Description of the neural network			Accuracy
	Layer	Activation function	Number of neurons	
1	1	Tanh	100	0.83
	2	RELU	25	
	3	Sigmoid	5	
2	1	RELU	100	0.89
	2	RELU	25	
	3	Sigmoid	5	
3	1	RELU	100	0.90
	2	RELU	50	
	3	RELU	50	
	4	Sigmoid	10	
4	1	RELU	100	0.90
	2	RELU	50	
	3	RELU	50	
	4	Sigmoid	10	
5	1	RELU	100	0.79
	2	RELU	50	
	3	RELU	50	
	4	RELU	10	
6	1	RELU	150	0.90
	2	RELU	100	
	3	RELU	100	
	4	Sigmoid	20	

As can be seen from Table 3, a further increase in the number of neurons and layers does not give a gain in the quality of attack classification. Thus, neural network number 4 gives the best quality of attack type classification with an accuracy of 0.90.

The type of attack is not the most important metric in assessing the integrity of data, it is much more important to understand whether the data is modified or not. To assess the presence of an attack, the following set of time series proximity metrics was used: determination coefficient (R2), mean absolute percent error (MAPE), Euclidean distance, and dynamic time transformation metric (DTW) [14]. Thus, the input data packet for the classifiers of the presence of an attack is as follows (table 4):

Table 4. Data package

R2	MAPE	Euclid distance	DTW	State class	Attack
0,96	6,83	3,8	7	dynamic	0
0,38	10,24	5	14	static	1
0,38	11,4	10	17	static	1

The classification quality of the classifier constructed using the Random Forest method was 0.89. It is possible to improve the quality of the classification using algorithms built on neural networks..

Table 5. Finding the best classifier for the presence of an attack

№	Description of the neural network	Accuracy

	Layer	Activation function	Layer	
1	1	ReLu	50	0.85
	2	ReLu	50	
	3	Out	1	
2	1	ReLu	100	0.90
	2	ReLu	50	
	3	Out	1	
3	1	ReLu	100	0.90
	2	ReLu	100	
	3	Out	1	
4	1	ReLu	50	0.98
	2	ReLu	50	
	3	ReLu	50	
	4	Out	1	
5	1	ReLu	100	0.98
	2	ReLu	50	
	3	ReLu	50	
	4	Sigmoid	10	

As can be seen from the table, further increase in the number of neurons and layers does not give a gain in the quality of establishing the fact of an attack. Thus, neural network number 4 gives the best classification quality when establishing the fact of an attack.

5. Conclusion

The article presents a way to improve the method of monitoring the integrity of data transmitted from the board of a mobile object, based on a comparison of two time series - obtained from the model of the mobile object and from the object itself. The improvement of the method consists in the justification of the number of state classes for each type of mobile objects. On the example of the operation of the ACS GTE, two classes of object states were distinguished: static and dynamic. This separation allows the classification process to explain a greater difference between two TTSs in a dynamic state class than between two TTSs in a static state class.

To further classify the type of possible attacks, mathematical models have been developed of the attacker's possible impact on the TVR. Four main types of actions have been formed, and on the basis of these data, a classifier of the type of attack of an attacker, built on a deep neural network, has been developed. The accuracy of determining the type of attack was 0.98.

To establish the fact of the presence or absence of an attack, a binary classifier was constructed, the input vector of which is the metrics of the proximity of TTS (determination coefficient, MAPE, Euclid distance and DTW), as well as the class of the state of the mobile object (static or dynamic). The attack presence classifier is also based on a deep learning neural network. The accuracy of determining the presence or absence of an attack was 0.98.

Acknowledgment
The work is partially supported by the RFBR grant # 20-08-00668.

References

[1]. Decision support in the task of ensuring information security of aviation telemetry systems Order of the Ministry of Transport of Russia dated February 13, 2013 No. 36 "On approval of requirements for tachographs installed on vehicles, categories and types of vehicles equipped with tachographs, rules of use, maintenance and control over the operation of tachographs installed on vehicles "

[2]. Order of the Ministry of Transport of the Russian Federation of July 31, 2009 N 128 "On Approval of the Federal Aviation Rules" Preparation and Performance of Flights in Civil Aviation of the Russian Federation "(with amendments and additions)
[3]. Gorbatov V S, Zhukov I Yu, Murashov O N 2014 Cryptographic protocol for authentication and generation of a common key for vehicle control devices *Saf.of Info. Tech* 5 24 27-
[4]. Chuyanov G A , Kosyanchuk V V, Selvesyuk N I, Kravchenko S V 2014 Directions for improving on-board equipment to improve the safety of aircraft flights *Izv. SFedU. Tech Sci.*.6 155
[5]. Bronnikov A M 2017 The efficiency of the technical operation of the onboard system of the aircraft, unattended during the service period *Sci. Bull. of MSTU GA*. 6 12
[6]. Guzairov M B Frid A I, Vulfin A M, Berkholts V V 2019 The concept of integrity of telemetric information about the state of an aircraft power plant monitoring *Proc.of 2019 ICOECS* (Ufa:Ufa aviation technical state university) 10.1109/ICOECS46375.2019.8950020
[7]. Vulfin A M, Frid A I 2011 A neural network model for analyzing technological time series within the framework of the Data Mining methodology *Inf. and Con.Sys*. 5 31-38
[8]. Robert L Thorndike 1953 Who Belongs in the Family *Psychometrika*. 18 4: 267-76.
[9]. Jolliffe I T 2002 Principal Component Analysis, Series: Springer Series in Statistics, 2nd ed., Springer, NY XXIX, 487 p. 28
[10]. Gashi I, Stankovic V, Leita C, Thonnard O 2009 An Experimental Study of Diversity with Off-the-shelf AntiVirus Engines *Proc of the IEEE Inter.Symp. on Net. Com and App*. (Cambridge, MA) 4-11.10.1109/NCA.2009.14
[11]. Hastie T, Tibshirani R, Friedman J 2001 *The Elements of Statistical Learning* Springer, *ISBN 0-387-95284-5*
[12]. Guzairov M B, Frid A I, Vulfin A M, Berholts V V 2020 Decision support in the task of ensuring information security of aviation telemetry systems *Wor. of Intern. Sym. Rel and Qual (accepted for publication)*
[13]. Kim, Y 2014 Convolutional neural networks for sentence classification *IEMNLP* 1746 -51
[14]. Ghazi Al-Naymat, Sanjay Chawla, Javid Taheri 2009 Sparse DTW: A novel approach to speed up Dynamic Time Warping *The 2009 Aus. D Min*. 101 117-127 arXiv:1201.2969

Duality of artificial intelligence technologies in assessing cyber security risk

N I Romancheva

The Moscow State Technical University of Civil Aviation, Kronshtadtsky blvd 20, 125499, Russia

E-mail n.Romancheva@mstuca.aero

Abstract. Issues of estimation of information technologies influence on information security were and remain issues of special importance. During the pandemic 2020 over one million effects on structures including CII were registered, where all of them were related to application of modern technologies including those based on artificial intelligence. Extensive penetration of intellectual intelligence systems and of "smart" (end-to-end) technologies urges for revision of approaches to cyber security risk assessment. Application of the latest technologies in cyber security risk assessment creates new options on the one hand, and represents new threats from another. Under cyber security risk assessment we shall understand an integral index of information safety achievement in terms that cannot be easily predicted, such as complicity, uncertainty, inability to foresee, lack of stability. On that basis this work aims to analyze influence of artificial intelligent technologies on information security indexes in risk assessment in terms of VUCA-world. This paper proposes an approach for creating a model of duality assessment for such technologies in cyber security e risk assessment with games theory applied.

Keywords: Duality; Artificial intelligence; Risks; Cyber security; Games theory.

1. Introduction

General trends in information security support in the state and public security are defined by the Information Security Doctrine of the Russian Federation adopted by the Decree of the President of the RF No. 6468 [1] dated 05 December 2016. The Action Plan for "information security" for the "Digital Economics of the Russian Federation" program was approved as well. Legal and regulatory issues of the cyber security are governed by several regulatory documents of the FSTEC of the RF. The Development Strategy [2] contains definitions of various modern information technologies including those applying artificial intelligence (AI). For regulatory and technical governance of various aspects of AI technology application, in July 2019 Russia established the technical committee 164 "Artificial Intelligence" (TC 164) that provided a set of standards related to AI, two of which were approved.

Existing theoretical, practical and legal-and-ethic norms of artificial intelligence application in corruption counterfeiting are extensively covered both in works of foreign authors [3, 4], and national studies [5, 6, 7]. In the works of V.S. Ovchinskiy and E. S. Larin the artificial intelligence is considered as a three-way technology that may be applied for civil, military and criminal purposes [8]. The work [9] notes that cyber risks layout is rapidly changing: the speed of innovations implementation is changing together with the speed of new cyber risks arising, response procedures established in regards to customary approaches to IT-solutions development get more complicated. Cisco report [10] underlines that "active research for threats should consist in revealing of the tactics, methods, and procedures (TMP) used by attackers. This is the most valuable asset, as the attackers are hard to change them." In some cases simple actions for active search for threats are based on studies or reports on recently revealed threats.

The work [10] considers a "pyramid of pain" method by David Bianco that describes approaches to information protection that strongly complicate network attacks. Each of six levels of the pyramid represents a different approach: from the simplest to the most complex. By moving upwards the pyramid, an attacker will use more resources on the network attack, that will complicate and increase a probability of being detected.

According to the same document, currently the common practice represents inclusion to similar studies of compromise indexes (CI) that could be used by other parties involved. These data points usually consist of IP-addresses, URL-addresses, hashtags of files, and other CI constituting a threat.

The term "cyber security" had different definitions for years. We will turn our attention to the term provided by JSC Kaspersky Lab: "this is set of methods and practices for protection against attacker's attacks on computers, servers, mobile devices, digital systems, networks, and data" [11].

The risk management idea is usually based on detection of trouble spots in network or systems, establishment of its importance, appointment of priorities and further adoption of respective measures for their elimination. It may include detection of sources of threats, and active research of threats may aid in obtaining information for risks assessment. With above all potential risks are considered (known and unknown).

Management of cyber risks assessment as a rule is performed based on monitoring of indexes and analyzing of their alteration dynamics. The work [12] offers an approach for cyber risks assessment based on a company's business processes analysis: assessment of effects on business and risk analysis, development of a response strategy for the risks, development of a strategy implementation plan.

The main problem of the modern cyber risks assessment consists in absence of complete formal description of parameters of the technologies applied, and their mutual effect on the final cyber risks assessment, application of methods and means that do not consider resistance to cyber accidents, and negative effects mitigation.

Analyzing existing and tested forms of risks assessment does not fully reflect the changed context in regards to society digitalization. Consideration of duality of the modern technologies applied, and of artificial intelligence in particular, in generating the information security system enhances and has a qualitative effect on the cyber risk assessment. Therefore, assessment of the duality effect of artificial intelligence technologies on cyber security risks assessment represents a critical task.

2. Informal statement of the problem

The purpose of this work consists in analysis of the dual impact of artificial intelligence technologies on the quality of cyber security risks assessment in modern conditions, and in generating a model of cyber risks assessment and management in VUCA-conditions. The research problem could be divided as follows: classification of the existing artificial intelligence (AI) technologies, and their application in the information security sphere; definition of AI technologies duality, and their effect on errors of cyber security risks assessment; ensuring set parameters of cyber security when selecting one or another artificial intelligence technology.

Therefore, the following issues shall be resolved to solve the set problem: to generate sets of existing cyber security risks assessment; to establish a type of reflection of sets of existing AI technologies on sets of risks assessment models corresponding to VUCA-security concept. A feature of this problem solution represents a necessity to define conditions and assessment of a probability of applied AI technology getting in an "abnormality" sphere. The term "abnormality" sphere will be applied to a zone of adverse effects of the applied technology on the cyber security indexes.

3. Methods

3.1. Classification of Existing Cyber security Risks Assessments

Understanding that information technologies carry threat to information security has resulted in coordination of criteria and establishment of the International Standard "Common Criteria for IT Security Evaluation, ISO/IEC 15408, by several states.

First of all, when using the artificial intelligence technologies for information security, focus is given on predicative analytics to assess risks of one or another threat arising, and a possibility to prevent an attack or other destructive actions.

Cyber security counterfeits three types of threats: a cyber crime is an action arranged by one or several attackers aimed to attack a system for breaking its work or receiving financial benefit; a cyber attack is actions aimed to information collection, mainly of politic type; cyber terrorism means actions aimed to destabilization of digital systems aimed to create fear or panics.

Modern information systems and computer networks consist of different components and represent a complex system. An issue of ability to preserve properties available by its components arises at an every stage of a complex system life cycle. The system property to preserve peculiar properties of its components is called compositionality [13], that is also referred to security properties.

Factors which negatively influence a complex system state, may include risks related to violation of protection of infrastructure, services, resources, information leaking by technical channels, not authorized and unintended impact on data and (or) other resources applied in the digital sphere. Thus, for example, intellectual intelligence application in various areas of industry, production, on the one hand, allows to adapt in new social and economic conditions, on the other hand, risk of qualitative replacement or sudden 5G-loss of large volume of information increases. Growth of deepfake technologies application, when AI is used to create a contact with replacement of faces and voices of famous people.

The risk management practice outlines several risk analyzing methods: statistics, expert assessment, analytics, method of analogue application, and risk accumulation consequences analyzing. Use of the statistic risk analysis method in utilization of AI technologies is based on historic data analysis, that in terms of uncertainty may deface risk assessment in general. The expert assessment method could be used at the first step of risk level definition until a well-organized registry of cyber risks related to application of one or another AI technology and consequences of their negative affect is available. The analogue application method for analyzing risk levels may be useful for threats registry formation, and is based on collection of information for previous incidents received from various sources. Supervised classification algorithms are widely spread recently in practice. However, complicity to control the tree size does not allow describing precisely complex areas of one class of objects.

3.2. The Risks Assessment Model with Consideration of AI Technologies Duality

AI technology represents a single process including such stages as information resource creation, ensuring this resource resistance to internal and external influence during transfer and processing. Following this consideration, we will deal with AI technology as with a complex system with multiple functional elements and relations between them, segregated from the environment in accordance with the required purpose for a certain moment of time with ability to analyze, self-educate and preserve resistance to destabilizing factors. Analysis and self-education properties allow the above technologies to show duality and, therefore, have both positive and negative effect on the information object that is a digital recourse. We will consider models, methods, and facilities for generation and utilization of an intellectual resource in the society as technology elements.

It is possible to outline technologies having a analysis mechanism, that unifies data receipt, storage and enhancement using processing of natural language to extract data from structured and non-structured data using AI, and multiple technologies based on comparison of visual information with some predefined classes based on an independent components analysis algorithm.

Problem setting is formulated as follows: in arbitrary time intervals, based on consequence of events and connections between them, to assess level of influence of the i-th parameter of the j-th AI technology belonging to the set of applied technologies on assessment of cyber security risks in accordance with the constructed model.

We will introduce a notion of "AI distrust index" that will mean a level of process destructurization leading to cyber threats implementation. Destructurization may be related to unintentional actions of the technologies (logics errors in the code, e.g., not detected during testing stages), and intentional actions (e.g. application of self-education algorithms for generation of new trouble points and threats). The set of acceptable risks of solutions has an uncertain origin and depends on VUCA-parameters: volatility, uncertainty, complexity, ambiguity. Impact of AI technologies parameters on the final target function may be both positive and negative: Therefore, the set problem may be considered as a problem of achieving minimum damage to information recourses and other correlated infrastructure system elements with applied AI technologies, i.e. absence of inverted maximum system protection result (1):

$$\sum_{i=1}^{k}\left(\sum_{j=1}^{n} p_{ij} - p_{ij}^{0}\right) \rightarrow \min \quad , \tag{1}$$

where p_{ij}^{0} is minimum specified damage of j-th resource with applied i-th technology, p_{ij} is actual damage of j-th resource at application of i-th technology.

We admit that if each AI technology element does not fall within the "abnormality area", the result of this technology application has no impact on the cyber security risks assessment.

The risk will be estimated through the likelihood of a risk event multiplied by the assessment of its consequences for the system operation. To assess such consequences, it is offered to rank AI technologies by the value of cyber security damage and by the value of measures related to consequences removal. At the first stage the problem of abnormality zone definition for each AI technology is defined by an expert method with further automation of this process.

Risk identification supposes a staged analysis of all technology elements involved in cyber security provision starting from the most to the least accessible for a threat.

The cyber security system assessment model may be represented as an integrated process in the form of combination of interrelated processes. Initial risk assessment, assessment of inverted risk probability and counter-measures efficiency, as well as analysis of residual initial and inverted risks should be grounds for this approach. For this we will introduce three initial sets of processes (of risk map formation, vulnerability map formation, formation of protection measures) into consideration, based on which a set of assessment areas is formed. The risk map includes both typical system risks and inverted influence risks of the used technologies. The vulnerability map is formed from a pool of fixed and declared vulnerabilities.

The threat level is proposed to calculate by vulnerability *(VL)*, based on the criticality and likelihood of threat implementation through this vulnerability. The threat level shows how the impact of this threat on one or another process performed by the system is critical with account of its implementation probability. The single threat mode uses equation (2):

$$VL = \frac{K}{100} * \frac{P(VL)}{100} \quad , \tag{2}$$

where K - criticality of threat implementation (specified quantitatively, i.e. in points assigned by experts, or, where applicable, as difference between BS failure in result of threat implementation, and probability of BS failure without threat implementation); P(V) is probability of threat implementation through such vulnerability related to AI technology application.

Vulnerability threat level values range from zero to one. Currently the existing vulnerabilities are classified in accordance with [14] as multifactorial. Accordingly, equation (2) may be presented as (3):

$$VL_f = \vartheta_i \sum_{i=1}^{i} (VL_i), \qquad (3)$$

where ϑ_i is weighted index of i-th vulnerability.

Risk R related to application of AI technology is proposed to calculate by equation (4):

$$R_T = 1/a * (VL_f * D), \qquad (4)$$

where: a is AI distrust index; VL_f is total level of threats; D is resource criticality (is specified qualitatively (be expert) or as quantitatively (as difference between failure probability without AI technology application, and failure probability with this technology applied).

We describe the protection measures formation process as a process of action program selection in terms of uncertainty and suppose the focus on an "expected value" of results of such actions. For its assessment mathematic expectations usually used.

Output data of risks assessment models represent binary forecasts for non-inclusion of applied technology T to the abnormality area. It allows to compare the simulation results with a selected working area tolerance limit (to ensure the specified parameters of cyber security). The risks assessment quality grows with extension of planning horizon This allows to reduce the time of risk management model formation in VUCA-conditions. Minimization of structural arrangement of technology elements between risk zones with account of VUCA-parameters and the abnormality zone (Figure 1) is used as a selection criterion.

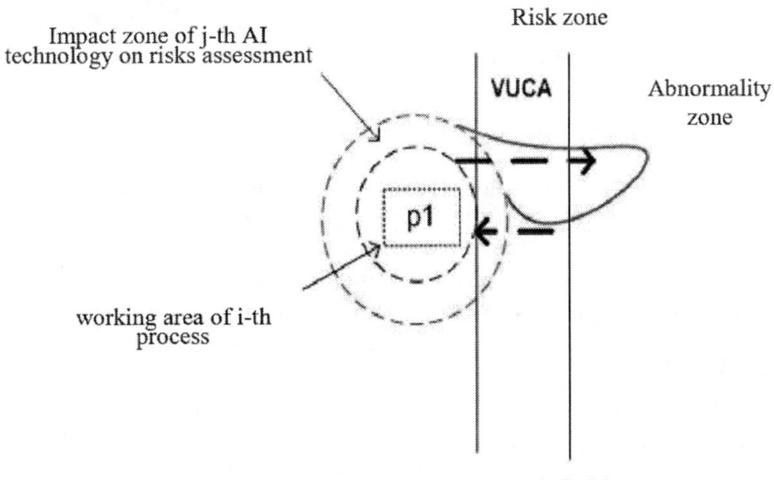

Figure 1. Structural arrangement definition

When assessing cyber security risks in conditions of uncertainty, it is required to specify VUCA working area, exceeding borders thereof shall be considered as a risk of exceeding the specified parameter and entering the abnormality zone.

Application of AI technologies as an instrument of analytics to assess cyber security risks is aimed to improve such processes as management processes. This includes placement and achievement of goals of long-term and short-term forecasting, transformation and detailed management of risks. Application of a balanced approach developed with consideration of priorities for protection from known and new threats allows to establish correlations between security properties and indexes. In terms of massive transition to remote work, a number of information channels requiring monitoring increases; in addition, there are events in data streams that need to be tracked, i.e. management of risks and their prediction is required. Threats analysis aimed to detection of dangerous actions is represented as a factor of re4adiness to threats including ones created in VUCA world. To ensure resilience to such threats, it is required to create conditions for restoration after cyber accidents, as well as negative consequences mitigation.

Therefore, applied advance technologies for proactive detection of cyber security threats should be balanced between the working parameters zone and VUCA - parameters zone. Process parameters entry to the abnormality zone indicates absence of flexibility properties in the applied AI technology, namely, effective measures to reduce risks and protection/recovery methods.

Ignoring a mutual effect of AI-based defenses and the threats created by the same technologies may lead to rather heavy consequences, increasing an attack space. Examples include chatbots that behave inappropriately, fake applications with malicious content embedded in components.

As it was mentioned earlier, the considered AI technologies consist of single elements, respectively; parameters of various elements of the technologies usually are not correlated. By the values of mathematical expectation of random variables m_{yi}, variance σ_{yi} corresponding to the period t, we can calculate the probability Pi that the values of the determining parameter σ_{yi} in the i-th section of random process $T_{vuca}(t)$ when using AI technology are within the risk zone tolerance and do not fall into the abnormality zone.

Assessment of risks of response to destructive processes created by AI technology elements includes such common stages as analysis, identification, description, measurement, and quantitative/ qualitative assessment.

Duality of AI technologies impact appears in, among others, in growth and in difficulty in understanding and assessment of risks of information loss or misrepresentation. For example, AI technologies applied in interception systems, on the one hand, increase coverage of channels, assist in general data review; on the other hand, cause data overloads and make their interpretation difficult in terms of reliability. New technologies allow to indicate important changes in processes, to pay attention to potential or currently existing threats in an events flow. From the other hand, all recent attacks on systems occur based on machine education algorithms.

The risks assessment model normally includes separate groups of risks known and mathematically described. Nevertheless, statistic analysis in general cannot follow changes in parameters of information flows, and their environment.

Algorithms of machine education applied for cyber security risks assessment separate and perform search of repeating patterns in information processes by separating a set of hidden interrelations between parameters, thus, forming a zone of abnormal behaviour and destructive characteristics.

In the abnormal behavior zone it is possible to indicate profiles of threats and vulnerabilities both traditional and related to AI application. Analysis of a complicated chain of events (information process phases) is performed for this purpose, and symptoms and processes, which lead to negative consequences, are identified.

To assess threats with consideration of duality of AI information technologies we will draw a table of risks analysis, perform risk calculation, and select a risk management method.

Increase in cyber attacks using AI technologies is primary related to application of machine education systems with an open code that allows attackers to avoid security systems. Classification of instruments applied for cyber security analysis, and their impact on changes in information resource properties is given in Table 1. This table uses the following designations: C - breach of confidentiality, I - breach of integrity, A - breach of accessibility. Changes of destructurization functions for process conditions depending on applied AI technologies are highlighted with color: green color - reduction, pink color - increase, orange color - ambiguous change, white color - change not detected.

Table 1. Classification of AI Technology Instruments and Their Impact on Information Properties

Instrument	Result	C	I	A
In-depth Traffic Analysis (NTA)	Availability of specific comprometation indexes			
Intellectual algorithms (X-Spider, Max-Patrol, PT Application Inspector)	Vulnerability and errors in web-applications			
In-depth technologic traffic analysis (PT, ISIM)	Detection of attacks related to hidden tunnel effect and extraction of data from APCS			
	Legal actions (development of attack vector)			
	Activity inside network			
IoT	Breach of IS policies			

To assess information technologies with AI application, the games theory [15] will be use Games theory application allows to simulate a conflict situations process (technologies duality) and take a decision based on the analysis result on practicability to apply one or another information technology. A game in a normal form is a triplet consisting of a set of players and their strategies, as well as player's payoff function, which assigns a payoff of this player to each set of strategies (situations). We will consider a case when a number of players is two. Attributes of AI technology will be considered as players having positive and, respectively, negative impact on results of its application. We will suppose that all players are rational, each player considers options available in his possession, forms notions in regards to unknown parameters, has clear preference, and choses own actions resulting from a certain optimization process. Players strategy is, respectively, minimization and maximization of the probability of destructive event occurrence risk. Winning means profit (based on attack vector development). We will employ a dominating strategy of player 2. To check that player 2 strategy dominates over player 1 strategy, it is necessary to consider behavior of these strategies against a certain standard. Consequent removal of dominating strategies of one of the players should lead to the balance. Choice of a unique balance from several equal outcomes is performed based on established criteria of options selection based on destructive functions analysis.

Input data for modelling are fixed-dimensionality attribute vectors, which contain numerical values, with support of a special variable value representing absence of the cyber security risk. A destructive event will be interpreted as a reverse spread of an error arising in the process leading to reduction or seizure of target function performance in result of reverse influence on the AI technology system.

Vectors of attributes of analyzed parameters falling into the abnormality zone are collected dynamically from results of processing of the risk assessment model parameters by the analytical module with further consideration of possibilities for risks implementation and developing measures for their mitigation. In result the module receives a possibility to define direction of AI technology adaptation to alteration of one or another parameter affecting cyber security risks.

4. Conclusion

- The paper provides fundamental principles of the offered approach to consideration of AI technologies duality.
- The set of existing cyber security risks assessments has been considered.
- The type of the set of existing AI technologies reflection to the set of risks assessment models has been established.

- The validity of the offered approach at the development stage is confirmed by expert assessments of specialists in the cyber security sphere provided in the public press that will be supported for its further testing.

References

[1] The general trends in information security support in state and public security are defined by Informational Safety Doctrine of the Russian Federation adopted by the Decree of the President of the RF No. 6468 [1] dated 05 December 2016

[2] Strategy for Information Society Development in the Russian Federation for 2017-2030 approved by the Decree of the President of the RF of 09.05.2017 No. 203 URL: http://kremlin.ru/acts/bank/41919 (reference date - 27.06.2020)

[3] Aiello M F 2018 Policing through Social Networking: Testing the Linkage between Digital and Physical Police Practices *The Police Journal* **91** 1 pp 89-101

[4] Webb A 2019 The Big Nine. How the Tech Titans & Their Thinking Machines Could Warp Humanity. Published March 5th 2019 by PublicAffairs (first published 2019).

[5] Borovskaya E V and Davydova N A 2018 Artificial Intelligence Fundamentals *Knowledge Laboratory* p 127

[6] Larina E S and Ovchinskyi V S 2008 Crime of the Future Is Already Here *Book World* p 512

[7] Larina E S and Ovchinskyi V S 2018 Artificial Intelligence. Big Data. Crime *Book World* p 416

[8] Ovchinskyi V S 2019 Digital World Criminology: Text Book for Magistrate *Norm Printing Press* p 352

[9] Cyber Risks Assessment Based on Company's Business Goals / JSC Deloitte&Touche CIS [Digital resource] URL: https://www2.deloitte.com/content/dam/Deloitte/ru/Documents/risk/russian/Ocenka-kiberriskov-na-osnove-biznes-celej-organizacii.pdf (reference date 02.07 2020)

[10] CISCO publications on information safety 2019 [Digital resource], URL: https://www.cisco.com/c/dam/global/ru_ru/assets/pdfs/cybersecurity-series-2019-threat-hunting-ru.pdf (reference date 02.07 2020)

[11] What Is Cyberdefense? [Digital resource] URL: https://www.kaspersky.ru/resource-center/definitions/what-is-cyber-security_(reference date 02.07 2020)

[12] Romancheva N I 2019 Conditions of Information Technologies Application in Digital Sphere *Transactions of International Conference. Innovation, Information Technologies and Communication Technologies* - Sochi pp 136-141

[13] Kosachev A S and Ponomarenko V N 2004 Analysis of Approaches to Verification of Safety and Mobility Functions *Triumf* p 101

[14] GOST R 56546-2015 National Standard of the Russian Federation Information Protection. Vulnerabilities in Information Systems Classification of Information System Vulnerabilities. Information Protection. Vulnerabilities in Information Systems. The Classification of Vulnerabilities in Information Systems OKC 35.020 Date of entry 2016-04-01.

[15] Smith M J 1982 Evolution and the Theory of Games *Cambridge University Press* p 10

Problems with the use of forensic techniques in the judicial investigation of the authenticity of an electronic digital signature.

V V Gordienko, L S Kryzhevich, R V Glagolev

Kursk State University, Kursk, Radisheva str., 33, 305000, Russia

E-mail: vika.gordienko.1973@ mail.ru

Abstract. This article aims to study the tactical foundations of conducting a forensic study of the authenticity of an electronic digital signature and documents that raise doubts about their reliability. The methodological foundations of identification, classification and search studies of "electronic traces" and signatures of the transformation made on the basis of algorithms with imitation and use of the latest methods are given. To identify "electronic traces" the functions of using hardware and technical means for counterfeiting an electronic signature are specified and methods of establishing to determine the kind and type of a particular device on which the documented information was manipulated. There are considered sets of features that allow identifying parts and details of documents subjected to an unsanctioned modification, as well as a specific list of correctly formulated questions, which should be put to the expert in the decision on the appointment of an examination to determine signs of forgery.

Keywords: electronic-digital signature, electronic trace, hash function, conversion algorithm, magnetic field, authentication, identification, illegal access, vulnerability, threat, knowledge, right of access, computer systems, hardware, certificate.

1. Introduction

Currently, cases of illegal alienation of property by means of forging an electronic digital signature have become more frequent. Transactions made using an electronic digital signature (EDS), which is analogous to this signature, as specified in the Federal law Federal Law of 06.04.2011 N 63-FL (red. From 08.06.2020) "On Electronic Signature", according to which an Electronic signature is equal to this signature of an individual and is applied by a specified group of users, and is implemented when registering transactions [1]. Among the works assigned to this group, it is impossible not to mention the article of RM Allalyev, VV Belyay. On the issue of digital rights in the Legislation of the Russian Federation, containing a comprehensive study of the problems of legal regulation of information technologies, based on recent Russian legal sources and their comparative analysis with foreign analogues. The authors considered the legal aspects of building an information society, the use of electronic documents and electronic signatures. analyzing this article, it should be noted that you can get a certificate of this signature in the name of each subject, provided that you have passport data and the insurance certificate of the pension Fund number [2]. It should be noted that the algorithm by which you can ensure your security in the information space does not currently exist, this is stated in the article Yadova N. E., Rusyaeva D. O. "Effective communication in e-commerce" [3]. drawing conclusions from which, it becomes clear that accredited certification centers issue qualified signatures in huge quantities. When applying to the Certification Center, a citizen is required to provide a passport and personally sign the documents required for issuance [3]. However, those employees who are authorized to check this person, their passport, and signature for authenticity dos'n always distinguish a fictitious passport from a real one. In addition, attackers can inject malware into a computer system that can output one type of document for visualization, and the user electronically signs another document that will be sent to the recipient. As a result of such implementation, a payment order generated by a malicious program will sent to a Bank or other institution. Ideally, there is a need to develop a digital signature that represents additional information attributed to the protected data. In order to understand the mechanism of conducting a forensic examination, you need to understand the essence and algorithm of forming this signature. Authors Jue-Ming, Chen-Hao, Zhang-Shan, Mei-Zhang, Qian-Wang. The article "Practical decoy-state quantum digital signature with optimized parameters" defines the legal status of computer information, which is divided into two types: documented information that has the form of a document and information that does not have a documented form. In order for information to have the status of a document as a type of information, it must have details that allow you to determine the authorship and authenticity of the data available in the document[4]. In electronic document management, according to the legislation, the confirming requirement is an electronic signature, both simple and enhanced, despite the fact that codes and passwords are used as analogues of a handwritten signature.An electronic digital signature is a block of electronic information attached to a certified document. This data allows you to identify the subject who certified the document and the encryption algorithm. As indicated in the work of R. J. Barceló, T. Vinje "Towards a european framework for digital signatures and encryption": The European commission takes a step forward for confidential and secure electronic communications. "the encryption Algorithm provides functions that determine:

1. the Integrity of an electronic document.
2. Authenticity of the transmitted data.
3. No changes in the document and no modified elements in it
4. Identification of authorship in the document (at the same time, it is unacceptable for the subject of the signatory to deny the fact of signing the document) [5].

The concepts and main provisions of the use of an electronic digital signature, as well as the procedure for its application, are discussed in international regulations, such as the UNCITRAL Model Law on Electronic Signatures with Guide to Enactment 2001. New York: UNITED NATIONS, 2002. In particular, the British act defines an electronic signature as "information in electronic form that is attached or logically associated with other electronic information and can be used as an identification method". the problem of its authenticity corresponds to the complexity of the digital signature key and is the only unique string of characters intended for creating an electronic digital signature, that is, for converting a document [6]. As noted in the work of Finnish researchers Pascale-L. Blyth Of Cyberliberation and Forbidden Fornication: Hidden Transcripts of Autonomous Mobility in Finland, where it is noted that the signature verification element is a certain string with characters corresponding to the digital signature key. The key allows you to decrypt and verify the authenticity of the signature. An accredited organization that registers EDS is issued a special document on paper or electronic media called a key certificate for identifying an electronic signature [7]. the key certificate certifies that the encryption key belongs to its owners. An electronic digital signature is generated using encryption algorithms, the same cryptographic tools allow you to verify the authenticity of the key, which is noted in the research given in the article Diffie W., Hellman M. E. New Directions in Cryptography the authors indicate the following types of electronic digital signatures: - simple, unskilled and qualified [8].

The use of a simple digital signature is well reflected in the article Rivest R., Shamir A., Adleman L. A method for obtaining digital signatures and public-key cryptosystems, in accordance with the proposed method, such a signature is formed using encryption, as well as passwords and other means and confirms the creation of an electronic digital signature by a specific subject. An unqualified digital signature must meet the following requirements: it is performed in the process of data conversion based on the key of an electronic digital signature and confirms the authentication of the subject certifying this document [9]. This aspect is also mentioned in Baum M. S. Secure electronic commerce — II: the ABA digital signature guidelines, where the requirement to use a non-qualified signature serves as a definition of authenticity in electronic document management used in electronic Commerce [10]. The scientific article Rivest R., Shamir A., Adleman L. " A method for obtaining digital signatures and public-key cryptosystems specifies a method that allows you to detect document modification or if an attacker made unauthorized changes to the document after signing. A qualified digital signature, in addition to having the features inherent in an unqualified digital signature, has additional features. It is fixed in a qualified certificate with an element confirming the key for verifying an electronic digital signature. These two types of unqualified and qualified signatures are also referred to as enhanced, despite the fact that the functions used to create an electronic digital signature must meet the requirements for the formation of an electronic digital signature [11]. A scientific and up-to-date approach to compliance with authentication rules based on cryptographic algorithms is given in the Tanmay study. Choudhury Post-quantum digital signature scheme based on multivariate cubic problem NibeditaKundu, where the proposed method determines the dependence of the digital signature on the data contained in the signed document and the existence of a key available only to the person signing it, is a mandatory requirement for the implementation of secure document flow [12]. It is also necessary to note the works of other foreign authors who offer methods for determining the legality of electronic document management, such as S. Aggarwal, N. Kumar. "Digital signatures", which provides General concepts of the use of signatures in the implementation of transactions using information technologies [13], as well as A. S. Kittur, A. R. Pais "Batch verification of Digital Signatures: Approaches and challenges" dedicated to the method of batch verification of digital signatures [14]. In addition, the Protection of e-Commerce using digital signatures is analyzed in the article "Securing electronic commerce with digital signatures" by A. Bequai, which offers a new approach to determining the requirements that documents of this type should have. Thus, Russian and foreign authors devote a huge array of research on the issue of personal identification when making transactions using information technology and determining the authenticity of the signature of the final document and making it legally sound. without the involvement of these sources, it would be impossible to write individual paragraphs of this scientific article.

2. Relevance and problem statement

The digital signature key is the only unique character string for creating a digital signature, that is, for converting a document. The signature verification element is a specific string with characters corresponding to the digital signature key. The key allows you to decrypt and authenticate the signature. An accredited organization that registers an electronic digital signature is issued a special document on paper or electronic media called a key certificate to identify the electronic signature. This key certificate verifies that the encryption key belongs to its owners. A digital signature is generated using encryption algorithms; the same cryptographic tools allow you to authenticate the key [2].

The Federal Law "On Electronic Signature" specifies the following types of electronic digital signatures:

1. Simple
2. Unqualified
3. Qualified

A simple digital signature is formed using encryption, as well as passwords and other means and confirms the creation of an electronic digital signature by a particular subject. An unqualified digital signature must meet the following requirements: it is carried out during the data conversion process based on the digital signature key and confirms the authentication of the entity certifying this document. [3]. Allows you to detect a modification of a document or if an attacker made unauthorized changes to the document after signing. A qualified digital signature, in addition to possessing features inherent in an unskilled digital signature, has additional features. It is fixed in a qualified certificate with an element confirming the key for verifying the electronic digital signature. These two types of unskilled and qualified signature are also called enhanced, while the functions used to create an electronic digital signature must comply with the requirements of the Federal Law. The dependence of a digital signature on the data contained in the document to be signed and the existence of a key available only to the person signing it is a mandatory requirement for the implementation of a secure document flow[4].

3. Problems formalizing the authentication of digital signatures

The essence of the identification mechanism is that the digital signature confirms that the signing entity intends to sign the electronic document and it was he who did this not by chance. The presence of a digital signature binds the fact of signing and allows you to record the time of such signing. As a result of these requirements, the person who signed the electronic signature is deprived of the opportunity to refuse to sign. The subject of research in undocumented electronic information is both the content of information and the material medium on which this information is contained, in addition, during the study of electronic documents, the main object of the study is the details, hardware and software used in their creation [5].

As with any forensic study, when implementing the authentication of an electronic digital signature, it is necessary to identify traces left by the attacker, and when studying documents created by the use of information technologies - electronic traces that allows the expert to conclude that unauthorized access occurred as an intrusion. In forensic science, there are several terms that denote these traces: virtual, computer, machine, digital, electronic and some other terms. The relevant terms have a relationship with an aspect of the study of such traces. The main means of information processing is an information system, therefore, this terminology conveys a special role as the main type of data transfer and, accordingly, the type of data processing, therefore, in the term "computer trace" the main role of the computer system as the dominant information processing function is transmitted [6]. The presence of "electronic traces" depicts the course of the scenario of actions of a technician functioning on the interaction of an array of electronic microcircuits, «digital traces»are able to display the form of presentation of information created and processed by transmission through digital channels. Used as synonyms "computer," "electronic," "digital traces," combine an important factor consisting of two elements of the information itself as a phenomenon of objective reality and the material medium on which this information is fixed. The filling of computer information is presented in the information system in digital form, adapted for storage and processing of the electronically calcined machine, and the electromagnetic field is its material carrier. The properties of the electromagnetic field, on which a certain amount of computer information is recorded, allows it to be located in a certain technical device (in a computer system or its external devices, in a removable medium, etc.), can be contained and transmitted by wireless communication or can exchange data via wired channels. The technical tool on which the information of interest is available in the expert study is studied as a secondary material medium of information. Such types of technical devices with precise spatial boundaries have a certain shape, for example, a computer, a removable storage medium. In the case of data exchange over a wired network, it is theoretically possible to allocate and establish such boundaries, but it is not possible to practically set and record the time of presence of certain information on a particular section of the network. A feature when computer information exists outside the technical device is that it does not have a real (object) form. Thus, the computer information medium can be both material and fixed on a hard disk or removable electronic media, and without having an object form when transmitting information via wireless channels in computer networks or via WI-FI, will nevertheless be subject to investigation [7].

The subject area of forensic research on the technical and software tools that create data digital signatures, as well as electronic and digital tools, includes the development of mechanisms, methods and recommendations for the recording, detection, seizure, storage and research of electronic traces, as well as the forensic study of these traces, as one of the complex and difficult methods of forensic examination, which is justified by the specific features of such objects. Information processed and transmitted by technical devices is not available for human perception because it is represented in the form of zeros and ones, while a certain amount of the presented information is not uniquely assigned to a specific material medium, and an electromagnetic field acting as a computer information medium cannot be individualized. A feature is such a fact that without mutual changes in the content of computer information, it can be segmentally separated, or the content of information can be separated from its material medium without mutual changes, which does not cause a change in its medium, or vice versa. It complicates the expertise and technical capabilities associated with the speed of processing and the ability of an attacker to modify and distort information by remote access, which is not controlled by persons who legally own information, which entails the possibility of restoring the original content of deleted and modified data. These factors are supplemented by technological advances aimed at improving computer technology, implementing new technologies for processing data related to the speed of processing information on the network and the inability to overcome protective equipment when investigating intrusion into a computer system. The tasks of the forensic research are divided into types that allow you to identify, diagnose, classify, reconstruct and search for these traces of modification. The tasks of identification and classification are combined due to the lack of tools among criminologists that allow the identification of a single material object within the framework of issues related to the use of these objects for individualization of the identification of other objects, therefore, when implementing the classification task, it is necessary to strive to reach a narrower classification group, which allows you to distinguish a separate individualizing feature [8].

All standard problems of identification and classification orientation are implemented for definition of type, a type, models and brands of the concrete hardware and software system applied to change of the digital signature. Solving such problems leads to the definition of the type of system or application software, and also allows you to narrow classification characteristics before determining the version of the program or the revision of the test file when inserting a signature into a modified document. In the same way, the types of computer equipment presented are determined up to the identification of types, brands and models, the determination of the sources of information presented on different material media, including the ability to determine the purpose of the equipment, the definition of users and providers of provider services, the time and place of non-authorized changes[9].

4. Conclusion

In order to carry out a forensic examination of the authenticity of an electronic digital signature, the following materials must be provided for examination: root certificates of an electronic digital signature issued by an authorized organization, electronic documents, the authenticity of which is questionable and the value of the key for verifying an electronic digital signature. The expert is asked the following questions to authenticate a digital signature:

- Determine whether the digital signature belongs to the owner of the signature key certificate.
- Have any changes been made to the signature key certificate that cast doubt on the legality of this key?
- Is the signature key certificate valid, is it invalid at the time the document is signed?
- Whether a certificate has been issued in a special authorized organization issuing digital signature key certificates.
- Can the issued key certificate be suspended or revoked at the time the document is certified?
- Are there any unauthorized changes to the digitally signed document?
- Is a genuine digital signature?
- Which type of digital signature was used to certify a specific document?
- Was the authority of the person who certified the document a given digital signature?

Thus: The certificate validity study is carried out by standard methods of document examination both in electronic form and on paper. This examination is a new type of research carried out within the framework of technical and computer expertise; however, taking into account the growth of electronic document management, it becomes one level of importance with handwriting and other expertise. In order for the findings of the examination to serve as evidence in the courts, it is necessary to combine qualitative and quantitative assessment methods in order to avoid reducing the level of expert error, to this end it is necessary to expand the fields of integrated research and continue further improvement of research methods. There is a need to revise the relationship between the establishment of the equality of the signature in the electronic digital version and the direct authorship in the document of doubtful authenticity, which should be taken into account in judicial practice.

In addition, the question related to the limits of the competence of experts involved in establishing the identity of the signature or changes in the document, the expert does not establish the fact of forgery itself - this is a problem within the competence of law enforcement officers, the expert indicates the identity or its absence, accordingly, if the authenticity does not correspond on paper, then the modifications can be established, but the signature made in electronic format sharply reduces the possibility of identification, because attackers can be entered into the program of a fake model, therefore it is necessary to distinguish the concepts of identity, authorship and ownership of the signature to the legitimate owner.

References

[1] On Electronic Signature *Federal Law of 06.04.2011 N 63-FL* (red. From 08.06.2020) URL: http://www.consultant.ru/cons/cgi/online.cgi?req=doc&base=LAW&n=354532&fld=134&dst=1000000001,0&rnd=0.5819060327627392#0006249100798215723

[2] Allalyev R M and Belyay V V 2020 To the question of digital rights in the legislation of the Russian Federation *Education and Law* №2 URL:https://www.elibrary.ru/item.asp?id=42838957 DOI: 10.24411/2076-1503-2020-10217

[3] Yadova N E and Rusyaeva D O 2018 Effective communication in e-commerce *In: Evropeiskii zhurnal sotsialnykh nauk* (European Journal of Social Sciences no. 7 (2)) pp 41-48.https://cyberleninka.ru/article/n/sovremennoe-sostoyanie-mirovogo-rynka-elektronnoy-kommertsii-i-perspektivy-ego-razvitiyaDOI: 10.18384/2310-6646-2019-2-116-125

[4] Jia-Ming, Chen- Hao, Zhang -Chun, Mei- Zhang, Qin-Wang 2019 Practical decoy-state quantum digital signature with optimized parameters *Physica A: Statistical Mechanics and its Applications (Electronic Materials* vol 535) 12234 https://doi.org/10.1016/j.physa.2019.122341

[5] Barceló R J and Vinje T. 1998 Towards a european framework for digital signatures and encryption *The european commission takes a step forward for confidential and secure electronic communications (Computer Law & Security Review Volume 14, Issue 2*) pp. 79-86. https://doi.org/10.1016/S0267-3649(97)82130-9

[6] Uncitral Model Law on Electronic Signatures with Guide to Enactment 2002 (*New York: united nations)* https://www.uncitral.org/pdf/english/texts/electcom/MLETR_ebook.pdf

[7] Pascale-L and Blyth 2019 Of Cyberliberation and Forbidden Fornication *Hidden Transcripts of Autonomous Mobility in Finland Journal Transportation Research Part D (Transport and Environment*) https://doi.org/10.1016/j.trd.2018.12.002

[8] Diffie W and Hellman M. E. 1976 New Directions in Cryptography *IEEE Trans. Inf. Theory / F. Kschischang* (Vol. 22, Iss. 6.) pp. 644—654. — ISSN 0018-9448 — doi:10.1109/TIT.1976.1055638https://doi.org/10.1016/j.jclepro.2019.06.265

[9] Rivest R, Shamir A and Adleman L 1978 A method for obtaining digital signatures and public-key cryptosystems *Commun. ACM —NYC*: ACM (Vol. 21, Iss. 2.) pp. 120—126. — ISSN 0001-0782; 1557-7317 — https://doi.org/10.1145/359340.359342

[10] Baum M S 1992 Secure electronic commerce *The ABA digital signature guidelines* №8 pp.217–222. https://doi.org/10.1016/S0267-3649(97)89803-2

[11] Wenbo M 2004 Modern Cryptography: Theory & Practice *Prentice Hall Professional Technical Reference* (New Jersey, 4) pp.308. ISBN 0-13-066943-1https://doi.org/10.1137/0217017

[12] Kumar S, Kundu N, Dheerendra, Tanmay M, Choudhury D. 2020 Post-quantum digital signature scheme based on multivariate cubic problem *Journal of Information Security and Applications*, 102512 https://doi.org/10.1016/j.jisa.2020.102512

[13] Aggarwaland S and Kumar N 2020 Digital signatures *Advances in Computers Available.* https://doi.org/10.1016/bs.adcom.2020.08.004

[14] Kitturand A S and A R Pais 2017 «Batch verification of Digital Signatures: *Approaches and challenges».Journal of Information Security and Applications:*Volume 37 pp.15-27 https://doi.org/10.1016/j.jisa.2017.09.005

[15] A. Bequai Securing electronic commerce with digital signatures 1999 (*Available online* 24) https://doi.org/10.1080/13600830701194711

Stability analysis of binary transposition algorithms during attacks on text data

L S Kryzhevich, V M Dovgal

Kursk State University, Kursk, Radisheva str., 33, 305000, Russia

E-mail: leonid@programist.ru

Abstract. Different approaches to using of transposition methods for data protection are examined in the article. The author has conducted comparative analysis of the approaches to increasing strength to protect the text from various attacks when its length and the data amount given per one symbol are raised. Basic criteria of evaluation of the offered methods are marked as well. The increase of the computational complexity in the described approaches is calculated.

Keywords: binary transpositions, binary data distribution, security evaluation, computing costs.

1. Introduction

Nowadays, volumes of communicated information increase each year. Especially, there are more and more amounts of personal and confidential data. From the one hand, it is connected with great development of messengers and social networks, which are subject to Article 23, Paragraph 2 of the Constitution of the Russian Federation and, as a consequence, to Article 138 of the Criminal Code of Russia that guarantee protection of our private correspondence. From the other hand, there is Federal Law N 152 FL "On Personal Data" where we can see that disclosure of personal information of a citizen carries a punishment. Thus, data collection and analysis and storage systems in automated systems or web resources [1] must follow vital security conditions, and all the information contained in a file storage should be enciphered [2].

Classical methods of cryptographic transformation include substitution ciphers and transposition ciphers of both stand-alone symbols and entire blocks of an original text.

Traditional cryptographic methods where transposition is used imply movement of stand-alone symbols of a secret message (see Figure 1) or replacement of higher least significant bits in data packets. It has been proved that they have a higher level of cryptographic strength than ciphers of block substitution [3]. We can look at their classical scheme.

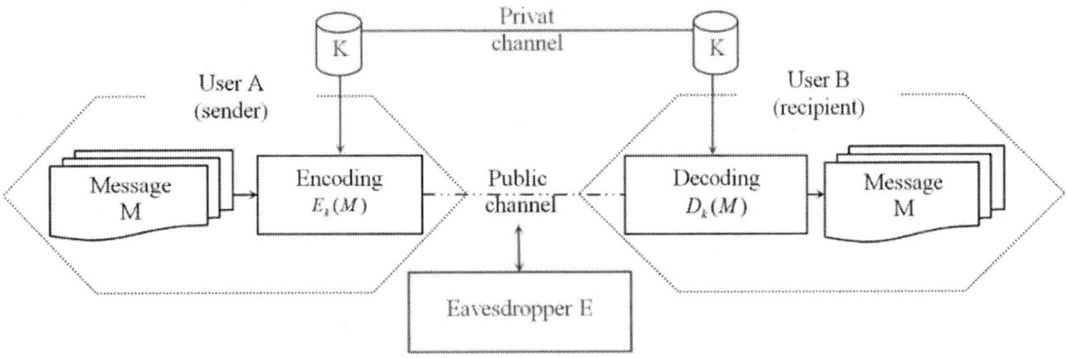

Figure 1. Message delivery with the help of transposition methods

The number of possible transpositions can be calculated according to the following formula:

$$\hat{P}(n) = \frac{n!}{k_1! \cdot k_2! \cdot \ldots \cdot k_l!} \quad (1)$$

where
n – the number of symbols of the text;
l – the number of signs in the alphabet;
k_i – the number of repetitions of the i-symbol of the alphabet;

A transposition process gives a great number of text modifications that have the same set of symbols. This process can be illustrated with the help of two examples[4]. We take a word that consists of 6 symbols (for instance, the word "приказ"). The number of combinations in this case is $6! = 720$ because there are no repeated letters in the word. However, none of these combinations (excepting "приказ" itself) has a lexical meaning in the Russian language. The next example relates to the phrase "секретное сообщение". This phrase consists of 19 symbols, some of which are repeated. It is necessary to pay attention to the fact that transpositions of the same symbols give the same words. Therefore, the number of transpositions is not $19! \approx 1{,}2 \cdot 10^{17}$, but, according to the formula (1), it is $\approx 2{,}1 \cdot 10^{14}$.

Thus, methods of transposition of stand-alone symbols of a text give plenty of senseless messages, which can be excluded according to the vocabulary. If the generated combinations represent words that have lexical meanings in the examined language, they can be sorted out according to the message context. If this is the case, it is easy to find out meanings of encrypted words and phrases [5].

The following *contradiction* appears from the above-stated facts: from the one hand, it is necessary to get a cryptographic method where symbols of an original text are evenly distributed and the number of combinations is an order of magnitude greater than transpositions of stand-alone symbols; from the other hand, it must give all possible variants of words, which will not be excluded according to vocabulary. Thus, the objective of our research, which is determined by the necessity of developing new approaches to solving the cryptographic problem, is rather topical.

In order to solve this problem, we should analyze such an approach where the plaintext message is transposition during its transmission with the help of a communication pathway. In such a case, it is necessary to pay attention to the fact that bit swapping is applied not to stand-alone blocks of the file, but to the whole file to avoid use of several sets of keys and generation of periodic repetition [6].

Thus, when a binary code is used, the length of the message increases eightfold if ASCII is applied or it increases 16 times if Unicode is applied. It is important to notice that the number of symbols in the alphabet drops to 2 (0 and 1). Later it leads to increase of repetitions of the alphabet signs themselves.

According to the formula (1), the number of combinations of the cryptotext, which has been received with the help of transposition of bits, can be calculated in the following way:

$$\hat{P}_8(n) = \frac{(8n)!}{k_0! \cdot k_1!}, \text{(ASCII)} \tag{2}$$

or

$$\hat{P}_{16}(n) = \frac{(16n)!}{k_0! \cdot k_1!}, \text{(Unicode)} \tag{3}$$

where

n – the number of the text symbols;

k_0 – the number of "0" symbols in the message code;

k_1 – the number of "1" symbols in the message code;

We can use the described approach to the above-stated examples and compare the results.

The phrase "*секретное сообщение*" from our first example has 152 binary bits in binary ASCII. In that case, the number of "0" symbols is k_0=81, and the number of "1" symbols is k_1=71.

If this occurs, the number of the ciphertext combinations according to the formula (2) can be calculated in the following way:

$$\hat{P}_8(n) = \frac{(8 \cdot 19)!}{71! \cdot 81!} = 2{,}7 \cdot 10^{44}.$$

This result $\hat{P}_8(n) = 2{,}7 \cdot 10^{44}$ is 30 orders of magnitude greater than the result $\hat{P}(n) = 2{,}1 \cdot 10^{14}$, which is received with the help of the transposition of the message symbols.

If a sixteen bit encoding is used, then the number of cryptotexts according to the formula (3) increases significantly:

$$\hat{P}_{16}(n) = \frac{(16 \cdot 6)!}{84! \cdot 220!} = 3.8 \cdot 10^{76}.$$

Thus, doubling the message capacity makes the number of combinations 32 orders of magnitude greater in comparison with $\hat{P}_8(n)$.

We can find out how much the number of transpositions decreases if the bit count swerves from the central value.

The maximum value of transpositions can be reached when the number of "0" symbols and the number of "1" symbols are equal. If we take into account our example where an eight bit encoding is used, these values are k_0=76 and k_1=76. It means that the number of transpositions can be calculated in the following way:

$$\hat{P}_8(n) = \frac{(8 \cdot 19)!}{76! \cdot 76!} = \frac{152!}{76! \cdot 76!} =$$

$$= \frac{2^{76} \cdot 152!!}{76!} = 3{,}8 \cdot 10^{44}.$$

We can pay attention to the fact that the departure of the correlation between the number of "0" symbols and the number of "1" symbols from the central value leads to a slight decrease of the number of the transposition combinations, which is less than one order of magnitude.

IOP Conf. Series: Materials Science and Engineering 1069 (2021) 012006 doi:10.1088/1757-899X/1069/1/012006

If a sixteen bit encoding is used, the result is formulated in the following way:

$$\hat{P}_{16}(n) = \frac{(16\cdot 19)!}{152!\cdot 152!} = \frac{304!}{152!\cdot 152!} =$$

$$= \frac{2^{152}\cdot 304!!}{152!} = 1.5\cdot 10^{90}$$

The difference is more significant here. It is about 14 orders of magnitude between $\hat{P}_{16}(84+220) \approx 3.8\cdot 10^{76}$ and $\hat{P}_{16}(152+150) \approx 1.5\cdot 10^{90}$.

We are going to examine an issue of changing the transpositions number when the symmetry of distribution deviates from the uniform value.

Bits can be grouped together in bytes at random. Therefore, any words of the used language that consist of the same set of bits can be formed. These combinations are not excluded according to the vocabulary. In that case, it is necessary to use an exhaustive search. Supercomputers are applied to sort out such great amounts of combinations.

There is a necessity to find a criterion of minimum computing costs. When this criterion is overcome with the help of the encryption algorithm, an eavesdropper will not be able to decode the message if he does not know the key.

At the moment, the most powerful supercomputer in the world is Summit (the USA) [7]. (Its productive capacities are 200 PFLOPS = $200\cdot 10^{15}$ operations per second). Let us suppose that one operation is a check of one combination (according to the most optimistic forecast). The necessary time that is spent on sorting out the variants of the above-stated example can be calculated in the following way:

$$T_{16}(n) = \frac{64{,}35\cdot 10^{27}}{200\cdot 10^{15}} = 3{,}22\cdot 10^{10}\,\text{сек}\,.$$

We know that a year consists of 31,6 million seconds. Thus, it will take more than 1000 years to finish the described process.

The criterion "*cryptographic strength over time*" is the strength level of the algorithm where the time that is necessary to sort out the variants exceeds the age of the Universe (the age of the Universe is 13,7 ± 0,2 billion years; it is a bit less than $1{,}4\cdot 10^{10}\,\text{лет} = 4{,}42\cdot 10^{17}\,\text{сек}$ ($1{,}4\cdot 10^{10}$ years = $4{,}42\cdot 10^{17}$ seconds) [8]).

The general computation power of all computing systems of the world is approaching to 10^{20} FLOPS [9]. It imposes an additional requirement on the level of the cryptographic strength of the algorithm [10]. Thus, if all computers of the world are attacked with a "Trojan horse" virus and later are used to look over all the transposition combinations, it must take so much time as at least the current age of the Universe is. As a sequence, the number of transpositions that are made with the help of this cryptographic algorithm should not be less than $4{,}42\cdot 10^{37}$.

In order to solve this problem, it is possible to use two directions of the research. They are an increase of the text length and an increase of the length of binary representation of the letters [11]. We are going to examine these two directions separately.

2. An increase of the text length.

The purpose of the method is to find the minimum text length that is enough to clear the threshold of *cryptographic strength over time.*

We should compute the number of transpositions in its binary representation. It can be calculated with the help of the formula (4):

$$P(n) = \frac{(k_0 + k_1)!}{k_0! \cdot k_1!} > 4{,}42 \cdot 10^{37} \quad (4)$$

According to the above-stated facts, $P(n) = \frac{(k_0 + k_1)!}{k_0! \cdot k_1!} > 4{,}42 \cdot 10^{37}$

The covered dependence (4) is portrayed in Figure 2.

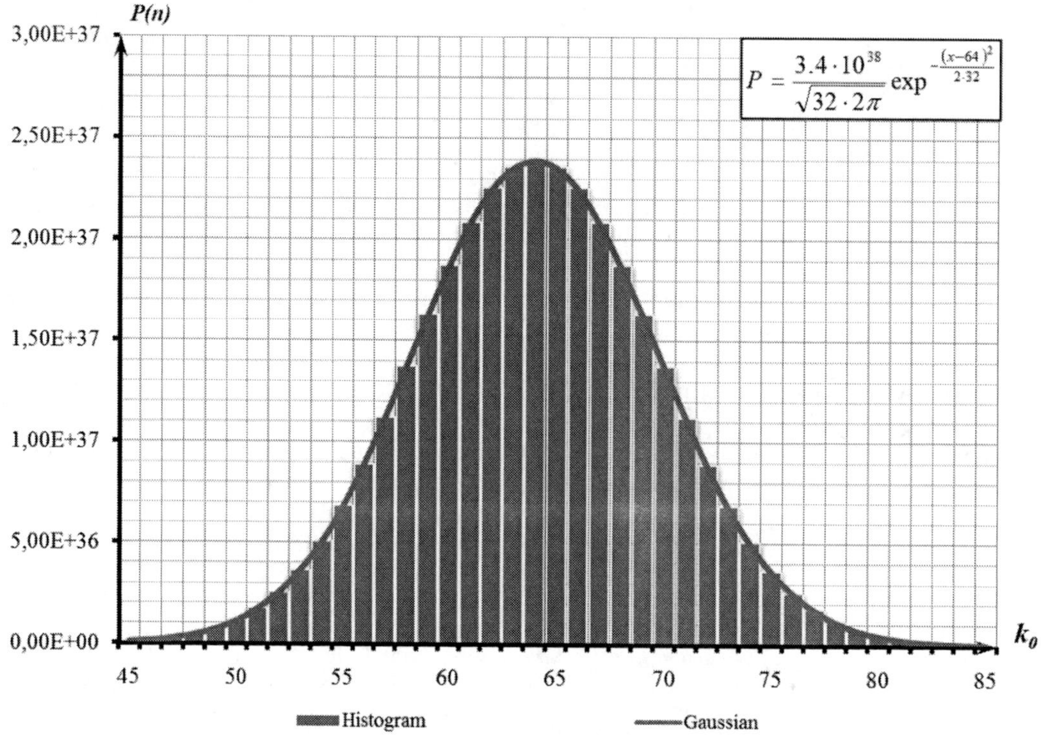

Figure 2. The histogram of the transpositions number for a word that consists of 16 letters when an eight bit encoding is used

When the text length n is fixed, the maximum value of the transpositions can be reached if $k_0 = \cdot k_1$.

It is necessary to get the lower bound of the number of n symbols in the original message. Thus, we have:

$$k_0 = \cdot k_1 \Rightarrow k_0 + \cdot k_1 = 2k \quad (5)$$

It means that the condition (4) can be written in the following way:

$\max\left(\frac{(k_0 + k_1)!}{k_0! \cdot k_1!}\right) = \frac{(k + k)!}{k! \cdot k!} = \frac{(2k)!}{(k!)^2} > 4{,}42 \cdot 10^{37}$ We should get limitations on the number $n_{\min} = \frac{k}{4}$ from the last condition and the condition (5):

$$\frac{(2k)!}{(k!)^2} = \frac{2^k \cdot (2k-1)!!}{k!} = 2^k \cdot \prod_{i=1}^{k} \frac{2 \cdot i - 1}{i},$$

After evaluation of the resulting expression, we have:

$$\frac{2^{2k-1}}{k} < 2^k \cdot \prod_{i=1}^{k} \frac{2 \cdot i - 1}{i} < 2^{2k}$$

According to the fact that $2^{125} < 4{,}42 \cdot 10^{37} < 2^{126}$, we receive the limitation $k \geq 65$.

Taking into account the condition about an integrality of the length of the bit sequence *n*, we can see that the threshold of "*cryptographic strength over time*" can be cleared when the length of the original message is more than 16 symbols if an eight bit encoding is used. If a sixteen bit encoding is used, the length of the original message should be more than 8 symbols respectively. It lets us use this method to keep even short passwords.

The next stage should be dedicated to the analysis of influence of the text length enlargement on the amount of time that is spent to look over the text [12]. We are going to find out how much the number of operations increases if the text is increased by 1 bit. Let us say that the number of "0" symbols has grown in the text encoding to raise certainty. In such a case, we have:

$$L = \frac{P(n+1)}{P(n)} = \frac{\frac{(n+1)!}{(k_0+1)! \cdot k_1!}}{\frac{n!}{k_0! \cdot k_1!}} = \frac{(k_0+k_1+1)!}{(k_0+k_1)!} \cdot \frac{k_0! \cdot k_1!}{(k_0+1)! \cdot k_1!} = \frac{k_0+k_1+1}{k_0+1} = 1 + \frac{k_1}{k_0+1}. \quad (6)$$

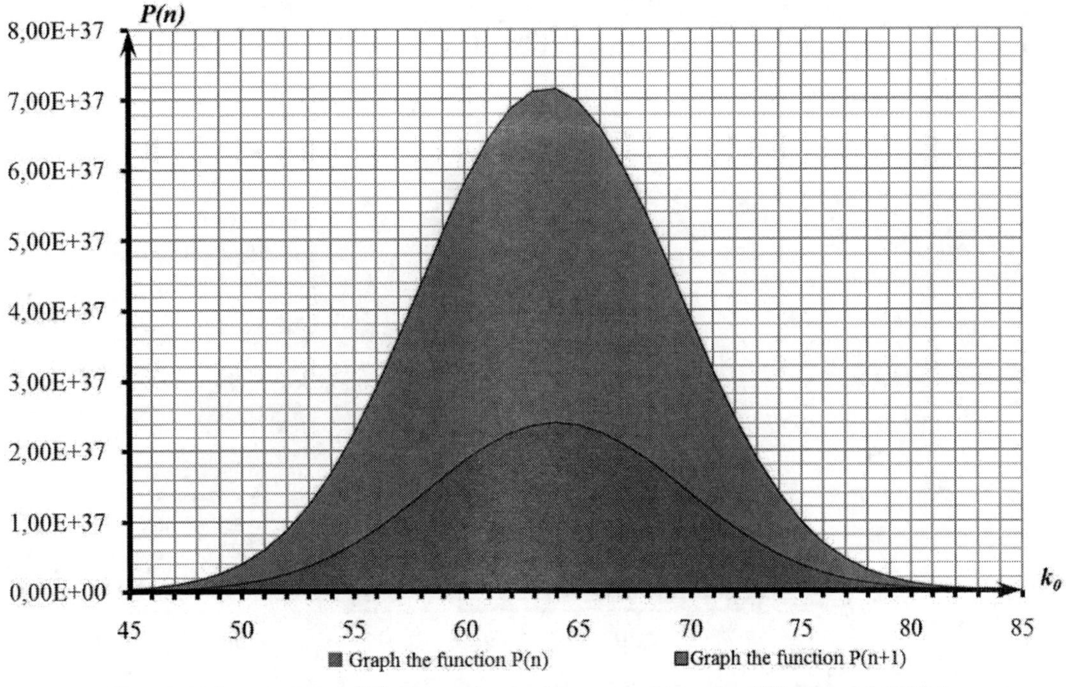

Figure 3. Comparison of the histograms of transpositions for 128 bits and 129 bits

Thus, we can see that the complexity increases in $L = 1 + \frac{k_1}{k_0 + 1}$ and depends on the asymmetry level of distribution of binary "0" and "1" in the text encoding and $L \in [1; n+1]$.

We should pay attention to the fact that the L index that is used in the formula (6) does not depend on a type of the chosen text encoding, but it depends on the deviation of k_0 and k_1 from the central value. This regularity is portrayed in Figure 4.

3.An increase of the length of the symbol encoding.

It is necessary to analyze the second issue, the purpose of which is to find out how the capacity increase influences the rise of complexity of computations [13].

Let us say that n – the number of the text symbols,

k_0 – the number of "0" symbols in the message code;

k_1 – the number of "1" symbols in the message code.

If this is the case, we can use the formula (2) to calculate the number of transpositions in binary representation of the text, the length of which is n symbols and it is expressed with the help of an eight bit encoding. It is, of course, meant in the formula (2) that $k_0 + \cdot k_1 = 8n$.

From the other hand, we can transform the formula (3) in order to calculate the number of transpositions in the same text, which is expressed with the help of a sixteen bit encoding:

$$P_{16}(n) = \frac{(16n)!}{(K_0)! \cdot (K_1)!} \cong \frac{(2 \cdot 8 \cdot n)!}{(2k_0)! \cdot (2k_1)!}, \tag{7}$$

where $K_0 + \cdot K_1 = 16n$.

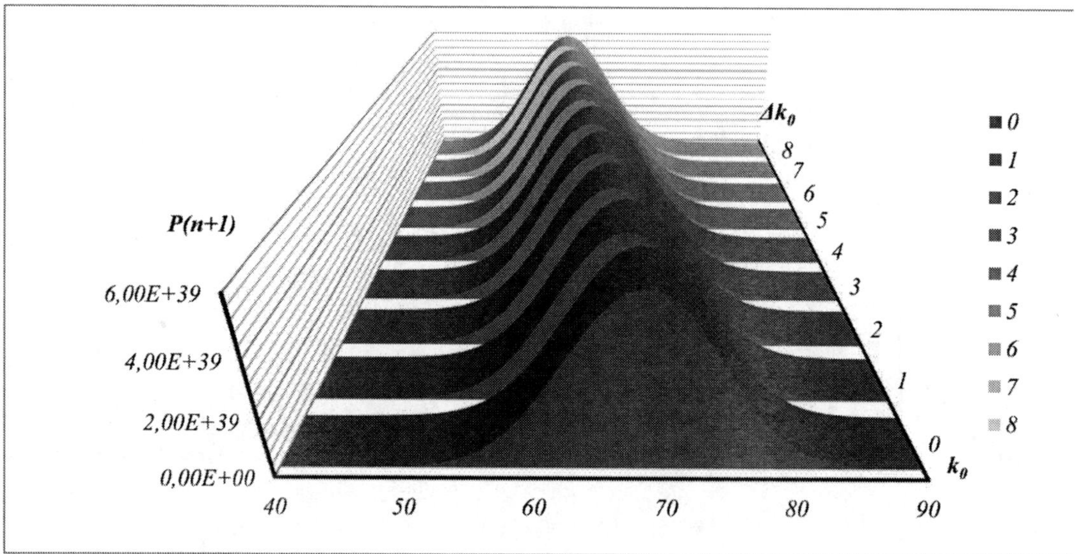

Figure 4. Comparison between the histograms of different transpositions for 129 bits

In order to find out how much the computing costs increase when the length of the text encoding is enlarged, we can divide the formula (2) by the formula (7):

$$\frac{P_{16}(n)}{P_8(n)} = \frac{\dfrac{(2\cdot 8\cdot n)!}{(2k_0)!\cdot(2k_1)!}}{\dfrac{(8\cdot n)!}{(k_0)!\cdot(k_1)!}} = \frac{(2\cdot 8\cdot n)!}{(8\cdot n)!}\cdot\frac{(k_0)!\cdot(k_1)!}{(2k_0)!\cdot(2k_1)!} =$$
$$= \frac{(16n-1)!!\cdot 2^{8n}}{2^{k_0}(2k_0-1)!!\cdot(2k_1-1)!!\cdot 2^{k_1}} \Rightarrow \frac{P_{16}(n)}{P_8(n)} = \frac{(16n-1)!!}{(2k_0-1)!!\cdot(2k_1-1)!!}; \qquad (8)$$

Thus, we have received the final formula that can be used to calculate the increase of the complexity of computations in order to look over the encoding combinations when a conversion key is changed from 8 bits to 16 bits (for example, when we face a switch from ASCII to UniCode).

This article has dealt with the analysis of different aspects of use of the binary transposition approach depending on the encoding method (8) of the text and its length. It has been shown that this approach produces several orders of magnitude of bit combinations more than traditional transposition methods do[14]. However, the method is especially effective when the text length is bigger than 65 bits. We have developed the criterion (6) that determines possible ways of increasing the cryptographic strength of the algorithm by adding the definite bits into the text. The prospects and advantages of this approach are described [15]. One of them is reducing possibilities of sorting out the deciphered parts of the text according to the context or vocabulary if words of a natural language are used. Another important aspect is impossibility to find a bit while encryption access passwords or keys that are generated with the use of pseudorandom sequences, even if an eavesdropper has a part of the plain text[16].

References

[1] Loik R 2016 Integration Trends of EU Internal Security and Law Enforcement: How Legal, Technological and Operational Advancements Matter Baltic Journal of European Studies 6(2) pp 3-27.

[2] Ho Ngoc Duy 2012 Algorithms of information processing in automated systems of an electronic document flow *Thesis for a Candidate Degree in Technical Sciences* p161.

[3] Singh S and Kleist V 2005 The Code Book: The Science of Secrecy from Ancient Egypt to Quantum Cryptography *IEEE Annals of the History of Computing* 24(2) pp 97-98.

[4] Shaska T and Ustimenkob V 2009 On the homogeneous algebraic graphs of large girthand their applications *Linear Algebra and its Applications* pp 1826-1837.

[5] Fix A S 2014 The history of cryptography and new ethic status of information *Online scientific journal "Philosophical problems of IT and Cyberspace"* (Volume 2 Issue 8) pp 65-76.

[6] Kant S, Verma N 2000 An effective source recognition algorithm: extraction of significant binary words *Pattern Recognition Letters* 21(11) pp 981-988.

[7] Abramov S M 2019 June 2019: Analysis of the Development of the Supercomputer Industry in Russia and Worldwide *Program Systems: Theory and Applications* 10(3) p 3-40

[8] Avelino A and Kirshner R 2016 the dimensionless age of the universe: a riddle for our time *The Astrophysical Journal* 828(1) 35
[9] Levin V K 2014 Supercomputer trends *Computational nanotechnology: The scientific journal concerning: Mathematics, Physics, Economics and economic sciences, Automation. Computer engineering* (Publishing house Yur-VAK) pp 35-38
[10] Moldovyan A A 2005 Synthesis methodology of algorithms of information security in computer systems based on controlled substitution-permutation networks *Abstract of a thesis for a Doctoral Degree in Technical Sciences* (Saint Petersburg: Printing office VAS) p 44.
[11] Laskari E C, Meletiouc G C, Stamatiou Y C, Vrahatis M N 2005 Evolutionary computation based cryptanalysis: A first study *Nonlinear Analysis: Theory, Methods & Applications* 63(5-7) pp 823-830.
[12] Doshina A D,. Mikhaylova A E , Karlova V V 2015 Cryptography. Basic methods and problems. Modern cryptography trends *Modern trends of technical sciences: proceedings of the IV International scientific conference* (Kazan: Buk) pp 10-13
[13] Chernyshev Yu O, Dubrov E O, Ryazanov A N, Sergeev A S 2013 Development of methods of cryptanalysis of transposition and substitution ciphers in information security systems on the basis of evolutionary and optimization methods *Reds: telecommunication devices and systems* Volume: 3 Issue: 3 (Moscow: Publisher: Russian scientific and technical society of radio engineering, electronics and communication of name A.S. Popov) pp 221-225.
[14] Laskari E C, Meletiou G C, Stamatiou Y C , Vrahatis M N 2007 Applying evolutionary computation methods for the cryptanalysis of Feistel ciphers *Applied Mathematics and Computation* 184 (1 SPEC. ISS.) pp 63-72
[15] Sotov L S 2011 The hardware implementation of permutation instructions of bits of the data *Heteromagnetic microelectronics* Issue: 10 (Saratov: OAO Research institute "Tantal") pp 25-50.
[16] Chávez M L 2005 Security services in fieldbuses: at what cost? *IFAC Proceedings Volumes (Electronic Materials* Volume 38 Issue 2) pp 166-171

Cloud processing of security events for VANET distributed sensors

I P Bolodurina[1], Y A Ushakov[1], D I Parfenov[1], J Cui[2]

[1]Orenburg State University, Orenburg, Pobedy str., 13, 460018, Russia
[2] Anhui University, Anhui, Hefei, 12487, China

E-mail unpk@mail.ru

Abstract. Machine-to-machine communication networks, which are related to human activities, are today one of the fastest growing IT sectors. The most vulnerable in terms of security is the Internet of Things and VANET segments. When devices interact, various types of targeted attacks are possible, as well as failures or random dangerous events. The handling and response to security events, in this case, must be fast. However, the specificity of VANET networks does not allow the use of classical intrusion prevention systems due to the constant movement between connection points, for example, in LTE or 5G networks. The article is devoted to the possibility of quickly deploying a cluster of security events analysis based on containerization on existing virtual machines that are closest to the event source. The analysis of the scaling speed, as well as the performance limits of event processing processes on various configurations of the data analysis cluster.

Keywords: cluster, Hadoop, Spark, big data, security.

1. Introduction

Around the world, the number of smart devices is increasing every year. The interaction of such network devices generates significant amounts of data transmitted. With the increase in the number of devices the volume of transmitted data the complexity of control over end devices are grows. As a consequence, the process of administering such heterogeneous networks requires more and more resources to monitor, identify, and prevent possible destructive information impacts on them. In this regard, the vehicular ad-hoc networks (VANETs) are the most vulnerable.

In VANET networks, data is transmitted both between vehicles (Vehicle To Vehicle, V2V) and between vehicles and base stations of the 5G network (Vehicle To Road), when vehicles move in their coverage area. The main purposes of such networks are road safety, assisted driving, distribution of alerts and warnings, navigation, provision of integrated road services. In this way, VANETs carry critical information that directly affects road safety [1]. Therefore, the development of methods to ensure confidentiality, integrity, and availability of data, as well as the design of means of protecting the data transmission medium in VANET networks is an important area of research.

The amount of data that needs to be processed and analyzed grows exponentially with the number of connected devices. This is especially true for data for security and attack detection. At the present stage of technology development, tools have appeared that allow storing, processing, and analyzing large amounts of data, and approaches to analysis have also begun to change.

The most popular big data analytics solutions are clusters and cloud structures [2]. Existing solutions for creating cloud systems, such as cloud resource management systems OpenNebula, OpenStack, CloudStack, Doku, etc. are universal, they do not take into account the specifics of working with big data - the possibility of parallelizing the network (alternative routing and switching), a large amount of memory consumed and calculations per cycle of analysis. This leads to the impossibility of deploying the structures responsible for processing big data in an arbitrary cloud.

Providers provide out-of-the-box cloud toolkits in BigData-as-a-Service format. In most cases, this is a Hadoop / Spark cluster with storage on HDFS / Hive / Gluster, NoSQL services like Redis / MongoDB [3].

There are several problems associated with the use of distributed computing for processing big data. Consumers often lack computing resources for big data processing. It is also necessary to ensure the planning of processes and flows serving individual parts of the algorithm or data structure, to plan the placement and transfer of data between nodes with isolated networks. Also, there is a need to predict the amount of computation, execution time, consumed processor, and memory resources when executing distributed applications and their subtasks. To efficiently use the resources of dedicated virtual machines, it is necessary to ensure the selection of the optimal pool of virtual machines for the tasks to be solved for processing big data. Also, there is the problem of isolating the work and deploying distributed applications in multi-user environments of public computing service providers - it is necessary to ensure the security of the calculations performed and the data being processed in an environment where tasks of different users can run on the same virtual or physical machine.

2. Review of big data processing methods

The most flexible solution when creating a cloud infrastructure is to use a platform that supports IaaS technology.

Cloud systems can plan computations at the level of virtual machines. With the advent of the era of cheap servers, the Apache Hadoop project began to develop, which made it possible to build clusters for storage and processing of data. But with the proliferation of cloud computing and the increase in data volumes to hundreds of terabytes, Hadoop's lack of flexibility in dealing with random sampling has led to the increased popularity of alternative methods. Apache Spark can use many components of the Hadoop cluster, but at the same time, it works much more efficiently with the analysis of arbitrary data samples, especially in distributed storage [4]. Also, a direct analysis of big data is available in neural network frameworks, which have also begun to integrate with Spark [5]. There is also a native Python Dask solution for distributed analysis, but the flexibility of the Spark cluster and full compatibility with the existing storage system, the ability to work with different schedules, and worker nodes makes it more versatile.

Not so long ago, elements of computing on GPU accelerators were added to common cluster platforms. Hadoop and Spark began to be used in various projects. An analyst does not have to write a new Hadoop program every time. Some extensions allow you to formulate queries in a simpler language, for example, Apache, Hive (queries in SQL language). Spark has a flexible programming model that allows you to store data in RAM for multiple processing [6].

For network operation, Hadoop uses the Master-Slave model, and there is work to optimize data transmission using software-defined networks [7]. For correct operation, it is necessary to have direct access from Slave to Master, which is not always possible to organize immediately.

Software-defined networks are actively used to solve various problems arising in cloud platforms: migration of virtual machines between different data centers [8], data flow control in parallel DBMS, data flow control in frameworks for the development of distributed scientific applications (Hadoop MapReduce, Hadoop -OFE, Orchestra, Coflow, OpenMPI), building locally managed MAC addresses for virtualizing the physical infrastructure of the data center [9], power management of network devices in the data center [10], etc.

Algorithms for routing data flows in software-defined networks when choosing paths do not take into account the need to provide QoS parameters for the laid and previously installed data flows [11]. The solution to this problem is important since requires proactive routing of data flows between nodes (containers, virtual machines) executing individual processes of neural network applications.

The main advantage of containerization technology compared to virtualization is the reduction of overhead. To organize the operation of full-fledged virtual machines, full emulation of the computer hardware by the hypervisor (for example, KVM, Xen, Hyper-V) is required, each virtual machine includes a complete copy of the guest OS. In the case of using container technology (for example, Linux-VServer, OpenVZ, LXC, Docker), there is a single common kernel of the computer OS, on top of which lightweight containers with processes of running programs run, there is no need to emulate the hardware - the physical means of the server are used [12].

The most important advantage of containers is the ease of deployment of applications and services, taking into account all the dependent components, files, and libraries [13]. Also, containers are launched much faster compared to virtual machines, because there is no need to load the OS, you just need to run the necessary applications [14].

As part of this work, it is planned to launch big data cloud applications in a set of containers mapped to the physical nodes of the data center.

3. Security events data collection and analysis scheme

The main requirements for the selection of tools are the ability to work in distributed systems and support for streaming data processing. Since a large amount of heterogeneous data is generated in real-time, which needs to be brought to a common form, processed and analyzed, the selected tools must support the ability to work with streaming data. The sources of data flows are authentication systems, active network equipment, IDS / ISP, server event logs, antiviruses, vulnerability scanners, and other information security protection and control systems.

The distributed security data processing system consists of two dependent parts:

- receiving and recording data;
- data processing and response.

Since event processing is required all the time, the delay between the arrival of data in storage and processing must meet the requirements for these events. To save history and restore the history of events, to create a description of incidents, and identify threats with a long pre/aftereffect, it is necessary to store a big data for a long time. The data entry scheme is shown in Figure 1.

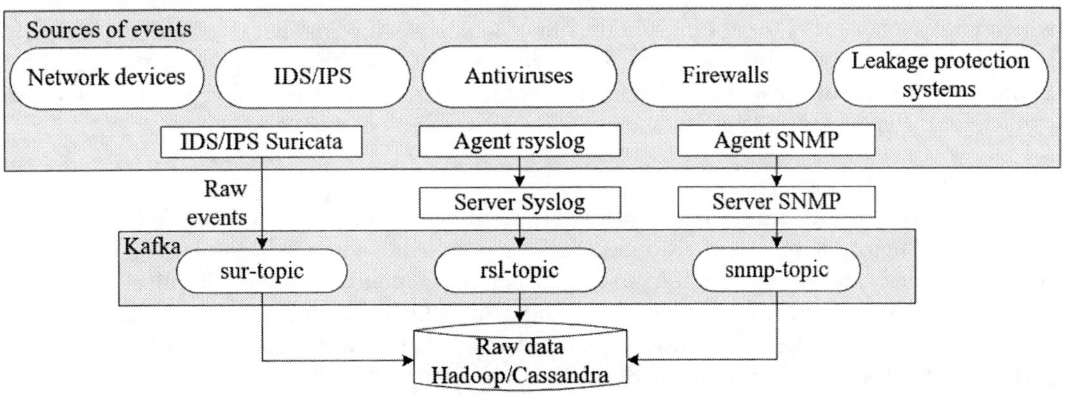

Figure 1. Security event data collection scheme.

Instead of Apache Kafka, you can also use Spark Streaming, which, according to tests, works more stable under high loads, but has its characteristics. The Hadoop NameNode is located where the node can be accessed from the outside; the node can be used for running internal Hadoop clients and for

IOP Conf. Series: Materials Science and Engineering 1069 (2021) 012007 doi:10.1088/1757-899X/1069/1/012007

external Hadoop services. To interact with external consumers directly, Hadoop uses the forwarded HDFS and WebHDFS ports, as well as the mode of collecting cluster addresses via DNS.

Reading data consists of the following steps (Figure 2):

- Request to the application, which will be accompanied by data reading
- Reverse proxying of the request to the container.
- Request for DataNode addresses with the required information
- Return the list of addresses.

Addressing the required address, reading information.

This architecture works transparently through the Swarm or K8n cluster overlay network and transmits data encrypted using configured IPSEC mechanisms.

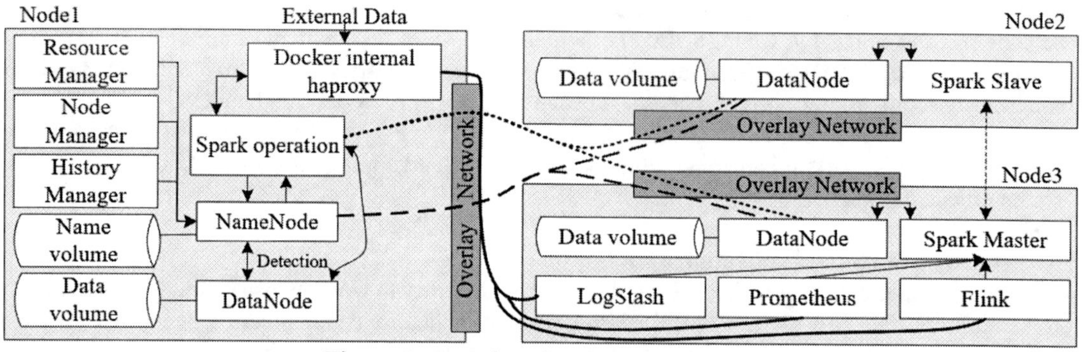

Figure 2. Container interaction scheme.

To use external processing systems you need to run Master and several Slaves. Their connection will be carried out according to the specified Master service name through the DNS of Docker.

Spark Streaming, an extension of the Apache Spark API, splits streams into small batches of time intervals. This mechanism is called DStream. DStream micro-packages contain resilient distributed datasets (RDDs). RDD sets support transformation and action operations. A feature of transformation operations is deferred execution: Spark remembers which transformation and on what data should be performed, and the result is returned when the action is called. DataSets can be used in place of RDDs. When using a DataSet, before performing the transformations, the optimizer is triggered to determine a more efficient way to obtain the result. Working with it occurs mainly using SQL queries [15].

The disadvantages of Spark include variable load on the network during data loading and processing, which makes it necessary to set limits on the input stream density. Another disadvantage of Spark is the lack of the ability to recover clusters from failures. To solve these problems, the Apache Kafka framework will additionally be used, which is capable of creating real-time data pipelines and streaming applications. Replicated and persistent storage reduces the risk of data loss.

4. Organization of data transfer and storage

Raw data enters the system from network devices, antivirus, firewalls, firewalls, etc. To receive data from these systems, there is Kafka Connect, a tool for scalable and reliable streaming of data between Apache Kafka and other systems. Kafka Connect can accept entire databases or collect metrics from all application servers into Kafka partitions, making data available for low latency streaming processing. Since the sources are different, a different Kafka section is created for each.

Events are handled by the Spark Streaming application. The hierarchical structure of events is determined by the way the event is received and its features. There are three large classes of events: events received using SNMP, events received using Syslog and other events. Each class can be subdivided into smaller subclasses; subclasses can be subdivided as well.

It becomes necessary to draw up rules according to which the correlation analysis will be carried out. In the system under development, it is planned to use only two types: templates and queries. A template is a set of rules with some fields that are filled with data from an event. The main difference between a request and a template is working with data streams. They allow you to combine, group, filter, deduplicate, sort, run events through sliding windows. The template and queries can specify the actions that the system should perform when the data matches the template. The simplest response is to alert the administrator of suspicious activity.

The data processing architecture is shown in Figure 3. Information from sources that do not cover SNMP and Syslog are collected using Kafka and fed to the data collection server. All received raw data is stored in the database, from where it is retrieved by the Spark application. It converts, filters, and aggregates events, and then performs correlation analysis. Correlation rules are taken from a separate store. All obtained results are saved in a separate database. If incidents are detected, a message is sent to the response module, which notifies the administrator.

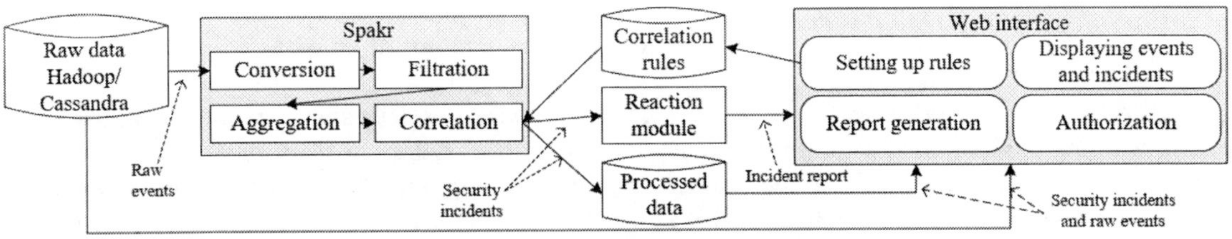

Figure 3. Data processing scheme.

Since data processing also includes notification of detected patterns and critical events, the diagram shows some of the modules for interacting with the interface.

5. Experiment with handling events in a container-based cluster

A Docker-based cluster was created for the experiment. Virtual machines in a private cloud based on OpenNebula are used as nodes; the nodes are connected by a 10Gbps network, and have common network storage with a separate connection to each node. Also, each node has local storage that is used to store data. To generate events, Metasploit attack generation, the method of mirroring traffic through SPAN, detection on IPS Suricata, and redirection to the cluster are used (Figure 4).

To test the scaling speed of a Docker-based cluster, all nodes use a dedicated repository located on the central node. This repository is used to download images to launch cluster members and is used to speed up container deployment. Also, the nodes have the usual variants of the cluster elements (Spark, Hadoop, Kafka).

The experiment consists of measuring the speed of automatic scaling of the cluster, as well as comparing the speed of Docker mode and the standard model of the cluster. In the absence of events, the Spark cluster can be reduced to one working container, but when events arrive, it is possible to scale up to the required number if the nodes with the operating system are already running.

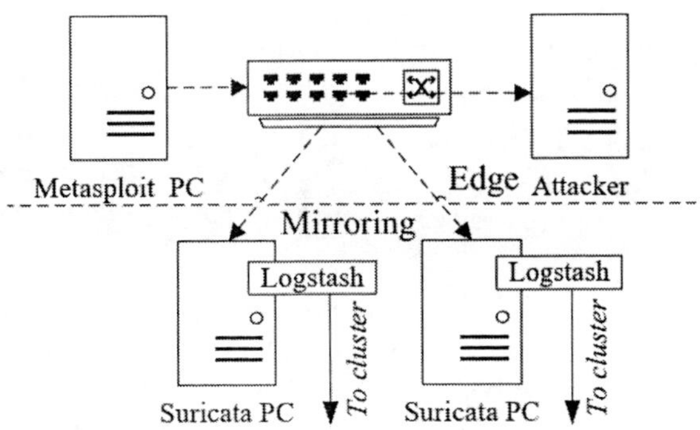

Figure 4. Generating events scheme.

The start times for a different number of Spark worker containers are shown in Figure 5. The time is estimated from receiving a start command to registering with the Spark Master control process. Up to 4 Spark worker processes will run on one physical node.

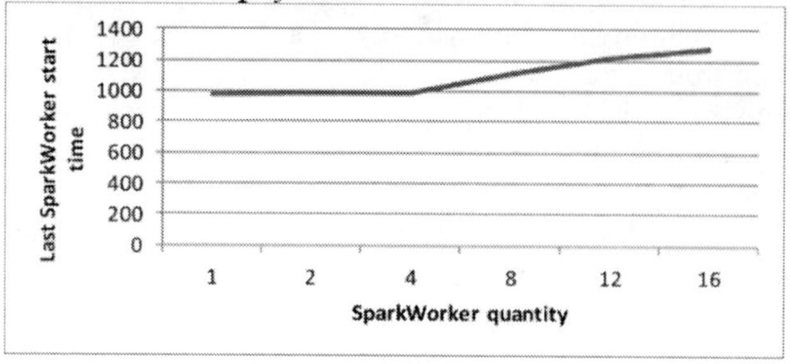

Figure 5. Scaling delay.

As you can see from the graph, the launch speed depends on the number of working containers launched but is within acceptable limits.

To estimate the recording rate and the rate of event processing, let's start generating events, as well as processing them in the cluster. Since latency is important in handling security events, the scheduler latency and event latency will be measured. As processing, aggregation, and grouping of events are used, which refers to the time window. This method simultaneously loads both the storage system and the kernel. Figure 6 shows the processing latency for a maximum number of containers and an inbound stream of events with the exponential time between events at a rate of 850 events / s.

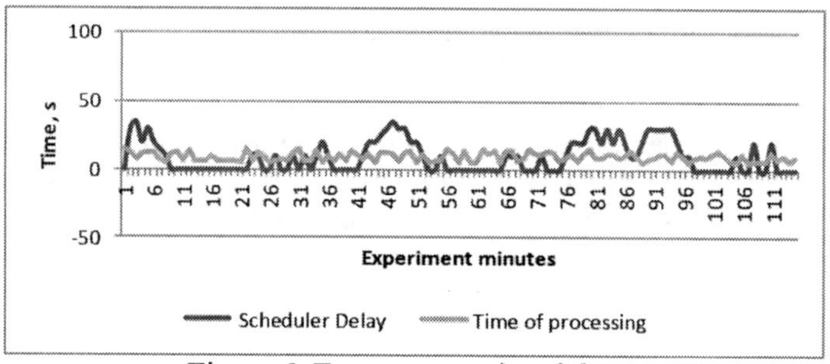

Figure 6. Event processing delays.

As a result of measurements, it can be seen that the failure of individual processes does not affect the overall processing process, increasing the processing time for individual jobs due to the redistribution of the job and waiting for a response from the old process. To estimate the maximum rate of event processing, let's measure the processing acceleration at various values of the number of worker processes.

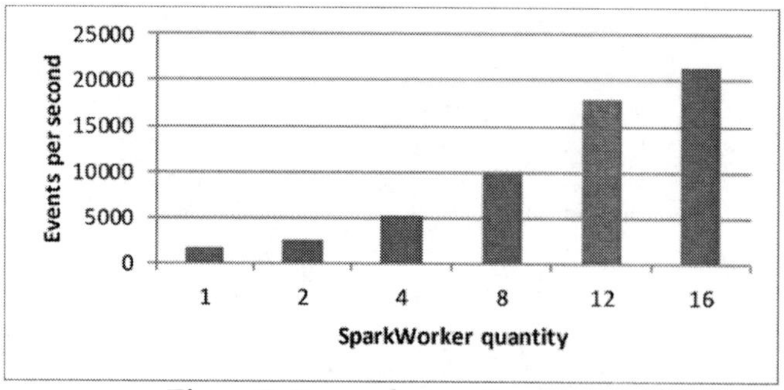

Figure 7. Processing performance.

Graph on Figure 7 shows non-linear acceleration, which is limited by the speed of concurrent storage access.

6. Conclusion

As a result of the research, a system for rapid deployment of a security event processing cluster was created. The use of containerization allowed us to create a system of fast scaling. The measurement result showed the efficiency of this approach, and measurements of its performance and delays allow us to conclude that it applies to the analysis of a security event.

Acknowledgments

The study was carried out with the financial support of the RFBR in the framework of scientific projects No. 18-07-01446, No. 20-57-53019, as well as a grant from the President of the Russian Federation Federations for state support of young Russian scientists - candidates of sciences (MK-860.2019.9) and by the grant of the President of the Russian Federation for state support of leading scientific schools of the Russian Federation (NSh-2502.2020.9).

References

[1] Cooper C, Franklin D, Ros M, Safaei F & Abolhasan M 2016 A comparative survey of VANET clustering techniques *IEEE Communications Surveys & Tutorials* **19(1)** 657-81
[2] Choi TM, Wallace SW & Wang Y. 2018 Big data analytics in operations management *Production and Operations Management* **27(10)** 1868-83
[3] Bhattacharjee A, Barve Y, Khare S, Bao S, Kang Z, Gokhale A & Damiano T 2019 Stratum: A bigdata-as-a-service for lifecycle management of iot analytics applications *2019 IEEE International Conference on Big Data* 1607-12
[4] Erraissi A, Belangour A & Tragha A 2017 Digging into Hadoop-based big data architectures *International Journal of Computer Science Issues* **14(6)** 52-9
[5] Johnsirani Venkatesan N, Nam CS & Shin DR 2019 Deep learning frameworks on apache spark: a review *IETE Technical Review* **36(2)** 164-77
[6] Mavridis I & Karatza H 2017 Performance evaluation of cloud-based log file analysis with Apache Hadoop and Apache Spark *Journal of Systems and Software* **125** 133-51
[7] Zhao S and Medhi D 2017 Application-Aware Network Design for Hadoop MapReduce Optimization Using Software-Defined Networking *IEEE Transactions on Network and Service Management* **14(4)** 804-16
[8] Boughzala B, Ben Ali R, Lemay M, Lemieux Y & Cherkaoui O 2011 OpenFlow supporting inter-domain virtual machine migration *Eighth International Conference on Wireless and Optical Communications Networks* 1-7
[9] Matias J, Jacob E, Sanchez D & Demchenko Y 2011 An OpenFlow Based Network Virtualization Framework for the Cloud *IEEE Third International Conference on Cloud Computing Technology and Science* 672-8
[10] Heller B, Seetharaman S, Mahadevan P, Yiakoumis Y, Sharma P, Banerjee S & McKeown N 2010 ElasticTree: Saving Energy in Data Center Networks *Proceedings of the 7th USENIX conference on Networked systems design and implementation* 1–16
[11] Ibanez G, Naous J, Rojas E, Rivera D, Schuymer BD & Dietz T 2011 A Small Data Center Network of ARP-Path Bridges made of Openflow Switches *The 36th IEEE Conference on Local Computer Networks*
[12] Zeng H, Wang B, Deng W & Zhang W 2017 Measurement and evaluation for docker container networking *2017 International Conference on Cyber-Enabled Distributed Computing and Knowledge Discovery* 105-8
[13] Zheng C & Thain D 2015 Integrating Containers into Workflows: A Case Study Using Makeflow, Work Queue, and Docker *VTDC '15: Proceedings of the 8th International Workshop on Virtualization Technologies in Distributed Computing* 31-8

[14] Huber N, von Quast M, Hauck M & Kounev S 2011 Evaluating and Modeling Virtualization Performance Overhead for Cloud Environments *CLOSER* 563-73.
[15] Nabi Z 2016 DStreams: Real-Time RDDs *Pro Spark Streaming* 29-50.

Approaches to the application of homomorphic encryption in sensor networks

L K Babenko, E A Tolomanenko

Southern Federal University, Taganrog, Chekhov str., 2, 347922, Russia

E-mail kat.tea@mail.ru

Abstract. This paper describes the basic principles of the sensor network. The article characterizes the most common types of attacks on wireless sensor networks. The most commonly used protocols of sensor networks are presented. Special attention is paid to schemes of partially homomorphic encryption and fully homomorphic encryption schemes. An approach to the implementation of sensor network protection using the joint use of a fully homomorphic encryption algorithm and a symmetric block encryption algorithm is proposed.

Keywords: sensor network, attacks on a sensor network, cryptography, homomorphic encryption

1. Introduction

The development of IT technologies is gaining momentum over time, cyber-physical systems are used almost everywhere: security systems, smart home systems, diagnostics of various equipment in production, collection of environmental indicators, healthcare, the military-industrial complex, etc. Wireless sensor networks are often the obvious basis for such systems. Sensor networks are a wireless network distributed in space, which consists of many small sensor devices - sensors. Sensors have the capabilities and functionality to receive and transmit information. All distributed sensors are used to record and track readings from their environment, or to collect all received data in one place. For example, touch sensors can collect information about environmental conditions, including temperature, humidity, noise level, pollution, wind speed, etc.

However, wireless sensor networks have a number of advantages and disadvantages. When it comes to value, the sensors are fault tolerant, scalable, mobile, able to withstand harsh environmental conditions, and are easy to use. Within the range of this network, you can access the device at any time. The disadvantages of this network are the low power consumption of the devices, low performance due to the small size of each sensor, and also the small amount of memory. Common methods of protecting computer networks are not suitable for protecting wireless sensor networks, so all this leads to the development of special approaches for protecting sensor networks that will be relevant specifically for them, for example, using homomorphic encryption schemes.

An example of building a sensor network is shown in Fig. 1. The nodes of this network are low-cost from an energy point of view, but at the same time this is the multifunctional sensors. Also, sensor networks often include base stations that receive information from a group of sensors associated with it.

IOP Conf. Series: Materials Science and Engineering 1069 (2021) 012008 doi:10.1088/1757-899X/1069/1/012008

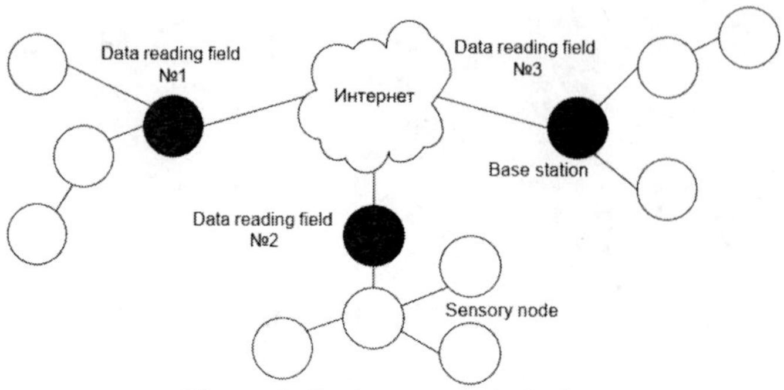

Figure 1. Sensor network device

The most famous protocols on the basis of which sensor networks are capable of functioning:

– MQTT is an open data exchange protocol in conditions of a small amount of data transferred and limited bandwidth;

– AMQP - used to exchange data at the application level, between system components through a broker that routes packets;

– MAVLink is a protocol that describes the interaction between ground systems and components of the network - various small unmanned aerial vehicles, ground vehicles, water vehicles, and drones.

2. Attacks on sensor networks

Sensor networks have a number of problems related to the security of the information they transmit. These problems arise due to the disadvantages listed above, which are the weak points of these networks. Since any attacker knows in advance that the nodes of the sensor network collect and transmit any information to each other, he can make attempts to invade such a network.

The most famous attacks on sensor networks are the following:

1. Denial of Service – Bringing a system to failure by transmitting a huge amount of data requests. Thus, the entire attacked transmission channel will be loaded, the nodes will become unavailable, and the resources of the sensors exposed to the attack will be depleted [1].
2. Creation of a fake route – Attackers register a new sensor in the sensor network implementing this attack, which is beneficial for surrounding sensors using fake routing data [2].
3. Creation of wormholes - In this attack, attackers work together to create a new fake node and transmit this data to the destination over a high-speed connection. Due to the high speed of the transmission channel, all nodes will start transmitting information through fake nodes, and the attacker will eventually receive all the data [3].
4. Passive collection of information - an attacker using additional technical means, in the form of a specialized antenna, collects information transmitted in the network, which he can then use to attack the network.

3. Types of homomorphic encryption

Homomorphic encryption is an encryption method that allows you to perform mathematical operations, namely addition and multiplication, with ciphertexts and obtain an encrypted result that will correspond to the result of mathematical operations on the plaintext. Homomorphic encryption allows you to perform various operations in an untrusted environment without revealing confidential data for each of the operations [4].

There are two types of cryptosystems: fully homomorphic and partially homomorphic. Fully homomorphic schemes support both addition and multiplication mathematical operations, that is, the properties of homomorphism are performed on these operations. A partially homomorphic cryptosystem allows performing only one of the above operations — either addition or multiplication.

Partially homomorphic cryptosystems are represented by the following schemes.

IOP Conf. Series: Materials Science and Engineering 1069 (2021) 012008 doi:10.1088/1757-899X/1069/1/012008

Cryptosystem of Goldwasser and Micali [5]. The partially homomorphic cryptosystem proposed by S. Goldwasser and S. Micali is the first of its kind proven stable public-key probabilistic system. The stability of this system is based on a computationally complex problem of quadratic residue. In the Goldwasser and Micali cryptosystem, sender A encrypts a plaintext bit as a random quadratic residue or non-residue modulo n. Receiver B decrypts the received ciphertext by checking the quadratic residues of the encrypted values using factorization of the number n. This cryptosystem includes encryption and deterministic decryption schemes, generation of encryption keys. The homomorphism properties of this scheme are performed over the addition operation:

$$E(M_1)\cdot E(M_2) = x^{M_1}y^2x^{M_2}y_2^2 = x^{(M_1+M_2)}(y_1y_2)^2 = E(M_1+M_2).$$

The computational complexity of the homomorphic encryption scheme is due to the fact that in order to encrypt a K –bit message, it is necessary to carry out operations in the amount $O(K(log_2N)^2)$. The original text will increase in log_2N times when encrypted, because each bit of the original text corresponds to a log_2N bit of the ciphertext. The semantic stability of this scheme, due to the resistance to an attack based on a selected source text, lies in the fact that an adversary with computational polynomially limited resources, as well as knowing the ciphertext and its length corresponding to a certain plaintext, will not extract any useful information.

Cryptosystem Benalo [6] is a modified Goldwasser and Micali scheme described above. The main difference is that this scheme allows one-time encryption of an entire block of data, in contrast to the Goldwasser and Micali scheme, where each bit of the plaintext is encrypted separately. In the Benalo cryptosystem, the homomorphism properties are also performed over the addition operation:

$$E(M_1)\cdot E(M_2) = (x^{M_1}u_1^r)(x^{M_2}u_2^r) = x^{M_1+M_2}(u_1u_2)^r = E(M_1+M_2 \; mod\; r)$$

where is r the block size. The computational complexity of this kriptoskhemy: $O(R(\log n)^2)$. Subject to the use of preliminary calculations, the time complexity of decryption $O(\sqrt{r})$. The robustness of Benalo's scheme is due to the computationally difficult task of high-degree residues: it is computationally difficult to find a plaintext, even with information about the block size, the magnitude of the modulus, with an unknown factorization and ciphertext.

PAYE's cryptosystem [7] invented by P. Paillier, it is a public-key cryptosystem, probabilistic, strength due to the complexity of factorization of a composite number, which is the product of two primes. The homomorphism of this scheme is also performed over the addition operation. In the PAYE cryptosystem, the multiplication of two ciphers will be decoded as the sum of their respective plaintext:

$$D(E(m_1,r_1) * E(m_2,r_2) modn^2) = m_1 + m_2 modn.$$

Multiplying the ciphertext by g^{m_2}, we get the sum of the corresponding plaintexts:

$$D(E(m_1,r_1) * g^{m_2} modn^2) = m_1 + m_2 modn,$$

where m_1 and m_2 are plaintext, (n, g) is a pair of numbers forming the public key, r_1 and r_2 are random numbers, E is the encryption function, and D is the decryption function.

Cryptosystem of Damgard and Yurik [8]. This system is interpreted as a generalization of the above-described PAYE cryptosystem for large modules for use in a wider range of tasks. The homomorphism property holds over the addition operation, the system is homomorphic with respect to addition to Z_{n^s}: $E(M_1)\cdot E(M_2) == E(M_1 + M_2 \; mod\; n^s)$ where n is a - module and s - natural number. As for the strength of the Damgard and Yurik cryptosystem, it is based on the assumption that it is difficult to calculate the root of the remainder modulo, as it is implemented in the above-described PAYE system.

Cryptosystem Okamoto – Uchiyama [9]. This system is probabilistic, based on a logarithmic function, which is defined over the multiplicative group and using large primes. Additive homomorphism of the system due to the fact that $M_1 + M_2 < p$ performed as follows:

$$E(M_1)\cdot E(M_2) = (x^{M_1}r_1^c)(x^{M_2}r_2^c) mod\; n = x^{M_1+M_2}(r_1r_2)^c mod\; n = E(M_1+M_2)$$

The stability of the Okamoto – Uchiyama system is based on the complexity of the factorization problem for the number n and requires $\mathrm{O}(\log_3 \mathrm{n})$ bitwise operations.

Fully homomorphic systems are a much more rare form of homomorphic encryption. As previously stated, Fully Homomorphic Encryption can support simultaneous addition and multiplication operations, allowing more computations to be performed on encrypted data. In 2009, an IBM employee,

Craig Gentry, proposed the first cryptographically strong asymmetric fully homomorphic encryption scheme based on ideal lattices [10]. In recent years, many works have appeared in which various modifications and optimizations of this scheme have been proposed. The homomorphic property of the simplified Gentry scheme is to obtain the sum of texts *m1* + *m2*, you only need to add the ciphertexts and perform the decryption operation:

$$D\big(E(m_1, pk) + E(m_2, pk)\big) = D\left(m_1 + m_2 + \sum_{i=0}^{n} C_i pk_i\right) = D(m_1 + m_2 + C_1' sk + C_2' n)$$
$$= m_1 + m_2$$

where *pk* and *sk* are the public and private keys, respectively.

In 2012, Gentry showed how you can perform self-correction of ciphers [11], and then showed how you can use a fully homomorphic encryption scheme in conjunction with the AES encryption algorithm [12]. The idea proposed by Gentry is as follows. For data encrypted first with a symmetric cipher and then with a homomorphic cipher on top, it is possible to decrypt with a symmetric system using the encryption key without removing the homomorphic encryption. To do this, it is necessary to represent all operations of the decryption algorithm in the form of simple Boolean functions, namely, those functions that allow the selected type of completely homomorphic cipher to be performed.

4. Development of a homomorphic encryption scheme using a symmetric algorithm

Based on Gentry's ideas and the functioning of sensor networks, as well as existing implementations and modifications of fully homomorphic encryption schemes, an approach to information protection in sensor networks was developed with the joint use of a fully homomorphic scheme and a symmetric encryption algorithm.

The Kuznechik encryption algorithm - GOST R 34.12-2015 [13] was chosen as a symmetric encryption algorithm. As you know, Kuznechik uses S-block substitution in his transformations, which is the most difficult. The replacement S-block, in turn, can be represented in the form of Zhegalkin polynomials, which will reduce the complexity of this task.

The development of a data transfer algorithm based on the selected encryption methods implies interaction between the sensor nodes of the network and the base station, where the most resource-intensive operations will be performed at the base station, which is not limited in the battery supply, and low-cost calculations and the exchange of processed data itself will be performed directly by the sensor nodes of the network. Thus, all complex encryption operations reproduced by the Kuznechik algorithm and the homomorphic Gentry scheme will be implemented at the base station. And the nodes will then be able to transmit homomorphically encrypted data without much impact on the energy and processor components.

In view of the fact that symmetric encryption algorithms are energy-intensive during the execution of encryption and decryption operations, it becomes necessary to develop a method for implementing the symmetric algorithm in the fastest possible way in order to reduce the computational and associated time costs of the sensor nodes and the base station.

5. Conclusion

This paper contains a description and application of sensor networks, which have become the most relevant main components in the construction of cyber-physical systems. The most common types of attacks on this wireless sensor network are considered. The most well-known partial homomorphic encryption schemes used in modern cryptographic systems have been described.

The analysis of existing fully homomorphic cryptosystems based on the idea of Gentry is carried out. An approach was developed for the joint use of a homomorphic encryption scheme and a symmetric cryptoalgorithm in order to protect data transmitted in sensor networks. Further work will be devoted to the working and development of an approach to the joint use of symmetric and homomorphic encryption algorithms.

6. Acknowledgment

The reported study was funded by RFBR, project number 19-37-90148.

References

[1] *What is a DoS attack?* (Binance Academy). URL: https://www.binance.vision/ru/security/what- is-a-dos-attack

[2] Sharmila S and Umamaheswari G 2011 Detection of Sinkhole Attack in Wireless Sensor Networks Using Message Digest Algorithms *International Conference on Process Automation, Control and Computing* (Coimbatore) pp 1-6

[3] Pandey and Rajshree 2014. Wormhole Attack in Wireless Sensor Network *International Journal of Computer Networks and Communications Security* vol 2 pp 22-26

[4] Sen J 2013 Homomorphic Encryption: Theory and Application *Theory and Practice of Cryptography and Network Security Protocols and Technologies* pp 31

[5] Goldwasser and Micali S 1982 Probabilistic encryption and how to play mental poker keeping secret all partial information *Proceedings of the 14th ACM Symposium on the Theory of Computing.* pp 365–377

[6] Benaloh J Verifiable secret-ballot elections URL: http://www.cs.yale.edu/publications/techreports/tr561.pdf

[7] Paillier P 1999 Public-key cryptosystems based on composite degree residuosity classes *Advances in Cryptology - EUROCRYPT'99* - Vol. 1592 of Lecture Notes in Computer Science. pp 223–238

[8] Damgard I, Jurik M A 2001 generalization, a simplification and some applications of Pailliers probabilistic public-key system *Vol. 1992 of Lecture Notes in Computer Science* pp 119-136

[9] Okamoto T and Uchiyama S 1998 A new public-key cryptosystem as secure as factoring *Advances in Cryptology - EUROCRYPT '98* - Vol. 1403 of Lecture Notes in Computer Science pp 308-318

[10] Gentry C Fully 2009 Homomorphic encryption using ideal lattices *Proceedings of the 41st ACM symposium on theory of computing (STOC)* (Bethesda, Maryland, just outside Washington, DC, USA) pp 169-178

[11] Brakerski Zvika Gentry Craig and Vaikuntanathan Vinod 2012 Fully homomorphic encryption without bootstrapping *In Innovations in Theoretical Computer Science (ITCS'12),* URL: http://eprint.iacr.org/2011/277

[12] Gentry C and Halevi S 2012 Smart NP Homomorphic Evaluation of the AES Circuit *CRYPTO 2012* pp 850-867

[13] GOST R 34.12–2015 Information technology. Cryptographic information protection. Block ciphers URL: https://tc26.ru/standarts/natsionalnye-standarty/gost-r-34-12-2015-informatsionnaya-tekhnologiya-kriptograficheskaya-zashchita-informatsii-blochnye-shifry.html

Graphical data processing in the Clinical Decision Making System for the Respiratory Diseases Diagnosis using ML methods

G R Shakhmametova[1], N I Yusupova[1], A A Evgrafov[1], R Kh Zulkarneev[2]

[1] Ufa State Aviation Technical University, Ufa, Karl Marx str., 12, 450000, Russia

[2] Bashkir State Medical University, Ufa, Pushkin str., 12, 450008, Russia

E-mail shakhgouzel@mail.ru

Abstract. This article considers the basic algorithm of graphic data extraction used in the developed system of clinical decision-making in diagnosis of respiratory diseases, methods for processing files in JPEG and DICOM formats, visualizing and preprocessing images as well as constructing neural network models based on a convolutional neural network that provide detection symptoms of pneumonia in patients based on radiographs. Data extraction and processing are performed in the "Python" programming language using additional libraries: "pydicom" for processing DICOM files. "Pillow" for visualization, "Keras" for building a convolutional neural network model. The relevance of the task is due to the large volumes of graphical data supplied to the input of the CDSS and necessary for its effective functioning. The novelty of this development lies in the application of a set of existing and development of new algorithms for extracting and processing graphic information.

Keywords: Clinical decision support system; Data processing; Processing of graphic information, CNN, Computer vision

1. Introduction

Active penetration of information technologies into medicine and healthcare has stimulated development of software systems with the main goal of improving quality of medical services by introducing modern intelligent technologies and machine learning technologies in the process of interaction between medical personnel and patients. One of the most effective means to achieve this goal is by using clinical decision support systems (CDSS). CDSS are systems designed to help doctors and other medical professionals in working with tasks related to clinical decision-making, focused on processing and managing information to formulate recommendations for patient treatment based on available data [1]. The use of CDSS ensures increased efficiency of diagnosis and treatment of patients, and also reduces the burden on the doctor when processing information.

In [2], the authors describe the development of the clinical decision support system architecture in the diagnosis of respiratory diseases. This system presupposes the presence of such functions as analysis of patient data in the form of text, graphic and sound information, formulation of a presumptive diagnosis based on the analysis and identified patterns as well as formation and issuance of medical recommendations to improve the effectiveness of patient treatment [3].

This article is devoted to the consideration, description and implementation of a basic algorithm used in:

- converting data from DICOM format to JPEG format;
- uploading images (results of X-ray examinations of patients) in the form of JPEG files;
- preprocessing of uploaded images;
- use of machine learning algorithms to predict the presence of a disease (pneumonia) in a patient.

The implemented algorithm is a component of the graphic data recognition unit of the developed clinical decision support system in the diagnosis of respiratory diseases.

2. State of art

Graphic information in medicine, as a rule, is represented by three types of images:

- radiographs;
- images of computed tomography;
- pictures of magnetic resonance imaging.

The most common type of medical imaging in the area under consideration (i.e., pulmonology) is a chest X-ray. To obtain images, X-ray devices allowing to project the structure of a person's internal organs onto a special film or paper using X-rays are used.

Chest X-rays are useful for diagnostics of such conditions as

- pneumonia;
- pneumothorax;
- interstitial lung disease;
- heart failure;
- bone fractures;
- hernia of the esophageal opening of the diaphragm [4].

Today, there are many solutions on the market that implement the functionality of medical image recognition.

For example, the Imsight Technology company, which produces computed tomographs, at the end of May 2020 presented a development in the field of artificial intelligence for CT diagnostics of lung diseases under the name of Imsight CT Analysis System [5], which establishes a preliminary diagnosis in 10 seconds.

The CheXNext algorithm [6], created at Stanford, examines a chest image and checks it for 14 pathologies. The results show that the algorithm detects 10 out of 14 pathologies faster and better than experienced radiologists.

One of the main Russian developments is the "Botkin.AI" company "Intellodzhik" [7]. Botkin.AI is used to diagnose and assess disease risks using artificial intelligence technologies and is a platform for analyzing medical images. The platform allows for analysis of the patient's chest based on computed tomography images, mammography screening, detection of lung cancer as well as pneumonia caused by Covid-19.

Another Russian development was carried out by the Care Mentor AI company [8], and is a neural network that is used to describe X-ray images. The technology allows you to get a description and a preliminary diagnosis within 3 seconds after taking a picture. This product has also been enhanced with the ability to detect Covid-19. The diagnostic accuracy is about 84%.

Nevertheless, the presented developments are independent projects that process exclusively graphic medical data and obtain results only on their basis, while the CDSS described in this article involves the complex use of information about the patient, i.e., text, graphic, or sound, which will significantly improve the efficiency of the system in diagnosing patients.

3. Main problems in recognition in CDSS

IOP Conf. Series: Materials Science and Engineering 1069 (2021) 012009 doi:10.1088/1757-899X/1069/1/012009

The main problem of creating a module for recognizing graphic data of the system being developed is the heterogeneity of the initial information, which consists of the following factors:

- radiographs can be performed on different X-ray machines with different settings and environmental conditions;
- the projections of the lungs captured in the image may be displaced relative to the center of the image due to insufficient accuracy of the device, as well as the physiological characteristics of patients;
- X-ray images can contain significantly noisy frames and frames of low quality;
- it is difficult to obtain a uniform and sufficient number of images from patients with various diseases, since the incidence of diseases can vary significantly.

It should also be noted that radiographs created with medical equipment are usually stored in DICOM format. DICOM (Digital Imaging and Communications in Medicine) is a medical industry standard for the creation, storage, transmission and visualization of digital medical images and documents of examined patients [9]. DICOM is based on the Open System Interconnection (OSI) standard developed by the International Organization for Standardization (ISO). This standard is supported by major medical device and medical software manufacturers.

4. Proposed solution

The problem of processing X-ray images in the developed DSSP may be approached by solving the problem of pattern recognition or by the task of training with a teacher.

This work solves the problem of binary classification (dividing patients into two classes - with and without detected pneumonia). The Chest X-Ray Images (Pneumonia) of the Kaggle platform [10] are used as a dataset for training and testing. The sizes of the data sets for training, validation and test samples are shown in Figure 1.

```
Description
Normal:
Train: 1342 pictures
Valid: 9 pictures
Test: 234 pictures
Total: 1585 pictures

Pneumonia:
Train: 3876 pictures
Valid: 9 pictures
Test: 390 pictures
Total: 4275 pictures
```

Figure 1. Number of images in the "Chest X-Ray" dataset

The current dataset is JPEG images that do not require additional conversion. If files in DICOM format are received at the block input, additional conversion is required using the "pydicom" library [11]. DICOM and JPEG file formats are automatically recognized by reading their own extensions.

Images are displayed using the Python Imaging Library (PIL) [12]. Examples of snapshots are shown in Figure 2.

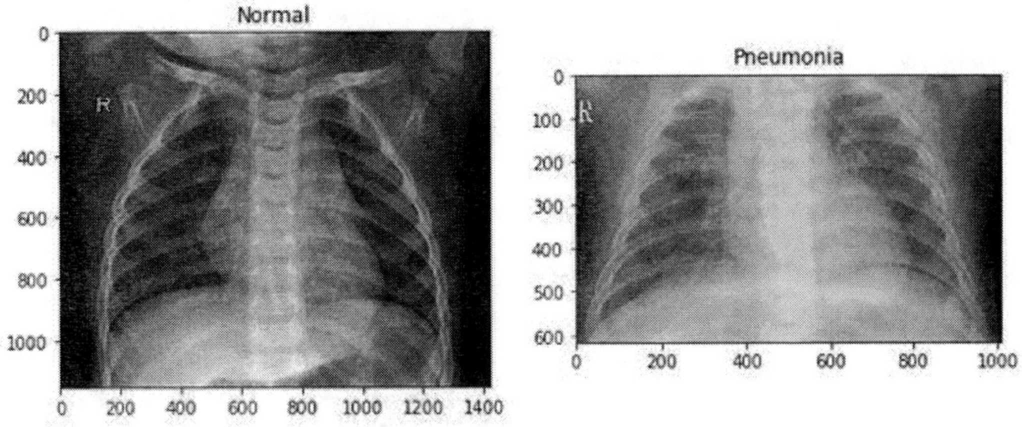

Figure 2. Examples of images of patients without pneumonia and with recognized pneumonia

As a framework for building a neural network, the "Keras"is used [13], since it is a fairly fast and efficient add-on over the deep learning libraries "TensorFlow", "Theano" and "Deeplearning4j".

To improve the generalizing ability of the future neural network, augmentation is performed as follows:

- horizontal flip of images is applied (horizontal_flip);
- the range of image pixel shift is set to 0.2 (shear_range);
- the range of random selection of image scaling (zoom_range) is set.

A convolutional neural network (CNN) is utilized as the architecture of a neural network, since networks of this type are known by their high ability to recognize patterns in images. In a convolutional neural network, the outputs of the intermediate layers form a matrix (image) or a set of matrices (multiple image layers). So, for example, three image layers (R-, G-, B-image channels) can be fed to the input of a convolutional neural network. The main types of layers in a convolutional neural network are convolutional layers, pooling layers and fully-connected layers [14].

The architecture of the created neural network is shown in Figure 3. The input of the neural network receives an image of 128x128 pixels in size. Next, the image goes into a bunch of Conv2D and MaxPooling2D layers.

Conv2D is a convolutional layer that applies convolution to input images and consists of customizable and trainable filters. The convolution kernel weights are trainable parameters. Another training weight is used as a constant shift.

MaxPooling2D is referred to as a sample-based sampling layer, which aims to downsample the input layer based on the maximum value of the subsampling window.

The Flatten layer transforms the resulting matrix into a one-dimensional vector, after which the final transformations are performed using fully connected layers (Dense).

The created implementation of the convolutional neural network is configured as follows:

- Algorithm for optimization of gradient descent - Adam;
- Loss function – Binary cross entropy;
- Quality metric – Accuracy.

The best results obtained after training the neural network and testing on test data:

- the accuracy value on the training set – 0.84;
- the accuracy value on the test set – 0.80;

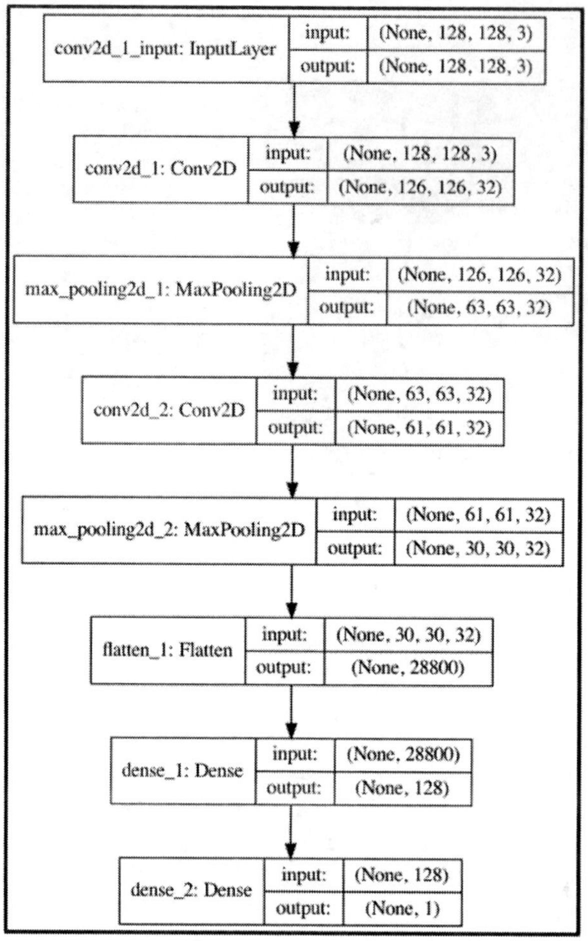

Figure 3. CNN Architecture for Pneumonia Recognition Module in CDSS

The accuracy values on the training and test samples differ insignificantly from each other, while the value on the test sample is consistently lower than the value on the training sample, whereupon it may be concluded that the resulting model is not overfitted and has a sufficiently high efficiency of X-ray pattern classification.

Further tasks to improve the quality of the model consist in increasing the number of images, expanding the library of images for diseases other than pneumonia, which will make it possible to move to the task of multiclass classification, and augmenting the capabilities of the DSS. The capabilities of the developed module for processing graphic information can also be expanded by building additional models for detecting diseases by evaluating images of computed and magnetic resonance imaging. The use of pre-trained neural network mechanisms such as ResNet, ImageNet is allowed.

The implementation of the developed algorithm for processing graphic information is located in the repository [15].

5. Conclusions

The article considers the algorithm for processing graphic information obtained from radiographs of patients with and without pneumonia, developed for a clinical decision support system in order to create a classifier that detects pneumonia in a patient:

- the algorithm has been developed that allows processing JPEG graphic files and medical standard DICOM files;
- functional for previewing the received radiographs has been developed;
- the architecture of a convolutional neural network has been developed, which most effectively performs the task of classifying patients;
- an assessment of the trained classifier was carried out, as a result of which the best results of the accuracy metric on the training set were 0.84, while on the test set they were 0.80;
- directions for further improving the quality of the classifier are highlighted.

The developed models and algorithms will make it possible to classify the presence or absence of pneumonia based on the obtained X-ray of the patient, then combine the information with the data extracted from the textual description of the symptoms and the tabular result of the patient's analyzes, and ensure its transfer to the decision-making module, whose functionality, in turn , will allow you to establish the most likely diagnosis for each patient.

6. Acknowledgment

The reported study was funded by RFBR according to the research projects No. 19-07-00780, 19-07-00709, 18-07-00193. The questions of determining the place of neural networks in the image recognition task were obtained as part of state task project No. FEUE-2020-0007.

References

[1] Shaikh F Dehmeshki J Bisdas S Roettger-Dupont D Kubassova O Aziz M and Awan O 2020 Artificial Intelligence-Based Clinical Decision Support Systems Using Advanced Medical Imaging and Radiomics *Current Problems in Diagnostic Radiology*

[2] Shakhmametova G, Zulkarneev R and Evgrafov A 2019 Clinical decision support system for the respiratory diseases diagnosis (Electronic Materials) *Proc. of the 7th Scientific Conference on Information Technologies for Intelligent Decision Making Support ITIDS 2019 (*Atlantis-Press)

[3] Shakhmametova G, Zulkarneev R and Evgrafov A 2018 Analytical Processing of Medical Data in the System of Diagnosis of Bronchopoietic Diseases *Proc. of the 7th Scientific Conference on Information Technologies for Intelligent Decision Making Support ITIDS 2019* vol 1 pp 256-261

[4] Novelline R 1997 *Squire's Fundamentals of Radiology* (Harvard University Press. 5th edition)

[5] Imsight CT Analysis System 2020 *TADVISER* (Electronic Materials)

[6] Aras M and Sengur A 2020 Deep learning approaches for COVID-19 detection based on chest X-ray images *Proc. of Expert Systems with Applications* vol 164

[7] Botkin.AI | AI-Powered Radiology Platform – Official company site (Electronic Materials)

[8] Vaishya R, Javaid M, Haleem Khan I and Haleem A 2020 Artificial Intelligence (AI) applications for COVID-19 pandemic *Diabetes & Metabolic Syndrome: Clinical Research & Reviews* vol 14 issue 4 pp 337-339

[9] Bulliard J Eggert S Ampranozi G Affolter R Sieberth T Thali M and Ebert L 2020 Preliminary testing of an augmented reality headset as a DICOM viewer during autopsy (Electronic Materials) *Forensic Imaging*

[10] Chest X-Ray Images (Pneumonia) Dataset, *Kaggle* (Electronic Materials)

[11] Shelmerdine S Singh M Norman W Jones R Sebire N and Arthurs O 2019 Automated data extraction and report analysis in computer-aided radiology audit: practice implications from post-mortem paediatric imaging *Clinical Radiology* vol 74 issue 9 pp 733.e11-733.e18

[12] Pillow – Official framework site (Electronic Materials)

[13] Nhu V Hoang N Nguyen H Thao Thi Ngo P Thanh Bui T Viet Hoa P Samui P and Tien Bui D 2020 Effectiveness assessment of Keras based deep learning with different robust optimization algorithms for shallow landslide susceptibility mapping at tropical area *CATENA* vol 188

[14] Guoqing L Xuzhao S Jiaojie L and Jiuyang W 2020 Diagonal-kernel convolutional neural

networks for image classification *Digital Signal Processing*
[15] bronchopulmonary_diag_system EvgrafovAlexander *GitHub* repository (Electronic Materials)

Evaluation of graphic data corruptions impact on artificial intelligence applications

O Molokovich[1], A Morozov[2], N Yusupova[3], K Janschek[4]

[1,3] Ufa State Aviation Technical University, Ufa, K. Marx str., 12, 450000, Russia

[2] University of Stuttgart, Stuttgart, Pfaffenwaldring str., 47, 70550, Germany

[4] Technische Universität Dresden, Dresden, Nöthnitzer str., 43, 01187, Germany

E-mail o.molokovich@ya.ru, andrey.morozov@ias.uni-stuttgart.de, yussupova@ugatu.ac.ru, klaus.janschek@tu-dresden.de

Abstract. The article is devoted to the problem of the reliability of applications based on artificial intelligence. The authors made an attempt to evaluate the impact of the graphic data distortion at the input of a convolutional neural network on the result of image classification. The experiment is based on the fault injection method. A series of independent tests were carried out for such distortions as Gaussian noise, salt and pepper, speckle and Poisson noise, as well as median blur, motion blur, scene brightness changing, rotation, rain, and snow. The results showed that Gaussian noise was the least critical distortion; environmental conditions (rain, snow, brightness) and image rotation up to 20 degrees are less critical than focus losing and motion blur, while the most critical distortion is speckle noise. It was verified that preprocessing the input data of the neural network improves the accuracy of image recognition.

Keywords: Reliability; Cyber-physical system; Fault injection method; Neural network; Artificial Intelligence

1. Introduction

The Cyber-physical system (CPS) is an system consisting of interconnected intelligent and heterogeneous components that interact with each other. Various components of a CPS are prone to faults. A malfunction is an abnormal condition that can manifest itself as an error propagating through the CPS, and ultimately leading to a failure of the system as a whole. This paper is devoted to assessing the impact of CPS faults on its intelligent components using the example of graphic data distortions at the input of a convolutional neural network for image recognition as an example of an artificial intelligence (AI) application.

Many researchers are engaged in issue of AI applications robustness. For example, Qi et al. [1] drew attention to the problem of data quality in machine learning and data mining. Hynes et al. [2] developed a data preparation and transformation tool for machine learning. Huang et al. [3] discuss how an attacker modifying data can cause a machine learning model to behave incorrectly. Li et al. [4] created the CleanML benchmark to investigate the effect of data cleansing in machine learning. Several tools have also been created for introducing faults into images [5,6,7,8].

2. Cyber-physical systems reliability problems analysis

CPS mechanisms are controlled by computer algorithms and AI systems. AI applications are highly dependent on the input data quality. Measurement errors can be the cause of CPS dysfunction.
There are CPSs that use a video camera to receive data from the outside world. The camera receives an image and transfers it to neural net, one of AI technologies. The network classifies the image and passes the result on to decide on the control action. An example of such a CPS is an autonomous driving system in which a camera is used to recognize traffic signs. Due to camera malfunctions or any network problems image distortions can occur. Then, the error extends to the driving scene recognition result, which may lead to erroneous vehicle behaviour. Image corruptions can be divided into two types:

1. Errors due to hardware malfunctions (Gaussian, shot, speckle, and Poisson noise, noise of salt and pepper, motion blur, defocusing).
2. Decreased quality due to environmental conditions (rain, snow, fog, brightness, image rotation).

3. Reliability evaluation method

In this paper, the error injection method is used to evaluate the reliability of AI applications. Faults are embedded in the data - images of the test dataset.

4. Classification problem interpreting

Video camera image recognition can be interpreted as one of the variants of the classification problem. The classical formulation of the pattern recognition problem is as follows [9]: M is given set of objects that need to be classified. The set is represented by subsets $K_1,\dots,K_l$, which are called classes: $M = K_1 \cup \dots \cup K_l$. The partitioning M is not fully specified, only some information I_0 about the classes $K_1,\dots,K_l$, is given. Similarly, the valid object S, whose belonging to classes $K_1,\dots,K_l$ is unknown, is determined by the values of some characteristics that make up $I(S)$. It is required according to information I_0 and description of object $I(S)$ to set the value of property $S \in K_j$, $j = 1,\dots,l$ for each j.
In the last decade, solutions based on a neural network approach have been successfully solving the problem of pattern recognition. The current research also used a neural network for image recognition.

5. Classification quality indicator

A natural assessment of the classification quality is the Mean Consequential Error:

$$MCE = \sum_{l=1}^{l} a(x_i) = y_i \quad , \tag{1}$$

where $a(x_i)$ – predicted class, y_i – original class, l – number of objects in the sample.

The metric shows the percentage of objects for which the algorithm gave correct answers. When used for unbalanced classes, the accuracy is calculated as a weighted sum, the weights are equal to the weights of the classes.

6. Case study neural network

To study the consequences of input data distortions pretrained open-source convolutional neural network (NN) was used [9,10]. The class-unbalanced test dataset network classification accuracy is 96%. It is based on a network designed by Vamsi Ramakrishnan and has two parts: localization network with spatial transformer and convolutional neural network for classification. The network consists of 15 layers. First, the image is fed into the localization network, which determines the transformation applied to the image by the spatial transformer. It leads to greater invariance to displacement, scaling, rotation, and general deformation [12]. Then the picture is fed into the neural network, which consists of several stacked convolutional layers. This part of the network is based on the architecture described by K. Simonyan and A. Zisserman [13], with each stack having the same structure: multiple convolutions followed by a PReLU activation function, and one Max Pool layer. By combining the outputs of the last three Max Pool layers before entering fully connected layers, it is possible to obtain the semantics of several extracted features from different levels of abstraction, which improves object detection [14,15].

7. Training dataset

Ramakrishnan network was developed for recognition of the German Traffic Sign Recognition Benchmark (GTSRB) [11]. It consists of over 50,000 images from 43 unbalanced class frequencies and is divided into three subsets for training, validation, and testing.

8. Fault injection tools overview

AVFI is a tool for entering faults into autonomous vehicle control systems. It allows one to study how various sensor malfunctions affect performance [8]. DeepRoad is a deep neural network framework, which automatically synthesizes driving scenes without using image transformation rules. It can create scenes with different weather conditions [7]. DpEmu is a software environment for emulating data problems, it is possible to introduce errors like rotation, snow, rain, delay of sensor signals, and others into various types of data sets [5]. DeepMutation is a system close to DpEmu that generates artificial faults for data and model implementation. The capabilities of these two tools are similar [6]. OpenCV is an open source cross-platform library that allows to process images in various ways [16].

9. Computer experiment

9.1. Input data

The input image set consists of 2950 class balanced images from the GTSRB test dataset. The instances distribution by class is a set of pairs (name of the picture, class number).

9.2. Methodology

The experiment of the evaluation of images distortions impact on the recognition result consists of an independent series sequence. Series algorithm is shown in Figure 1. The main steps are:

1. Select the distortion type and determine the range its power values.
2. For each range value:
   A. Inject distortion into images.
   B. Classify
   C. Calculate image recognition accuracy.

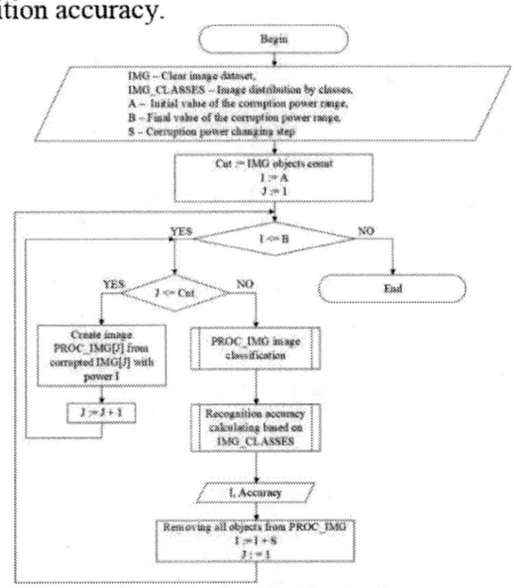

Figure 1. Series block diagram

9.3. Software

The computer experiment was implemented using Python. DpEmu, and OpenCV was chosen as distortion injection tools. Both libraries are also written in Python. The environment was deployed using the Ubuntu-18.04 operating system.

9.4. Data corruption

Blurring is a filter that defines the output pixel value as the weighted sum of the output pixel values (formula 2):

$$g(i,j) = \sum_{k,l} f(i+k, j+l)h(k,l) \quad (2)$$

where *h(k, l)* – filter coefficients, thus filter is a visualization of coefficients sliding across the image.

Gaussian Blur is a filter that uses a Gaussian function or normal distribution to compute transform applied to each pixel in an image (formula 3):

$$G(x,y) = \frac{1}{2\pi\sigma^2} e^{-\frac{x^2+y^2}{2\sigma^2}} \quad (3)$$

where x – distance from the origin in the horizontal axis, y – distance from the origin in the vertical axis, σ – the standard deviation of the normal distribution.

The formula result in two dimensions is concentric circles with a normal distribution from the center point. The values from this distribution are used to construct a convolution matrix that is applied to the original image. The new value of each pixel is set as a weighted average of the neighboring pixels. The original pixel takes on a heavyweight, the neighboring pixels get smaller.

Gaussian blur implementation using OpenCV is presented below:

output_image = cv2.GaussianBlur(input_image, (ksize.width, ksize.height), 0),

where output_image – distorted image, input_image - clear image, width and height of the sliding window are determined by formulas 4, 5:

$$ksize.width = 2*n-1 \quad (4)$$

$$ksize.heigth = 2*m-1 \quad (5)$$

where *m, n* $\in N$ – distortion parameters. Blurred image example is shown in Figure 2, Figure 3 shows the original image.

Median blur assigns to each pixel the average value of neighbouring pixels from some square neighborhood:

output_image = cv2.medianBlur(input_image, ksize),

where output_image – distorted image, input_image – clear image, window size is determined by the formula 6:

$$ksize = 2*k-1 \quad (6)$$

where $k \in N$ – distortion parameters.

Motion blur is also implemented using OpenCV:

output_image = cv2. filter2D(input_image, (ksize.width, ksize.height), 0),

where output_image – distorted image, input_image – clear image, sliding window width and height are determined by formulas 7, 8:

$$ksize.width = 2*n-1 \quad (7)$$

$$ksize.heigth = 2*m-1 \quad (8)$$

where *m, n* $\in N$ – distortion parameters. Processed image example is shown in Figure 4.

Speckle noise is implemented as the sum of the pixel value and its product by the distortion parameter:

output_image = input_image + input_image * n,

where output_image – distorted image, input_image – clear image, $n \in N$ – distortion parameter. Image example in Figure 5.

Gaussian noise is implemented as the sum of the point value and the value of a random variable *X*, obeying the normal distribution law (formula 9):

$$X = f(x) = \frac{1}{\sigma\sqrt{2\pi}} e^{-\frac{(x-\mu)^2}{2\sigma^2}} \quad (9)$$

where μ – expected value, σ – standard deviation. Image example in Figure 6.

Poisson noise is implemented as a change in the values of the components (0-255) of red, green and blue colours at H points of the picture. Random variable H obeys the Poisson distribution law (formula 10):

$$H = \frac{\lambda^k}{k!} e^{-\lambda} , \tag{10}$$

where: k – events number, λ – observed interval.

Salt and pepper noise is implemented as a change of the red, green, blue components (0-255) at random picture points. The distortion parameter is distorted pixels' percentage. Image example in Figure 7.

Figure 2. Gaussian Blur applied image

Figure 3. Clear image

Figure 4. Image processed with Median Blur

Figure 5. Image with speckle noise applied

Figure 6. Gaussian noise applied image

Image rotation is implemented using the DpEmu. Distortion parameter is rotation angle. Sample image in Figure 8.

Snowy and rainy scenes are also generated using dpEmu. In these cases, the parameter is the probability of a snowflake or raindrop in percent. Examples in figures 9 (for snow) and 10 (for rain). Changing brightness is another distortion under consideration. Image example in the Figure 11.

Figure 7. Salt & pepper noise

Figure 8. Rotated image

Figure 9. Snow-covered scene

Figure 10. Rain scene

Figure 11. Brightness change

10. Results

In total, 332 experiments were carried out to recognize a set of 2950 images. In each experiment, different distortions were applied to the pictures with different intensities. The graph of the recognition accuracy change depending on the filter and the strength of distortion is shown in Figure 12.

At the second iteration, the Gaussian blurred image recognition accuracy increases slightly. This can be explained by the realism of the GTSRB dataset. Blur reduces noise in the original images and contributes to better classification. However, the Median blur does not improve accuracy. Motion blur dramatically reduces the quality metric from 0.901 to 0.870. Then the accuracy almost returns to the original value and falls unevenly below. Slight picture blurring does not hinder recognition so much. But with an increase in the distortion power (by 5 iterations), the accuracy drops by almost 12 percent. The graph shows the influence of Gaussian noise on the metric depending on the mathematical expectation of its occurrence probability (σ = 0.3). We can say that the distortion is not critical to the classification process. At the same time, Poisson noise almost uniformly decreases the accuracy over the entire range of variation in the percentage of noisy points. Although the parameter for salt and pepper noise is pixel value, its effect is similar to Poisson noise. Perhaps a similar effect arises from a similar filter implementation. Speckle noise critically reduces the accuracy to 0.574% already at the first iteration, then the rate of decrease in accuracy slows down. The graph of classification accuracy changes of images with a small image rotation angle indicates the network resistance to image rotation. We can say that is due to the specifics of the training set (the meaning of some traffic signs does not change even if they are turned over) and the localization network. It determines which transformations need to be applied to the picture, then increases the invariance to turns. Excessive brightness reduces the efficiency of the network already at the first iteration. With an increase in the parameter value, the picture becomes more and more like a white canvas. Snowflakes create a "missing" area in the image, as a white spot appears instead of scene detail. Accuracy drops rapidly across the entire range. Jumps on the chart can be associated with the superposition of snowflakes on top of each other.

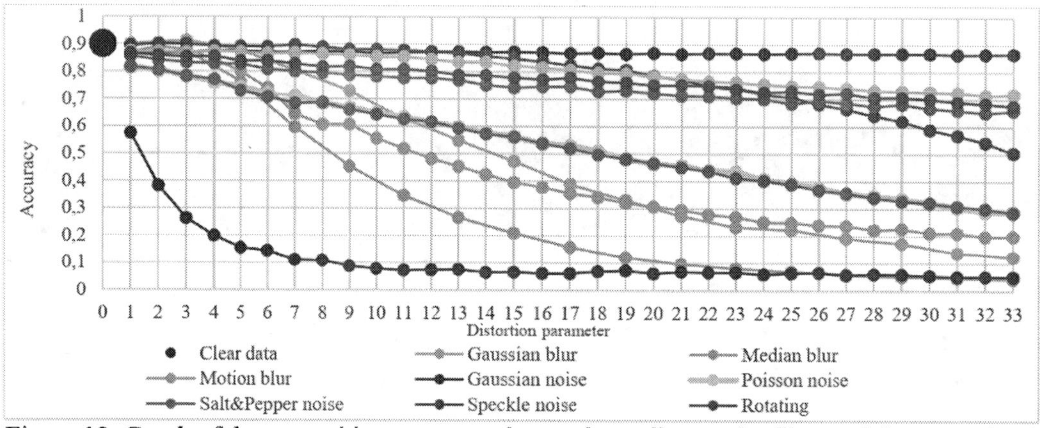

Figure 12. Graph of the recognition accuracy change depending on the filter and the strength of the distortion

Rain affects accuracy in a similar way to snow. This is probably due to the rain implementation in dpEmu. The weather condition impact evaluation will be more objective if the distortion is implemented in a different way.

In general, we can say that:

- Gaussian noise turned out to be the least critical distortion;
- environmental conditions (rain, snow, brightness) and image rotation up to 20 degrees turned out to be more critical;
- focus losing of and motion blur most strongly affect the accuracy;
- the most critical distortion is speckle noise.

The considered network turned out to be quite resistant to basic distortions applied with little force. In some cases, the small number of data errors in some cases slightly increased the classification accuracy. However, we shouldn't make any mistake that errors have a beneficial effect on network results. Due to the limitations of the hardware used, the experiment was conducted on a quarter of the GTSRB test data.

The results of the experiment highlighted the importance of pre-processing AI applications input data. Due to the peculiarities of the considered neural network architecture, namely the localization network in the first part, it turned out to be resistant to small rotations of the picture.

11. Conclusion

The article attempts to evaluate the impact of image distortions on artificial intelligence applications used in security-critical CPS.

The results of the experiment allowed us to conclude that:

1. The training dataset plays an important role, especially if it contains corrupted data.
2. Network resilience is highly architecture dependent.

3. Similar methods of injecting different distortions can lead to similar classification results.
4. It is necessary to carry out a similar experiment for different neural networks. This will allow identifying architectural features that contribute to improving network resilience.
5. Fault injection tools should be expanded. It will reduce the dependence of the results on the fault injection implementation and will increase research objectivity.
6. A lot of experiments carried out, perhaps, will allow deriving some recommendations for creating a fault resilient neural network.

12. Acknowledgment

The work is supported by the DAAD «Ostpartnerschaften» funds.

References

[1] Z. Qi, H. Wang, J. Li, and H. Gao 2018 Impacts of Dirty Data: and Experimental Evaluation
[2] N. Hynes, D. Sculley, G. Brain, M. T. Google Brain, and M. Terry 2017 The Data Linter: Lightweight, Automated Sanity Checking for ML Data Sets NIPS MLSys Workshop
[3] L. Huang, A. D. Joseph, B. Nelson, B. I. Rubinstein, and J. D. Tygar 2011 Adversarial machine learning Proc. of the 4th ACM workshop on Security and artificial intelligence - AISec '11 (New York, USA: ACM Press)
[4] P. Li, X. Rao, J. Blase, Y. Zhang, X. Chu, and C. Zhang 2019 CleanML: A Benchmark for Joint Data Cleaning and Machine Learning [Experiments and Analysis] 4
[5] K. Jukka, Nurminen, T. Halvari, J. Harviainen, J. Myllari, A. Roysko, J. Silvennoinen and Tommi Mikkonen 2011 Software Framework for Data Fault Injection to Test Machine Learning Systems" Proc. of 30th Annual IEEE International Symposium on Software Reliability Engineering (ISSRE 2019): 4th IEEE International Workshop on Reliability and Security Data Analysis
[6] L. Ma, F. Zhang, J. Sun, M. Xue, B. Li, F. Juefei-Xu, C. Xie, L. Li, Y. Liu, J. Zhao and Y. Wang. 2018 DeepMutation: Mutation Testing of Deep Learning Systems" Proc. International Symposium on Software Reliability Engineering, ISSRE, vol. 2018-October IEEE Computer Society, 11 2018
[7] M. Zhang, Y. Zhang, L. Zhang, C. Liu, and S. Khurshid 2018 DeepRoad:GAN-based metamorphic testing and input validation framework for au-tonomous driving systems" Proc. 33rd ACM/IEEEInt. Conf. AutomatedSoftw. Eng.
[8] S. Jha, S. S. Banerjee, J. Cyriac, Z. T. Kalbarczyk, and R. K. Iyer 2018 AVFI: Fault Injection for Autonomous Vehicles Proc. 48th Annual IEEE/IFIP Int. Conf. on Dependable Systems and Networks Workshops
[9] V. Ramakrishnan 2017 Traffic Sign Recognition
[10] M. Beyer 2019 Evaluation of the Effects of Common Hardware Faults on the Accuracy of Safety-Critical AI Components
[11] J. Stallkamp, M. Schlipsing, J. Salmen, and C. Igel 2012 Man vs. computer: Benchmarking machine learning algorithms for traffic sign recognition Neural Networks (2012) issn: 0893-6080
[12] M. Jaderberg, K. Simonyan, A. Zisserman and Koray Kavukcuoglu 2015 Spatial Transformer Networks CoRR abs/1506.02025 (2015). arXiv: 1506 . 02025
[13] K. Simonyan and A. Zisserman 2014 Very Deep Convolutional Networks for Large-Scale Image Recognition CoRR abs/1506.02025 (2015). arXiv: 1506 . 02025
[14] S. Bell, C Lawrence Zitnick, K. Bala and R. Girshick 2016 Inside-outside net: Detecting objects in context with skip pooling and recurrent neural networks Proc. of the IEEE conference on computer vision and pattern recognition (USA, Las Vegas)
[15] B. Hariharan, P. Arbelaez, R. Girshick and J.Malik 2015 Hypercolumns for object segmentation and Fine-grained localization Proc. of the IEEE conference on computer vision and pattern recognition (USA, Boston)
[16] OpenCV official documentation

Study of the characteristics of the synchronization algorithm for a quantum key distribution system based on comparing the number of samples from an adjacent pair of time windows with a threshold level

Y K Mironov[1], P D Mironova[1], K E Rumyantsev[1]

[1] Department of Information Security of Telecommunication Systems, Southern Federal University, Taganrog, str. Chekhov, 2, 347922, Russia

E-mail tmiyap117@gmail.com

Abstract. The analysis of time and probabilistic characteristics is carried out for a two-stage synchronization algorithm in a quantum key distribution (QKD) system with automatic compensation of polarization distortions. Formulas are obtained for calculating the missing probabilities a signal pair of windows, false alarm and successful synchronization. The dependences of the missing probability a signal pair on the average number of photons and noise pulses are investigated during the analysis of a pair of windows for different threshold levels. The numerical material obtained in the course of the study indicates the possibility of using for express calculations the probability of detecting a signal pair of windows formulas oriented to the normal distribution of the number of recorded pulses at the output of a single-photon avalanche photodiode.

Keywords: Quantum key distribution; Autocompensation system; Synchronization; Time parameters; Probabilistic characteristics; Fiber-optic line

1. Introduction

Systems of quantum key distribution (QKD) provide increased security of transmitted information. For the stable operation of the QKD system, accurate synchronization of user stations is required with minimal time. The issues of security of systems for quantum key distribution and station synchronization were studied in the works of domestic and foreign scientists [1-11].

In [12], the influence of the atmosphere on the accuracy of synchronization in the quantum key distribution in free space (QKD) was investigated. It is noted that fluctuations in the intensity of the synchronizing light due to atmospheric disturbances make a much larger contribution for the synchronization accuracy than others. In [13], the effect of random noise is studied, which is constantly present in synchronization in a practical CV-QKD (CV – continuous-variable) system, on system safety, including its effect on data sampling and parameter estimation.

In [14-15], synchronization algorithms imply the division of a frame into windows. The duration of the time window, equal to the duration of the pulse repetition period, significantly exceeds the duration of the synchronizing pulse. Time frames are analyzed sequentially. The signal window is the window where the largest number of photons is registered.

Another synchronization method also involves dividing a frame into windows [16]. However, the window duration is comparable to the duration of the synchronizing pulse. If the window durations and the synchronizing pulse are equal, the probability of finding the pulse at the border of adjacent windows is high. Therefore, at the first stage of synchronization, it is necessary to analyze windows in pairs.

The modified algorithm for searching for a photon pulse by analyzing a pair of adjacent windows, each of which is equal to the pulse duration, provides a higher probability of detecting a pair of windows compared to algorithms with an average number of registered photons in a pulse of more than 6 [17].

In the proposed algorithm for searching for a photon pulse, the decision to accept an adjacent pair of windows as a signal one follows when the total number of photons from an adjacent pair of time windows exceeds a given threshold level. Here, a photon pulse means an optical pulse, where the average number of photons during the observation time does not exceed 1. The tests continue until the threshold level kth is exceeded or the time resources are exhausted.

The purpose of the research is a complex analysis of probabilistic and time characteristics of detecting an adjacent pair of time windows containing a photon pulse for an autocompensating fiber-optic quantum key distribution system.

2. The missing probability a signal pair

In the proposed algorithm, the detection process begins with dividing a frame equal in duration to the pulse repetition period T_s into time windows. The duration of the windows is commensurate with the duration of the synchronizing pulse τ_s. When the durations of the synchronizing pulse and frames are comparable, the latter must be analyzed in pairs. Sequential interrogation of adjacent time windows is performed until the number of photons (hereinafter photons) and/or dark current pulses (DCP) registered in them exceeds a given threshold level k_{th}. If the result of the threshold test is positive, the corresponding pair of adjacent windows is taken as a signal pair, and then the process of accuracy the time position of the synchronizing pulse in the adjacent pair of time windows begins.

Note that, when speaking about registered photons, we take into account the fact that not all received photons are converted into single-photon pulses (SPP), which represent the responses of a single-photon avalanche photodiode (SAPD). Indeed, the quantum efficiency of the SAPD photocathode is always less than 1.

Let the average number of photons in a photon pulse be $\overline{n_s}$, and the average number of noise pulses (DCP) in the time window – $\overline{n_b}$.

The probability of detection in the signal pair $p_{th.s}$ at a given threshold level k_{th} when orientation on the description of the statistical properties of photodetection by the Poisson model is

$$p_{th.s} = \sum_{k=k_{th}}^{\infty} \frac{\overline{n_{sb}}^k}{k!} \cdot exp(-\overline{n_{sb}}), \tag{1}$$

where $\overline{n_{sb}} = \overline{n_s} + 2 \cdot \overline{n_b}$ – average total number of photons and DCP in a signal pair of windows.

If the Gaussian model is used, then

$$p_{th.s} = \frac{1}{2} - \frac{1}{2} \cdot erf\left(\frac{k_{th} - \overline{n_{sb}}}{\sqrt{2 \cdot \overline{n_{sb}}}}\right), \tag{2}$$

where $erf(x) = \frac{2}{\sqrt{\pi}} \cdot \int_0^x exp(-t^2) \cdot \mathrm{dt}$ – error function.

The threshold value is selected from the requirement of the admissible probability of a false alarm $p_{th.b}$, i.e. exceeding the threshold level k_{th} by the total number of $k=n_i+n_{i+1}$ noise pulses recorded in an adjacent pair of noise time windows.

The false alarm probability $p_{th.b}$ is related to the threshold level k_{th} by the ratio

$$p_{th.b} = \sum_{k=k_{th}}^{\infty} Pos(k, 2 \cdot \overline{n_b}), \tag{3}$$

when orientation on the Poisson model, and the relation

$$p_{th.b} = \frac{1}{2} - \frac{1}{2} \cdot erf\left(\frac{k_{th} - \overline{n_b}}{2\sqrt{\overline{n_b}}}\right), \tag{4}$$

when orientation on the Gaussian model.

For $\overline{n_{sb}} \geq 9$ (multiphoton or current synchronization mode), the threshold level k_{th} for a given probability $p_{th.b}$ is determined using the inverse function $\mathrm{erf}^{-1}(x)$ to $\mathrm{erf}(x)$:

$$k_{th} = 2 \cdot \overline{n_b} + \sqrt{4 \cdot \overline{n_b}} \cdot erf^{-1}(1 - 2 \cdot p_{th.b}). \quad (5)$$

Figure 1 shows the dependences of the missing probability a signal pair $p_{th.sb}$=1-$p_{th.s}$ on the average number of signal SPP and noise pulses $\overline{n_{sb}}$ for threshold levels k_{th}, equal to 2 (solid line), 4 (dashed line) and 8 (dotted line). The calculations were performed according to formula (1) with orientation to the Poisson model (step functions) and (2) with orientation to the Gaussian model (continuous functions).

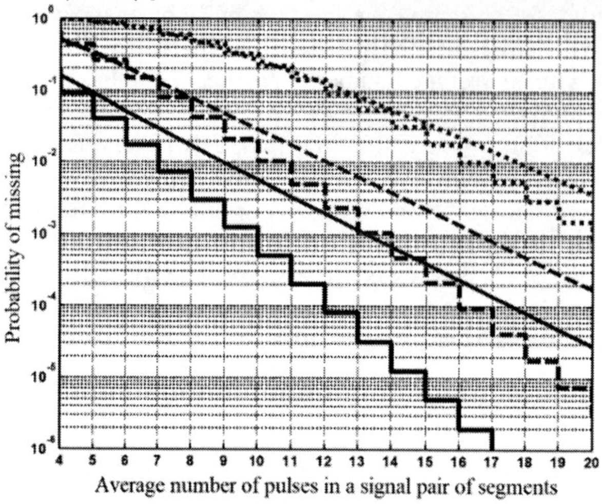

Figure 1. Dependences of the missing probability a signal pair on the average number of photons and noise pulses in a signal pair of time windows

It can be seen that the missing probability a signal pair, calculated when orienting to a normal distribution, gives an over value. Moreover, the probabilities differ by a factor of 2 for the average number of photons and noise pulses $\overline{n_{sb}}$=5 and the threshold level k_{th}=2. The difference exceeds 4 times with an average number of pulses $\overline{n_{sb}} \geq 7$.

Note that the difference decreases with increasing threshold level. So, for example, if with the average total number of photons and noise pulses $\overline{n_{sb}}$=3 the difference in the missing probability is 1.42 times at the threshold level k_{th}=2, then already only 1 % at k_{th}=8.

An acceptable difference in the missing probabilities (about 10...20 %) is provided under the condition $k_{th} \cdot p_{th.sb}$>1. From the formula

$$p_{th.s} = \frac{1}{2} \cdot \left[1 - erf\left(\frac{k_{th} - \overline{n_{sb}}}{\sqrt{2\overline{n_{sb}}}}\right)\right] = \frac{1}{2} \cdot erfc\left(\frac{k_{th} - \overline{n_{sb}}}{\sqrt{2\overline{n_{sb}}}}\right), \quad (6)$$

for normal approximation with $k_{th} = \overline{n_{sb}}$ we find $p_{th.s}$=0.5. Therefore, according to the normal approximation, the threshold level k_{th} should exceed the value of the average total number of photons and noise pulses $\overline{n_{sb}}$ to ensure the detection probability in the signal pair is more than 0.5. The same conclusion is also valid when orientation on the Poisson model.

The missing probability a signal pair of 0.1 is provided at the threshold level k_{th}=2 and the average number of photons and noise pulses in a pair of time windows $\overline{n_{sb}} \approx 4$ for the Poisson model and $\overline{n_{sb}} \approx 4.8$ for the Gaussian model. The difference in the average number of photons and noise pulses in the signal pair increases from 20% to 32 and 41% while ensuring the probability of missing, respectively, to 0.01 and 0.001. The numerical material indicates the possibility of using formula (2) or (6) for express calculations of the probability of detecting a signal pair of windows.

It is important to note that the missing probability a signal pair is determined not only by the signal-to-noise ratio, but also by the average number of noise pulses in a pair of time windows (Figure 2).

3. Synchronization probability

The synchronization process continues until the registered number of photons and noise pulses in the signal pair of windows exceeds the value of the set threshold. During synchronization, errors are possible due to a false alarm in a pair of windows preceding the signal pair in time. In addition, it is possible to miss a signal pair if the number of pulses registered in it does not exceed the set threshold.

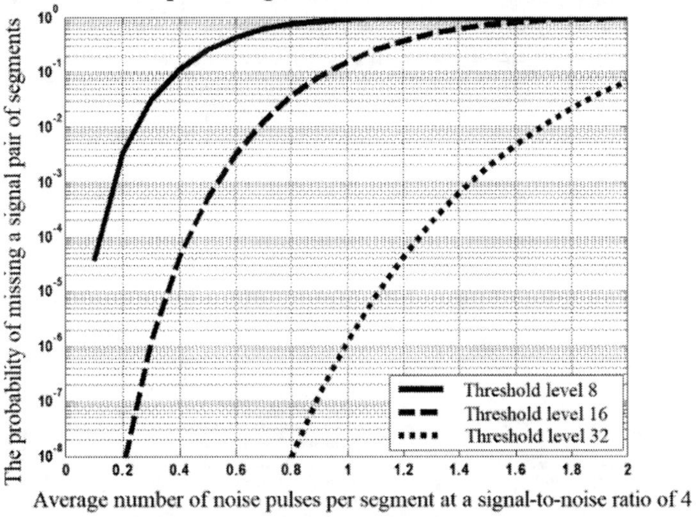

Figure 2. Working characteristic of detection

Therefore, to estimate the efficiency of synchronization, the synchronization probability is introduced, which will mean the probability of detecting a signal pair of windows, in which the synchronization signal is located, for a given number of time frames.

Let the sync pulse be located in the i-th pair of windows. The total number of paired windows (1, 2), (2, 3), …, (N_w-1, N_w), (N_w, 1) in the time frame is equal to N_w. It is considered that the sync pulse can be equally likely to be present in any pair of time windows.

The conditional probability of synchronization during the analysis of the first time frame j_T=1 is

$$p_{sync}\{i, j_T = 1\} = p_{sync1}\{i\} = (1 - p_{th.b})^{i-1} \cdot p_{th.s}.$$

Averaging this probability over all possible pairs of windows (sync pulse positions) gives the conditional synchronization probability during the analysis of the first time frame

$$p_D\{j_T = 1\} = p_{D1} = \frac{1}{N_w} \cdot \sum_{i=1}^{N_w} p_{sync1}\{i\} = \frac{1}{N_w} \cdot \sum_{i=1}^{N_w} (1 - p_{th.b})^{i-1} \cdot p_{th.s},$$

from where

$$p_{D1} = \frac{p_{th.s}}{N_w} \cdot \left[\frac{1-(1-p_{th.b})^{N_w}}{p_{th.b}}\right]. \quad (7)$$

Figure 3 shows the graphs of the dependences of the conditional probability of synchronization in the process of analyzing the first frame on the average number of signal SPP over the duration of the sync pulse. When plotting the graphs, it is assumed that the average number of noise pulses in the window is 2. The choice of the threshold level 10 provides the probability of a false alarm when analyzing a pair of noise windows for a sync pulse duration of 0.0081.

The graphs show that with an increase in the average number of photons over the duration of the sync pulse, the probability of synchronization increases during the analysis of the first frame. However, this growth is slowing down, tending to a certain limiting value. To find this limiting value, we use the power series expansion function $(1 - p_{th.b})^{N_w} = 1 - N_w \cdot p_{th.b} + \frac{(N_w \cdot p_{th.b})^2}{2!} + \frac{(N_w \cdot p_{th.b})^3}{3!} - \cdots$.

Then, expression (7), limited by three terms, transforms to the form

$$p_{D1.lim1} = p_{th.s} \cdot \left[1 - \frac{N_w \cdot p_{th.b}}{2}\right]. \quad (8)$$

Calculations by formula (8) give results that differ from calculations by formula (7) by less than 10% when the number of time windows is 80. Moreover, the difference decreases rapidly with decreasing number of windows (only 1% at N_w=20).

More accurate results are given using the formula

$$p_{D1.lim2} = p_{th.s} \cdot \left[1 - \frac{N_w \cdot p_{th.b}}{2} + \frac{(N_w \cdot p_{th.b})^2}{6}\right], \quad (9)$$

is obtained from (7) under the restriction of 4 terms in the expansion in a power series. The difference relative to the calculations using the exact formula is 1% even with 80 windows.

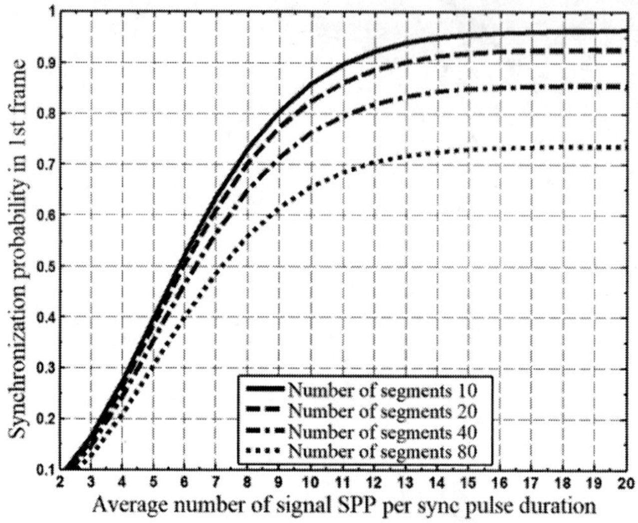

Figure 3. Dependences of the probability of synchronization in the analysis of the first frame on the average number of signal SPP for the duration of the sync pulse

Note that the calculation according to the formula (8) gives an underestimated value, and according to the formula (9) – an overestimated value. Using these features, it is possible to formulate requirements for the probability of a false alarm in a noise pair $p_{th.b}$. Naturally, while ensuring the limiting probability of synchronization when analyzing the first frame, the difference in the values calculated by formulas (8) and (9) will be minimal.

Let the distinction be K_{pD1} times. Then $K_{pD1} = p_{D1.lim2}/p_{D1.lim1}$. Solving the quadratic equation we find $N_w \cdot p_{th.b} = -1.5 \cdot (K_{pD1} - 1) + \sqrt{2.25 \cdot (K_{pD1} - 1)^2 + 6 \cdot (K_{pD1} - 1)}$.

Therefore, the choice of the probability of a false alarm in a noise pair $p_{th.b}$ must always satisfy the condition

$$p_{th.b} \le p_{th.b.max} = \frac{\sqrt{2.25(K_{pD1}-1)^2 + 6(K_{pD1}-1)} - 1.5(K_{pD1}-1)}{N_w}. \quad (10)$$

It follows from (10) that the maximum permissible probability of a false alarm in a noise pair is inversely proportional to the number of time windows. In the particular case with K_{pD1}=1.1, we have $N_w \cdot p_{th.b.max} \approx 0.64$.

Having performed similar calculations, we find the ratios for calculating the limiting permissible probability of a false alarm in a noise pair $p_{th.b.max}$ for a given permissible deviation of K_{pD1} times from the limiting synchronization probability $p_{D1.lim2}$:

$$N_w \cdot p_{th.b.max} \approx \begin{cases} 0.84 \; at \; K_{pD1} = 1.20; \\ 0.64 \; at \; K_{pD1} = 1.10; \\ 0.48 \; at \; K_{pD1} = 1.05; \\ 0.32 \; at \; K_{pD1} = 1.02; \\ 0.23 \; at \; K_{pD1} = 1.01. \end{cases} \quad (11)$$

A decrease in the deviation of the real probability of synchronization from the limiting value implies the presentation of more stringent requirements for the probability of a false alarm in a noise pair. For example, the requirement to reduce the deviation from 20% to 10% will require a decrease in the probability of a false alarm by at least 1.3 times, to 5% – by 1.75 times, and to 1% – by almost 4 times.

According to the adopted synchronization algorithm, if no sync pulse is detected during the first frame, then it is necessary to repeat the analysis of the second time frame. The conditional probability of detecting a sync pulse during the second frame is $p_D\{j_T = 2\} = p_{D2} = (1 - p_{D1}) \cdot p_{D1}$.

Using the method of mathematical induction, we find the conditional probability of detecting a sync pulse in the j_T-th time frame $p_D\{j_T\} = (1 - p_{D1})^{j_T-1} \cdot p_{D1}$.

Successful synchronization with the assumption of search in N_T time frames is possible with an unconditional probability (hereinafter the probability)

$$P_D = \sum_{j_T=1}^{N_T} p_D\{j_T\} = \sum_{j_T=1}^{N_T}(1 - p_{D1})^{j_T-1} \cdot p_{D1} = \frac{p_{D1}}{1-p_{D1}} \cdot \sum_{j_T=1}^{N_T}(1 - p_{D1})^{j_T}.$$

Using the properties of a geometric progression, we find

$$P_D = 1 - (1 - p_{D1})^{N_T}. \tag{12}$$

Note that as $N_T \to \infty$, the probability of successful synchronization tends to 1.

From (12) it follows that to provide a given synchronization probability P_D with a known probability p_{D1}, an analysis of at least $N_{T.min}$ time frames is required. Thus,

$$N_T \geq N_{T.min} = \left\lceil \frac{lg(1-P_D)}{lg(1-p_{D1})} \right\rceil, \tag{13}$$

where ⌈E⌉ – smallest integer greater than E.

The average number of time frames for successful synchronization is determined by the formula:

$$\overline{N_T} = \sum_{j_T=1}^{N_T} j_T \cdot (1 - p_{D1})^{j_T-1} \cdot p_{D1}.$$

In the limiting case, at $N_T \to \infty$, we find the limiting-maximum value of the average number of time frames for successful synchronization:

$$\overline{N_T} = \frac{1}{p_{D1}} = \frac{N_T}{p_{th.s}} \cdot \left[\frac{p_{th.b}}{1-(1-p_{th.b})^{N_T}}\right].$$

Therefore, the average time of entry into synchronism

$$\overline{T_{sync}} = T_s \cdot \overline{N_T} = \frac{N_T \cdot T_s}{p_{th.s}} \cdot \left[\frac{p_{th.b}}{1-(1-p_{th.b})^{N_T}}\right]. \tag{14}$$

If the probability of false alarms $p_{th.b}$ is close to one, then the average time taken to detect a sync pulse can be very large $\overline{T_{sync}} \approx N_w \cdot T_s / p_{th.s}$. On the contrary, if $N_w \cdot p_{th.b} \ll 1$, то $(1-p_{th.b})^{N_w} \approx 1 - N_w \cdot p_{th.b}$. Then $\overline{T_{sync}} \approx 1/p_{th.s}$. Taking into account condition (14), it can be argued that the average time spent on detecting a sync pulse will always exceed the value

$$\overline{T_{sync}} = T_s \cdot \overline{N_T} = N_{T.min} \cdot \frac{T_s}{p_{th.s}} \cdot \left[\frac{p_{th.b}}{1-(1-p_{th.b})^{N_T}}\right] = \frac{lg(1-P_D)}{lg(1-p_{D1})} \cdot \frac{T_s}{p_{th.s}} \cdot \left[\frac{p_{th.b}}{1-(1-p_{th.b})^{N_T}}\right]. \tag{15}$$

The formulas obtained make it possible to develop a technique for designing equipment for detecting a sync signal based on an analysis of the sum of counts from an adjacent pair of windows with a threshold level in the course of further research.

4. Conclusion

Formulas for calculating the missing probabilities a signal pair, successful synchronization and a false alarm have been determined. The dependences of the missing probability a signal pair on the average number of photons and noise pulses are investigated for various threshold levels. The missing probability a signal pair, calculated when orienting to the normal distribution, gives an overestimated value, compared to the missing probability a signal pair, calculated when orienting to the Poisson distribution. The dependences of the conditional synchronization probability during the analysis of the first frame on the average number of photons over the duration of the sync pulse are investigated. With an increase in the average number of photons over the duration of the sync pulse, the probability of synchronization increases during the analysis of the first frame. The numerical material obtained in the course of the study indicates the possibility of using formulas oriented to the normal distribution for express calculations of the probability of detecting a signal pair of windows.

Relationships are obtained for calculating the limiting permissible probability of a false alarm in a noise pair for a given permissible deviation from the limiting synchronization probability. A decrease in the deviation of the real probability of synchronization from the limiting value implies the presentation of more stringent requirements for the probability of a false alarm in a noise pair.

The resulting formulas and dependencies made it possible to form relationships for calculating the average time spent on detecting a sync pulse.

5. Acknowledgment

The reported study was funded by RFBR, project number 20-37-90040.

References

[1] Pljonkin A and Rumjantsev K 2016 Single-photon synchronization mode of quantum key distribution system *Proceeding of the International Conference on Computational Techniques in In-formation and Communication Technology* pp 531–534 DOI: 10.1109/ICCTICT.2016.7514637

[2] Salvail L, Peev M, Diamanti E, Alleaume R, Lütkenhaus N and Länger T 2010 Security of trusted repeater quantum key distribution networks *Journal of Computer Security* 18 (1) pp 61–87

[3] Gobby C, Yuan Z and Shields A 2004 Quantum key distribution over 122 km of standard telecom fiber *Applied Physics Letters* 84 (19) pp 3762–3764 DOI: 10.1063/1.1738173

[4] Cabello A 2000 Quantum key distribution without alternative measurements *Physical Review A* 61 (5) 052312 pp 1–4 DOI: doi.org/10.1103/PhysRevA.61.052312

[5] Hiskett P A, Rosenberg D, Peterson C G, Hughes R J, Nam S, Lita A E, Miller A J and Nordholt J E 2006 Long-distance quantum key distribution in optical fibre *New Journal of Physics* 8 pp 1–7 DOI: 10.1088/1367-2630/8/9/193

[6] Patel K A, Dynes J F, Choi I, Sharpe A W, Dixon A R, Yuan Z L, Penty R V and Shields A J 2012 Coexistence of high-bit-rate quantum key distribution and data on optical fiber *Physical Review X* 2 (4) pp 1–8 DOI: 10.1103/PhysRevX.2.041010

[7] Lindsey W C 1972 *Synchronization Systems in Communication and Control* (Prentice-Hall, Englewood Cliffs, New Jersey)

[8] Gobby G, Yuan Z L and Shields A J 2004 Unconditionally secure quantum key distribution over 50 km of standard telecom fibre *Electronics Letters* 40 pp 1603–1605 DOI: 10.1049/el:20045038

[9] Gottesman D, Lo H K, Lutkenhaus N and Preskill J 2004 Security of quantum key distribution with imperfect devices *Proceedings of International Symposium On Information Theory* p 136

[10] Fröhlich B, Lucamarini M, Dynes J F, Comandar L C, Tam W W S, Plews A, Sharpe A W, Yuan Z and Shields A J 2017 Long-distance quantum key distribution secure against coherent attacks *Optica* 4 pp 163–167 DOI: 10.1364/OPTICA.4.000163

[11] Kurochkin V L 2012 Eksperimental'nye issledovaniya v oblasti kvantovoj kriptografii [Experimental research in the field of quantum cryptography] *Fotonika [Photonika]* 5 pp 54–66

[12] Wu Qing-Lin, Han Zheng-Fu, Miao Er-Long, Liu Yun, Dai Yi-Min and Guo Guang-Can 2007 Synchronization of free-space quantum key distribution *Optics Communications* 275 (2) pp 486–490

[13] Xie Cailang, Guo Ying, Liao Qin, Zhao Wei, Huang Duan, Zhang Ling and Zeng Guihua 2018 Practical security analysis of continuous-variable quantum key distribution with jitter in clock synchronization *Physics Letters A* 382 (12) pp 811–817

[14] Rumiantsev K and Rudinsky E 2017 Parameters of the two-stage synchronization algorithm for the quantum key distribution system *Proceedings of the 10th International Conference on Security of Information and Networks (SIN-2017)* pp 140–150 DOI: 10.1145/3136825.3136888

[15] Mironov Y K and Rumyantsev K E 2020 Single-Photon Algorithm for Synchronizing the System of Quantum Key Distribution with Polling Sections of a Fiber Optic Line *Futuristic Trends in Networks and Computing Technologies* pp 87–97 DOI: https://doi.org/10.1007/978-981-15-4451-4_8

[16] Gagliardi R M and Karp Sh 1976 *Optical communication* p 432

[17] Mironov Y K and Mironova P D 2020 Veroyatnost' obnaruzheniya signal'nogo okna v algoritme poiska fotonnogo impul'sa s razbieniem vremennogo intervala na vremennye okna [Probability of detecting a signal window in the algorithm for searching for a photon pulse with a division of the time interval into time windows] *Fundamental'nye i prikladnye aspekty komp'yuternyh tekhnologij i informacionnoj bezopasnosti [Fundamental and applied aspects of computer technology and information security]* pp 135–138

Semi-structured information in the field of artificial intelligence and information security: processing results

N.I. Yusupova[1], O.N. Smetanina[1], M.M. Gayanova[1], N.P. Komendantova[2]

[1]Ufa State Aviation Technical University, Ufa, K. Marks str., 12, 450000, Russia

[2]International Institute for Applied Systems Analysis, Laxenburg, Schlossplatz, 1, A-2361, Austria

E-mail maya.gayanova@gmail.com

Abstract. This article is devoted to the semantic analysis of weakly structured information in the field of "Artificial intelligence and information security". The methodology of this research included two stages and is based on the meta-analysis of existing studies. The received results allow development of further methodological recommendations on semi-structured information and artificial intelligence.

Keywords: Semi-structured information, Artificial Intelligence, Information Security, Semantic analysis.

1. Introduction

The processing of semi-structured information is an urgent task, many scientists are engaged in it at present day. There are no general solutions in this area, the authors make specific decisions depending on the subject area. This article discusses some of the results of semantic analysis of semi-structured information in the field of knowledge "Artificial Intelligence and Information Security".

2. Research methodology

The research includes two stages. The first stage is a preliminary analysis of the information of the selected subject area based on data from a known platform, the purpose of which is to highlight the most significant publications. The second stage is a semantic analysis of the most significant publications to highlight the most interesting scientific results.

The scientific electronic library eLIBRARY.ru was the source for empirical data collection [1]. eLIBRARY.ru is a Russian scientific electronic library integrated with the Russian Science Citation Index (RSCI), which currently contains over 34 million articles. The API integrated into eLIBRARY.ru allows us to get a list of publications and metadata about publications.

On the main page of eLIBRARY.ru there is a built-in search function and an advanced search in the library, where a specific query is entered. Further, the system asks where to look for information (in the title of the publication, in the annotation, in keywords, in the names of the authors' organizations, in the lists of cited literature and in the full texts of publications) and in what types of publications (articles in journals, books, conference materials, deposited manuscripts, dissertations, reports and patents).

We can also specify search topics, specific authors or journals, which will refine the search result. Then we can set the search parameters: search based on morphology, search for similar text, search in publications that have full text on eLIBRARY.ru, etc. Further, we can set a specific year or an interval by years for the search and sorting type: by years or by relevance.

The system generates a list of publications for the request, which can be saved to a collection with the appropriate name. In personal selections, we can analyze publications according to various parameters, which are quantitative characteristics of publications that are issued by the system: the total number of publications, the number of articles in journals included in the Web of Science or Scopus citation bases, included in the RSCI or RSCI core. Further, it gives a weighted impact factor of the journals in which articles were published, the total number of authors, the average number of publications per author, the total number of citations, the average number of citations per article, the number of articles cited at least once, the number of self-citations and the Hirsch index.

Here we can also get data for statistical reports on the types of distribution of publications by topic, by keywords, by journals, by organizations, by authors, by years, by the number of co-authors, by the number of citations, citing publications on the topic, citing publications by keywords, citing publications by journals, citing publications by organizations, citing publications by authors, citing publications by years, citations by topic, citations by journals, citations by organizations, citations by authors, citations by years of citing publications, citations by years of cited publications.

Quantitative indicators are taken from the above-mentioned procedure. On their basis, indicators specific to this area are built and directions for further immersion are determined in order to identify hidden patterns.

The second phase, associated with immersion in the semantics of the body of scientific texts, suggests the involvement of knowledge experts in the given field of science. It should be noted that if the analysis is carried out automatically, only with the built-in tools of eLIBRARY.ru, without the participation of an expert, articles that are not related to the given field of science can be included in the selection, and the terms are only mentioned. When a knowledge expert is connected, non-material articles are rejected, and the number of articles in the collection decreases, and only articles that present significant results remain.

3. Results of the first stage

For a query in eLIBRARY.ru [1], four combinations of keywords from the field of knowledge were selected: "Artificial Intelligence and Information Security" (AI&IS), "Machine Learning and Cybersecurity" (ML&C), "Data Mining and Cybersecurity" (DM&C), "Machine Learning and Information Security "(ML&IS), "Data Mining and Information Security" (DM&IS). The search was carried out only in the title of the publication, in the annotation and in keywords, which significantly reduced the number of publications in the collection.

The obtained generalized quantitative data obtained from eLIBRARY.ru, which is the most important characteristics of publications on these topics, is summarized in Table 1.

Table 1. Key characteristics of publications on the analyzed topics

	Research Topics			
Quantitative characteristics	ML&C	DM&C	ML&IS	DM&IS
Total publications	27	28	107	419
Articles in journals	16	19	61	281
Articles in journals included in Web of Science or Scopus	1	1	6	21
Articles in journals included in the RSCI core	3	6	21	44
Articles in RSCI journals	2	6	19	34
The weighted impact factor of the journals in which the articles were published	0,410	0,685	0,412	0,428
Total number of authors	69	73	263	987
Average number of publications per author	0,39	0,38	0,41	0,42

Total number of citations	30	117	192	1452
Average number of citations per article	1,11	4,18	1,79	3,47
Number of articles cited at least once	9	12	33	189
Total number of self-citations	0	0	8	25
Hirsch index	3	5	6	18

The conclusion from the Table 1 is that the largest number of publications, the largest number of articles in journals, articles in journals included in the Web of Science or Scopus citation bases, the total number of authors, the total number of citations, the average number of citations per article, the number of articles cited at least once, the number of self-citations and the Hirsch index is in the field of "Data Mining and Information Security".

Further, according to data from eLIBRARY.ru, a graph was built reflecting the articles appearance dynamics from 2011 to 2020 (figure 1).

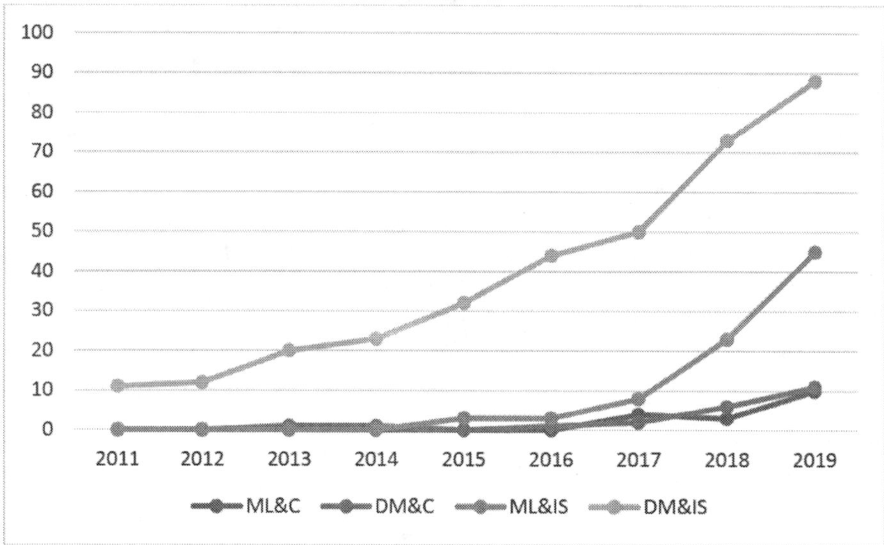

Figure 1. Articles appearance dynamics

From Fig. 2 it is easy to see that since 2011 there has been a steady increase in the number of articles over the years, which shows that this area of knowledge is relevant and the interest of researchers in this topic is growing. 2020 is not included as it is not yet complete.

Further, an analysis of the performance of the most active researchers by topic was carried out (Figures 2-5).

IOP Conf. Series: Materials Science and Engineering 1069 (2021) 012012 doi:10.1088/1757-899X/1069/1/012012

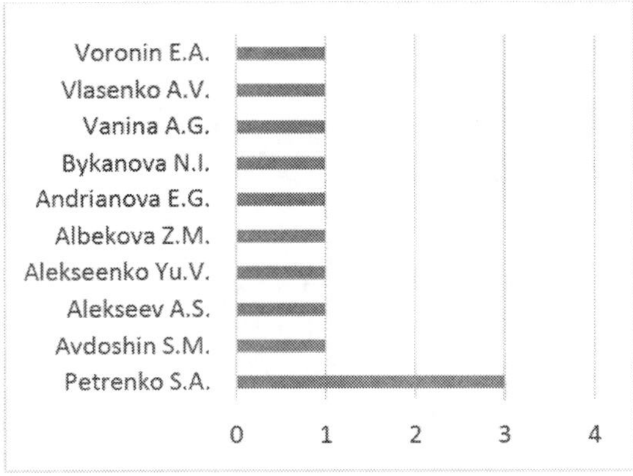

Figure 2. Comparative analysis of performance on the topic "Machine Learning and Cybersecurity"

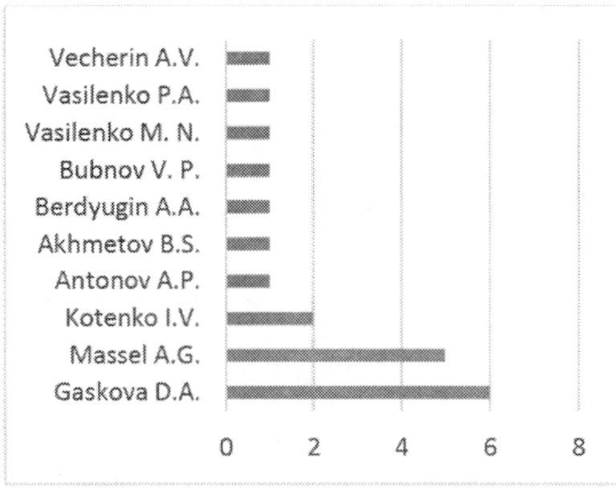

Figure 3. Comparative analysis of performance on the topic "Data Mining and Cybersecurity"

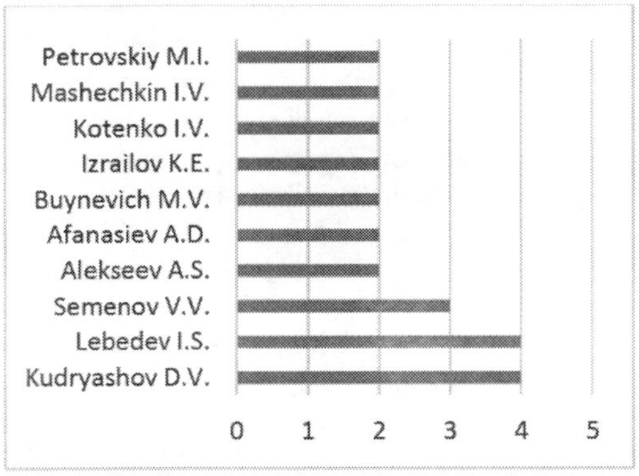

Figure 4. Comparative analysis of performance on the topic « Machine Learning and Information Security»

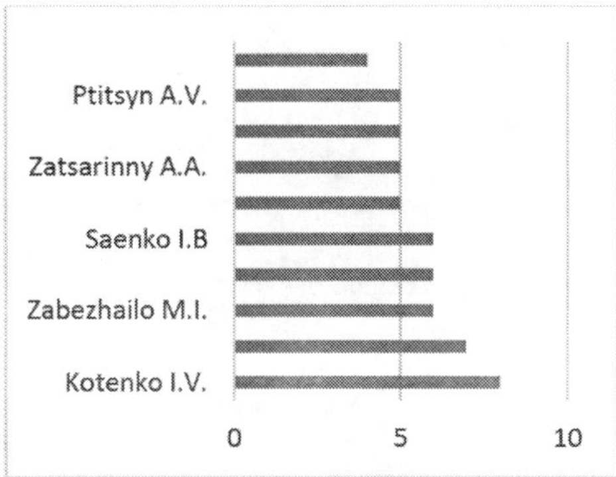

Figure 5. Comparative analysis of performance on the topic «Data Mining and Information Security»

Although the query was based on keyword combinations from the Artificial Intelligence and Information Security domain, there is not much intersection between authors by subject. Only Professor Kotenko I.V. from SPIIRAS, St. Petersburg, is one of the ten most active researchers on three topics and takes the first position in the topic "Data Mining and Information Security". Alekseev A.S. also meets twice. from Don State Technical University, in the topics "Machine Learning and Cybersecurity" and "Machine Learning and Information Security".

Further analysis of the publication activity of research organizations and scientific schools was carried out according to data from eLIBRARY.ru. The results of such analysis is shown in Figures 6-9.

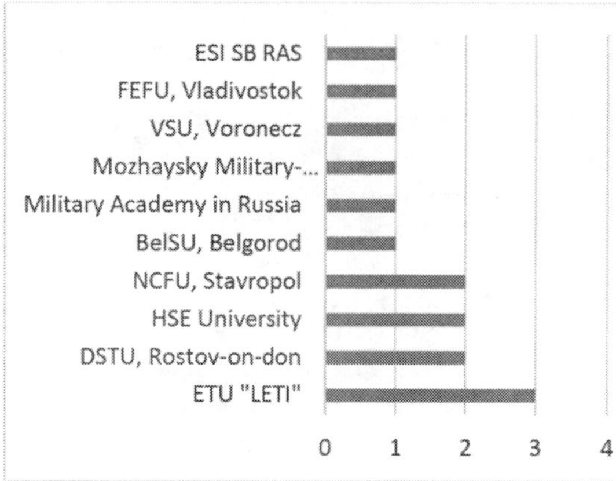

Figure 6. Performance analysis of the most active organizations by topic «Machine Learning and Cybersecurity»

Figure 7. Performance analysis of the most active organizations by topic «Data Mining and Cybersecurity»

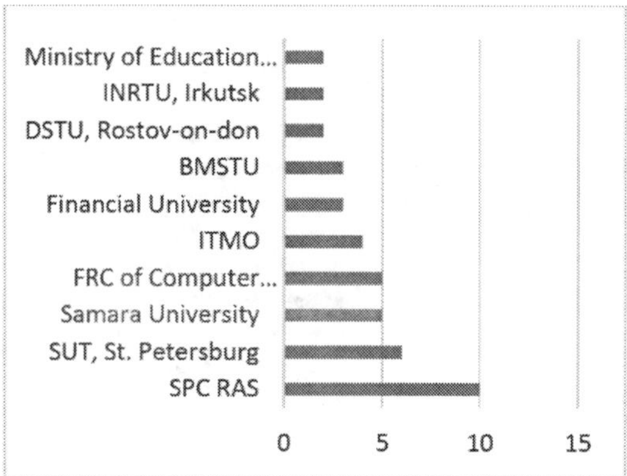

Figure 8. Performance analysis of the most active organizations by topic « Machine Learning and Information Security»

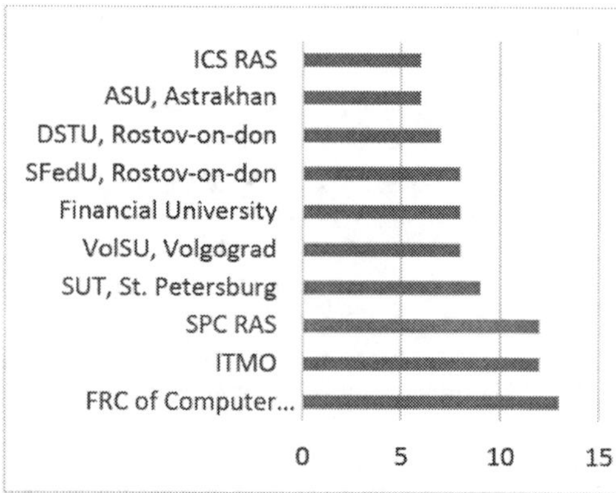

Figure 9. Performance analysis of the most active organizations by topic «Data Mining and Information Security»

There is a stronger intersection of organizations here. The undisputed leader is the St. Petersburg Federal Research Center of the Russian Academy of Sciences, which is present in three out of four topics and takes the first position in the topic "Machine Learning and Information Security". We can also note the Bonch-Bruevich Saint-Petersburg State University of Telecommunications, the Federal Research Center Computer Science and Control of the Russian Academy of Sciences and the National Research University ITMO, noted in two topics and occupying leading positions in them.

Also of interest are the journals in which the authors publish the results of their research. eLIBRARY.ru allows you to extract data on publication activity by journals; the diagrams based on these data are presented in Figures 10-13.

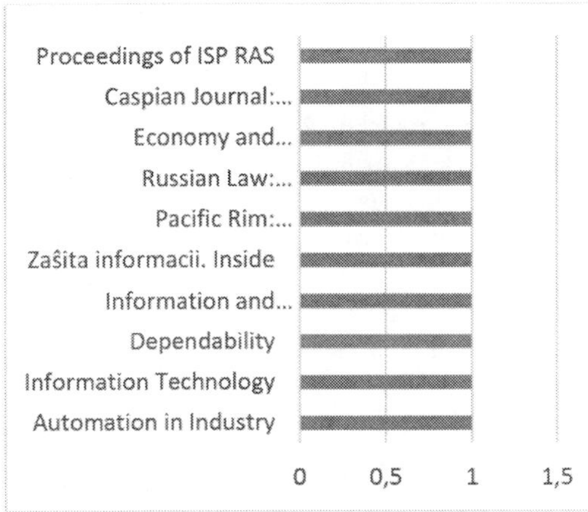

Figure 10. Analysis of journal activity by topic «Machine Learning and Cybersecurity»

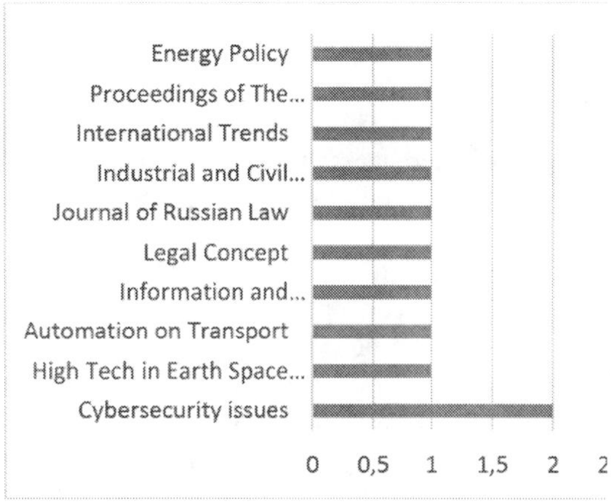

Figure 11. Analysis of journal activity by topic «Data Mining and Cybersecurity»

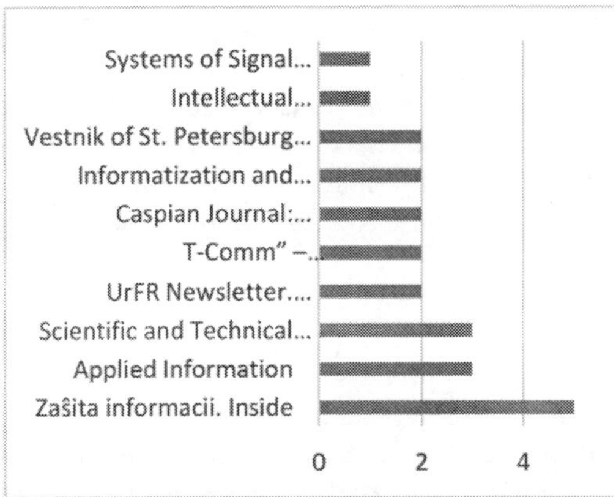

Figure 12. Analysis of journal activity by topic «Data Mining and Information Security»

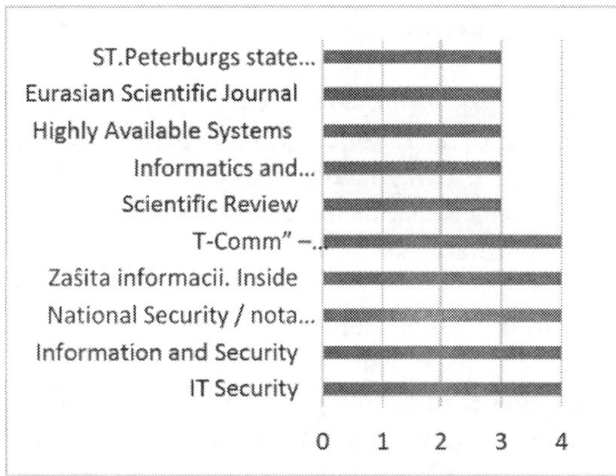

Figure 13. Analysis of journal activity by topic «Data Mining and Information Security»

It can be seen that 4 combinations of keywords do not overlap in active journals. The only exception is the journal "Zaŝita informacii. Inside ", it published the results of research on three topics: "Machine Learning and Cybersecurity", "Machine Learning and Information Security" and "Intelligent Analysis and Information Security". Also worth mentioning is the «Caspian Journal: Management and High Technologies» magazine, which is mentioned twice.

4. Results of the second stage

Based on the preliminary processing of information at the first stage, the most popular publications of active authors were found. Analysis of the content of these articles allows us to highlight the most interesting results in the given field of science.

In the papers of Petrenko S.A. from ETU LETI [2, 3] approaches to ensuring the security of information systems of banks and e-commerce using multifunctional BI-platforms are considered. The analysis of key problems in the organization of big data is carried out and problems of implementation, operation and maintenance of BI systems from the point of view of ensuring information security are identified. The cybersecurity ontology (meta-ontology) was proposed as a way to represent knowledge about the qualitative characteristics and quantitative laws of information warfare.

In the papers of S.M. Avdoshin from the Higher School of Economics [4], an analysis of successful cyber-attacks on the blockchain systems industry has been carried out. A list of the main types of successful thefts by hackers is presented.

The results of research by D.A. Gaskova and A.G. Massel from the Melentiev Energy Systems Institute of Siberian Branch of the Russian Academy of Sciences, Irkutsk, are given in [5, 6, 7]. In [5], a risk-oriented decision support approach was formulated in identifying critical objects in the energy sector from the point of view of cyber threats. The novelty of the research lies in the development and application of a fractal stratified model of the relationship between risks, energy facilities and technologies.

In [6], the concepts of Smart Grid and digital energy were discussed, and the main information technologies proposed for their implementation were analyzed. It was proposed to include intelligent technologies in this list to support strategic decisions on energy development.

In [7], a risk-oriented approach was described for analyzing threats and assessing the risk of cybersecurity breaches at energy facilities, taking into account damage caused by damage or destruction of the facility, using quantitative and qualitative parameters.

The papers of the most active author, Professor Igor Kotenko from SPIIRAS are given in [8-12]. In [8], the analysis of existing methods of attacker behavior, characteristics of the attacker's profile and their application for predicting future steps of an attack was given. The analysis made it possible to identify the main advantages and limitations of approaches to predicting attacks and building an attacker's profile, existing problems and prospects in this area.

The [9] examined an approach to online decision-making in telecommunication systems based on the use of hierarchical fuzzy situational networks, which allows making effective decisions in the context of dynamically changing external factors and presents the mathematical foundations for the creation and application of hierarchical fuzzy situational networks for the implementation of fuzzy logical inference in telecommunication networks.

In [10], an approach to the analysis of cybersecurity data based on a combination of a set of machine learning methods and big data technologies for network attacks and anomaly detection was presented. The approach was characterized by several levels of data processing, including extraction and decomposition of datasets, compression of feature vectors, training, and classification.

The paper [11] discussed the development and application of a semantic model for security assessment. The proposed model was presented as an ontology of metrics based on the relationship between security-related data sources, the primary characteristics of the underlying security data, and the objectives of the security assessment.

In [12], methods and techniques for managing the safety of complex heterogeneous systems with an emphasis on the correlation of events and safety assessment were considered. The proposed approach is based on a comprehensive analysis of large heterogeneous security data for event correlation, including syntactic and semantic analysis of security events and information.

The works of I.S. Lebedev from the St. Petersburg Federal Research Center of the Russian Academy of Sciences are presented in [13, 14]. In [13], the issues of ensuring the cybersecurity of autonomous unmanned objects were considered. The prerequisites were identified that determine the need for external monitoring systems. An approach to the analysis of the state of cybersecurity of an autonomous object based on classification methods was proposed and allows one to identify the current state using the processing of digitized acoustic information.

The article [14] presents methods for creating signatures of executable files based on the frequency distributions of their informative characteristics, which can be used to identify programs. A new method for identifying executable files was proposed and the results of experiments on their identification using a statistical criterion were presented.

Works by A.A. Grusho from Federal Research Center Computer Science and Control of the Russian Academy of Sciences are presented in [15-17]. In [15], the problem of revealing weakly expressed anomalies in the model of generalized systems was considered.

In [16], security policy rules for a multilevel access system of a distributed information system were constructed, and on the basis of the infrastructure associated with metadata, the possibility of implementing this security policy in a distributed information computing system was shown.

In [17], an approach was formulated to the study of some types of fraud in the digital economy using causal relationships.

5. Conclusion

Thus, the proposed method of processing semi-structured scientific information allows to get an insight of the most interesting scientific results in the subject area of interest in recent years. The analysis of publications based on data from the eLIBRARY.ru platform is a preliminary stage in the study of the corpus of scientific texts on the selected topic. It should be noted that when working with eLIBRARY.ru data, articles are automatically selected by keywords, without taking into account the specifics of the type and specifics of scientific research. To select the necessary material from the selection, you have to resort to a knowledge expert in the given field is science.

Carrying out the analysis "manually" is a rather cumbersome procedure, so it is advisable to develop an information system to support researchers with generalized data based on the analysis of the corpus of scientific texts. Such a system should provide information on the most interesting scientific results obtained recently.

6. Acknowledgment

The results of the research presented in this article were supported by Grants RFBR 19-07-00780.

The study is conducted with financial support from the Ministry of Education and Science of the Russian Federation as part of the basic part of the state assignment to higher education educational institutions # FEUE-2020-0007.The work is partially supported by the RFBR grant # 18-01-00796.

References

[1] Moskaleva O, Pislyakov V, Sterligov I, Akoev M and Shabanova S 2018 Russian Index of Science Citation: Overview and review *Scientometrics* **116** 449–62

[2]. Petrenko S and Makoveichuk K 2020 Development of BI-Platforms for Cybersecurity Predictive Analytics pp 273–88

[3] Petrenko S, Makoveichuk K and Olifirov A 2020 New Methods of the Cybersecurity Knowledge Management Analytics pp 296–310

[4] Lazarenko A and Avdoshin S 2019 Financial Risks of the Blockchain Industry: A Survey of Cyberattacks *Proceedings of the Future Technologies Conference (FTC) 2018* ed K Arai, R Bhatia and S Kapoor (Cham: Springer International Publishing) pp 368–84

[5] Massel A and Gaskova D 2020 Identification of Critical Objects in Reliance on Cyber Threats in the Energy Sector *Acta Polytech. Hungarica* **17** 61–73

[6] Massel L and Massel A 2018 Intelligent support tools for strategic decision-making on Smart Grid development ed N Voropai, F-J Lin, G Chang, A Kler and R Lis *E3S Web Conf.* **69** 02009

[7] Daria G and Massel A 2018 Intelligent System for Risk Identification of Cybersecurity Violations in Energy Facility 2018 3rd Russian-Pacific Conference on Computer Technology and Applications (RPC) (IEEE) pp 1–5

[8] Doynikova E, Novikova E and Kotenko I 2020 Attacker Behaviour Forecasting Using Methods of Intelligent Data Analysis: A Comparative Review and Prospects Information 11 168

[9] Kotenko I, Saenko I and Ageev S 2019 Hierarchical fuzzy situational networks for online decision-making: Application to telecommunication systems *Knowledge-Based Syst.* **185** 104935

[10] Kotenko I, Saenko I and Branitskiy A 2020 Machine Learning and Big Data Processing for Cybersecurity Data Analysis *Data Science in Cybersecurity and Cyberthreat Intelligence* (Springer) pp 61–85

[11] Doynikova, E., Fedorchenko, A., Kotenko, I., 2020. A semantic model for security evaluation of

information systems. Journal of Cyber Security and Mobility 9, 301–330. doi:10.13052/JCSM2245-1439.925
[12] Kotenko I, Fedorchenko A and Doynikova E 2020 Data Analytics for Security Management of Complex Heterogeneous Systems: Event Correlation and Security Assessment Tasks *Advances in Cyber Security Analytics and Decision Systems* (Springer) pp 79–116
[13] Semenov V, Sukhoparov M and Lebedev I 2019 Approach to Side Channel-Based Cybersecurity Monitoring for Autonomous Unmanned Objects *International Conference on Interactive Collaborative Robotics* pp 278–86
[14] Salakhutdinova K, Lebedev I, Krivtsova I, Bazhayev N, Sukhoparov M, Smimov P, Markelov D, Davvdov A, Pecherkin S, Kolcherin D, Shaparenko Y and Iskanderov Y 2017 A Frequency Approach to Creation of Executable File Signatures for their Identification *2017 IEEE 11th International Conference on Application of Information and Communication Technologies (AICT)* (IEEE) pp 1–7
[15] Grusho A, Grusho N and Timonina E 2020 Method of Several Information Spaces for Identification of Anomalies *International Symposium on Intelligent and Distributed Computing* pp 515–20
[16] Grusho A A, Grusho N A and Timonina E E 2019 Using Metadata to Implement Multilevel Security Policy requirements *Informatics Appl.* **13** 85–9
[17] Grusho A A, ZABEZHAILO M I, Grusho N A and Timonina E E 2019 Architectural Decisions in the Problem of Identification of Fraud in the Analysis of Information Flows in Digital Economy *Informatics Appl.* **13** 22–8

Artificial intelligence tools for analyzing emotionally colored information from customer reviews in the service sector

N I Yusupova[1], D R Bogdanova[1], N P Komendantova[2]

[1]Ufa State Aviation Technical University, Ufa, Karl Marx str., 12, 450008, Russia

[2] International Institute for Applied Systems Analysis, Laxenburg, Schlossplatz, 1, A-2361, Austria

E-mail dianochka.bogdanova@gmail.com

Abstract. On the Internet, you can find millions of service reviews. This information can be useful to support decision-making for both potential customers and for managing a service business It is especially important to consider the emotional tone of such reviews, because emotions can attract or repel potential customers. The article discusses the actual task of the automatic processing of customer reviews and identifying emotionally colored information in it. In this paper we provide the definition of emotions and describe the results of the analysis of their classification. Further on, we identify six main emotions from the customer reviews. As our methodological basis we use a well-known approach for accounting emotionally colored information in customer reviews. This approach made it possible to set the task of automatic analysis of customer reviews, which we applied in this paper.

Keywords: Artificial intelligence tools; emotionally colored information; customer reviews; sentiment analysis

1. Introduction

Nowadays several specialized platforms and websites exist that contain reviews on a variety of products and services. This information contains the feelings and emotions that a person has experienced as a result of using these products or services. Modern artificial intelligence and affective computing tools make it possible to extract useful data from this emotionally colored information that decision makers can use when promoting these products and services.

Emotions – is an integral part of human thinking. They accompany any thought, any activity, any action. Emotions are a key element for human intelligence emulating. Humanity is on the verge of merging the real and virtual worlds. We can tell this by observing the fact that people spend more and more of their time in virtual reality such as social media, professional communities, where they exchange opinions about certain events, goods, services, etc.

Intelligence is a multifaceted concept, and emotional intelligence is its important component. Today, the development of information technology makes it possible to develop systems that account the user's emotions and are able to emulate them. And automatic analysis of emotionally colored information from user reviews is its important aspect. The solution of this problem is impossible without the use of the artificial intelligence tools.

The modern scientific community is faced with the task of developing models and methods for recognizing and accounting emotionally colored information in intelligent decision support systems. The first part of this article discusses the definition of emotions and approaches to their classification. The second part contains the formulation of the problem of recognizing and accounting emotions and opinions from customer reviews on the Internet in natural language. The third part analyzes the existing approaches to emotionally colored information accounting on the Internet. And the last part offers solutions and tools for analyzing emotionally colored information in the decision-making process.

2. Emotion classification in service sector customer reviews

According to Wikipedia, emotion is a medium-term mental process that reflects a subjective evaluative attitude towards existing or possible situations and the objective world. Emotions are studied and classified by many scientists in psychology and affective computing fields. There are several approaches to their classification, and we describe some of them below.

One of these approaches is based on the idea of a discreteness of emotions. In Tomkins' theory of discrete emotions [1], eight emotions are considered, such as: surprise, interest, joy, anger, fear, disgust, shame and suffering. These emotions are considered - culturally recognizable, basic, illogically determined emotional responses, in which both expression and recognition are the same for all people, regardless of their ethnic and cultural backgrounds [2].

In his cross-cultural study Paul Ekman identified six main emotions (anger, disgust, fear, happiness, sadness, and surprise), which are expressed to varying degrees through certain characteristics [3].

Another approach to the emotion classification is based on the assumption of the existence of a multidimensional space of emotions. Usually, there are one or more parameters, such as arousal, valence, and intensity, that are usually considered as measurements of emotional space. Within this approach, there is the Watson and Tellegen PANA model [4]. Russell's Circumplex model can also be attributed here [5]. The Pleasure-Arousal-Dominance (PAD) emotional state model also uses a three-dimensional approach to measure emotions [6]. Plutchik's model is the most famous three-dimensional hybrid of both basic and complex categories, in which emotions of different intensities can be combined into emotional dyadic. He presented a wheel of eight emotions: joy, trust, fear, surprise, sadness, disgust, anger, and anticipation in twenty-four primary, secondary and tertiary dyadic [7].

In the service sector, the role of emotions is increasing. The fact that emotion is a mental process of medium duration becomes important here. It means that we experience emotion exactly at the moment of service consumption. That is why it is important to be able to correctly assess emotions and take them into account in the process of providing services.

The brighter emotions are experienced by person, the more this person is motivated to share emotionally colored information about a particular service provided to him. This information can now be found in customer reviews on social media and specialized platforms. This information is extremely important and useful for decision makers in the management and promotion of certain goods and services.

3. Statement of the task

To improve the effectiveness of decision making in complex systems, it is necessary to take into account emotionally colored information from customer reviews. The task is to extract and account the emotions of the authors from the reviews of the services rendered using the example of a tourist complex.

To solve this problem, this study will consider six basic emotions - joy, surprise, sadness, fear, anger and disgust, as shown in Figure 1.

As a source of emotionally colored information processed, we will consider the texts of reviews about services from specialized Internet portals. The analysis uses text mining techniques and the benefits of natural language processing (NLP).

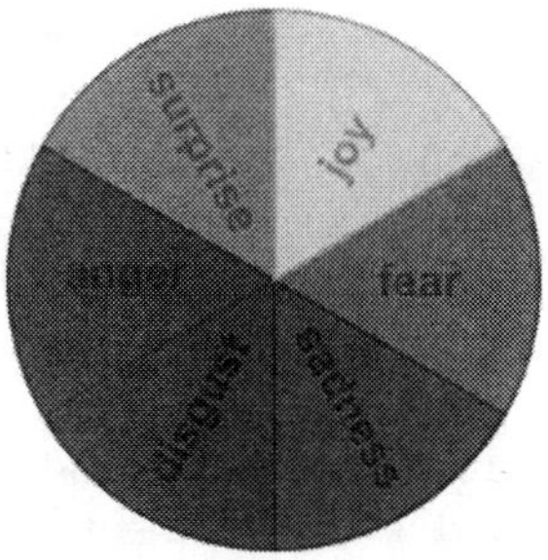

Figure 1. Considered emotions

4. Existing approaches

Many scientists are engaged in the recognition of emotionally colored information from user activities. The existing approaches can be roughly divided into several groups. We can select reviews analysis of sentiment, analysis of the sentiment polarity, analysis of the context. The approaches also can be grouped according to the type of source of information being processed - unimodal and multimodal.

The automatic fixation of public sentiments on social issues, political movements, marketing campaigns and preferences for goods and services has generated growing interest in the scientific and business communities. This led to the emergence of new areas of affective computing and text mining. And it led to the development of methods of information searching, multimodal signal processing, extraction of people's sentiment from the ever-growing volume of social data on the Internet [8].

Text sentiment analysis consists of extracting information about opinions, feelings and emotions transmitted by authors in relation to topics they interested in. It is often applied to opinion mining and emotion mining. Natural language processing and machine learning are used to determine the author's attitude to the subject described in the text, followed by opinion mining. Similar technologies are used to make an intelligence analysis authors' emotion and classify them in relation to events or topics. Methods of the text analysis of emotions are used in a variety of tasks, such as obtaining information about customer satisfaction, helping to select educational materials in online education, recommending products and services based on customer emotions, and even predicting mental disorders. In earlier works of mood analysis, emotions were not given enough attention [9].

Modern information technologies allow everyone to create and share their opinions and ideas in the generated Internet content with millions of people around the world. Most of the content posted and consumed on the Internet is multimodal. Today there are many devices that allow to create such content, and the amount of generated information is growing very fast. It is becoming increasingly difficult for researchers to evaluate and process this multimodal content stream, and even harder to organize or comprehend it. Receiving information from photos and videos is a critical need for mood analysis because product and service reviews are gradually shifting from unimodal too multimodal. Some researchers have proposed the usage of deep convolutional neural networks to extract features from visual and textual modalities. This provides a significant improvement in multimodal emotion recognition and mood analysis on various data sets [10].

Social media today are platforms that give people a possibility to share their opinions on various issues and topics with their friends, family, relatives and other members of the community. To do this, people use texts, images, audio and video messages and posts. But despite the presence of other forms of communication, text is still one of the most common methods of communication on social networks. The use of sentiment analysis methods for analyzing messages on the Internet allows to determine moods and emotions by posts and their comments and give user impact assessments based on various user parameters. It is also possible to create generalized and personalized recommendations for users, based on their activity in the posts of social networks [11].

The analysis of feelings and emotions are related tasks, and, therefore, are more effectively solved together. Multimodal input signals convey diverse and differing information and usually have different contributions to decision making. There is an intermodal method that works at the context level for simultaneously predicting moods and expressed emotions in statements. This approach works well on the CMU-MOSEI dataset for multimodal analysis of moods and emotions and shows a better result in comparison with the solution of only one of the considering problems [12].

The task of determining how different cultures respond to a crisis is another interesting challenge that uses the analysis of opinions and emotions. What feelings prevail in societies and political agenda in dealing with the crisis situation. The decisions made in this case are conditioned by events, social pressure and may not reflect the wishes of the population of a particular country. And while some are happy with this, others may show resentment. Social media is a mirror for public sentiment. Despite geographic proximity, many neighboring countries respond differently to the same threat. For example, in 2020, the leaderships of Denmark and Sweden made different decisions regarding the new threat. In 2020, the actions of the leadership of Denmark and Sweden to counter the danger did not coincide, and both countries populations supported these decisions mostly unanimously. The purpose of the research is to analyze the reaction of people of different cultures to the crisis, as well as the attitude of people to subsequent actions taken in different countries. To solve this problem, deep long-term short-term memory (LSTM) models are used on data extracted from social media posts and trained on the sentiment140 dataset. A distinctive feature of this study was the inclusion of smileys when analyzing the tone and emotion of messages. [13].

The application of sentiment analysis and emotionally colored information has a variety of applications, also including in the financial services sector. Online peer-to-peer (P2P) lending platforms are gaining popularity for providing finance and loans to small and micro-entrepreneurs. By applying an assessment of emotionally charged information, borrowers in a P2P financial platform can increase the probability that their microloan applications will be funded. The text of the application is analyzed, and the set of characteristics of the description of the microcredit is assessed. The influence of this description on the success of financing is also revealed. Sentiment mining methods are used to analyze and extract emotion from an unstructured P2P credit dataset. Borrowers can increase the chances of financing success by improving the quality of credit descriptions. They can make them more attractive to potential lenders by investing in them certain emotions [14].

5. Proposed solutions and tools for analyzing emotionally colored information in the decision-making process

To solve the problem of the emotionally colored information accounting in decision-making in the service sector, it must be understood that different customer emotions have different importance for a decision-maker. Negative customer emotions are critical for the stable operation of the enterprise. This can cause the rejection of the service by current customers and reduce the number of customers in the future.

In this paper we propose to analyze customer reviews in several stages. The stages of customer reviews analysis is presented in more detail in the Figure 2. A functional model for solving the problem of analyzing emotionally colored information from customer reviews in the decision-making process was developed in the SADT methodology.

At the first stage - collecting reviews from Internet resources. At the second stage - cleaning, preprocessing, and analyzing the collected reviews. It includes reviews markup, lemmatization and coding the reviews in vector form. At the third stage, the sentiment analysis is carried out using with machine learning tools and data from open-source corpus from sentiment-marked words. At the fourth stage, the analysis of emotions shown by customers is carried out in the allocated classes of positive, negative, and neutral reviews by using pattern recognition methods and a fuzzy search tool. It is important for the decision maker to know which aspect of the service provided is caused by the emotions that the customer have experienced. Therefore, at the fifth stage, an aspect analysis of classified reviews is carried out, and the emotional sentiments of individual statements about aspects are determined using the formation of dictionaries of provided services aspects. Qualitative research of customer reviews is carried out using the construction of models based on decision trees, where the emotional tone of the response is a variable dependent on the presence or absence of emotionally colored information on certain aspects of services. It allows you to identify patterns that affect customer satisfaction and form the rules for decision-making.

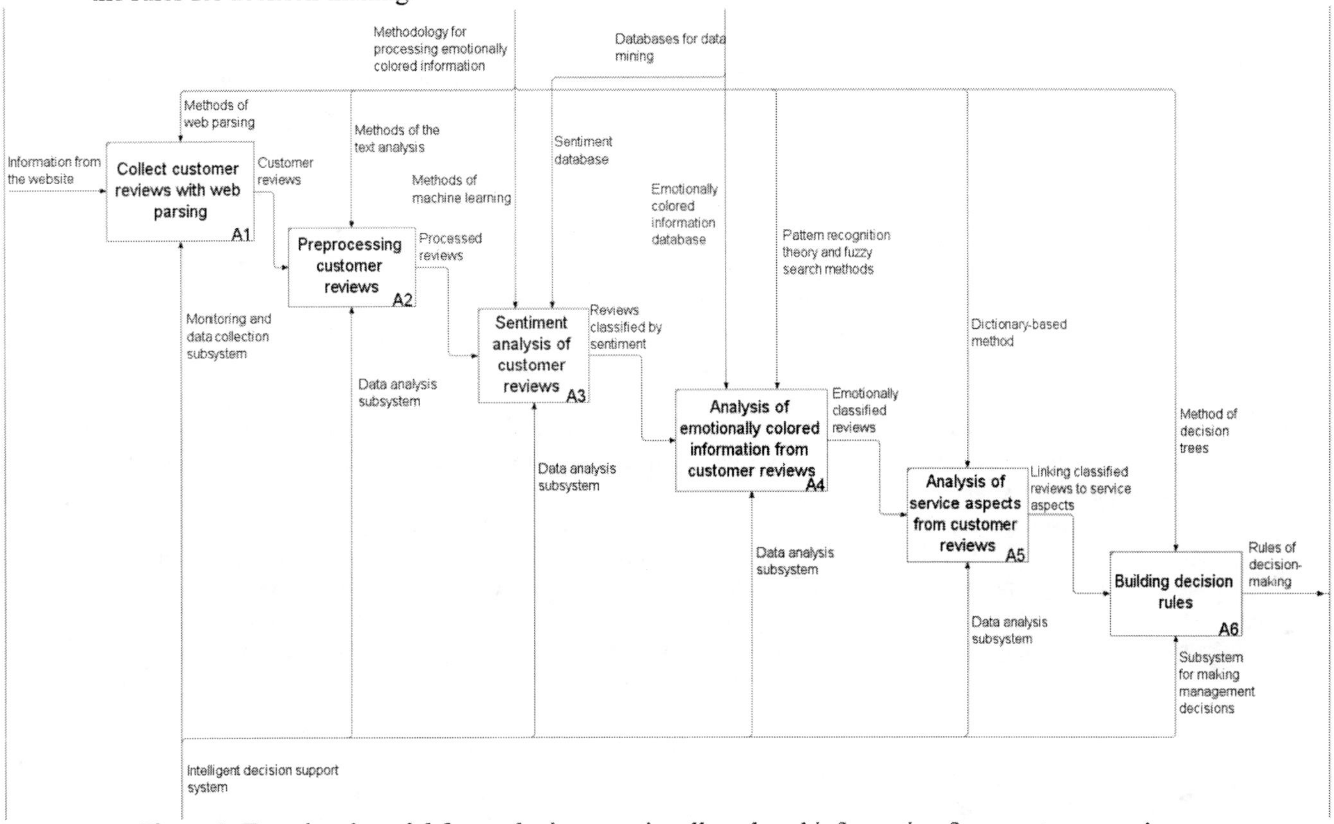

Figure 2. Functional model for analyzing emotionally colored information from customer reviews in the decision-making process

Today, one cannot do analysis without artificial intelligence tools at different stages of solving the problem of analyzing emotionally colored information from customer reviews in the decision-making process. The applied methods are shown in Figure 3.

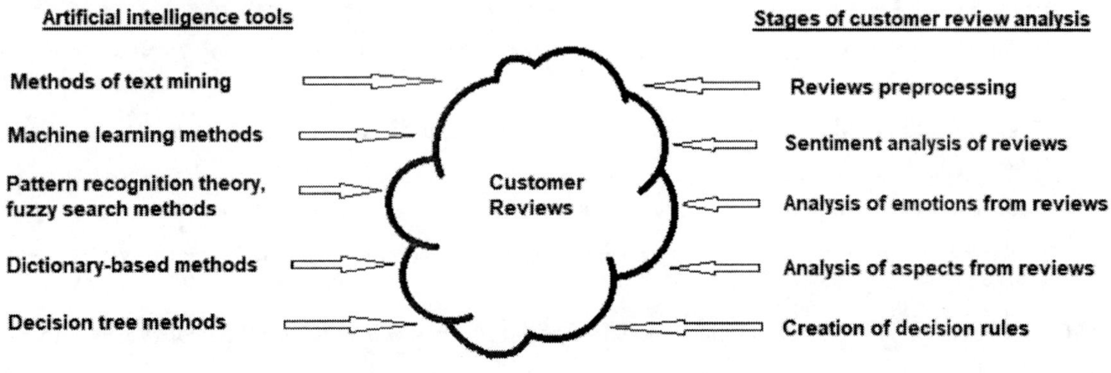

Figure 3. Artificial intelligence tools used to analyze emotionally colored information from customer reviews in the decision-making process

The developed model for analyzing emotionally colored information from customer reviews of the service sector allows us to define requirements for the applied methods of customer reviews analysis.

To date, the stage of determining the sentiment of customer reviews based on the machine learning tools and the Naive Bayesian classifier has been implemented. This stage is presented in more detail in this article [15].

6. Conclusion

On the Internet, you can find millions of service reviews. This information can be useful to support decision-making for both potential customers and for managing a service business. It is especially important to consider the emotional tone of such reviews because emotions can attract or repel potential customers. In this article the actual task of the automatic processing of customer reviews and identifying emotionally colored information in it was discussed.

Reviews on the Internet contain a lot of information, which makes it impossible to process them without use of the artificial intelligence tools. Current studies should show which of these tools are effective for processing customer reviews in natural language texts.

The article considers the actual problem of automatic customer reviews processing and the identification of emotionally colored information in them using artificial intelligence tools.

The definition of emotions was given, and an analysis of their classification is carried out, six main emotions are selected in customer reviews. The well-known approaches to accounting emotionally colored information in customer reviews, which made it possible to set the task of automatic analysis of customer reviews, had been analyzed.

A model analyzing emotionally colored information from customer reviews of the service sector was developed. This made it possible to define requirements for the methods used to analyze customer reviews. The application of artificial intelligence tools at different stages of the proposed approach to the analysis of customer reviews was considered. Further research will focus on the implementation of the presented stages of customer review analysis. To date, the stage of determining the sentiment of customer reviews based on the machine learning tools and the Naive Bayesian classifier has been implemented. In the future, it is planned to implement the stage of analyzing emotionally colored information using the theory of pattern recognition and methods of fuzzy search in dictionaries.

7. Acknowledgment

The results of the research presented in this article were supported by Grants RFBR 19-07-00709 and 18-07-00193.

The study is conducted with financial support from the Ministry of Education and Science of the Russian Federation as part of the basic part of the state assignment to higher education educational institutions # FEUE-2020-0007.

References

[1] Tomkins S S 1962 Affect Imagery Consciousness: The Positive Affects *vol. 1* (New York, NY, USA: Springer)

[2] Colombetti G 2009 From affect programs to dynamical discrete emotions *Phil. Psychol (vol 22 no 4)* pp 407–425

[3] Ekman P 1992 An argument for basic emotions *Cognition Emotion (vol 6 nos 3–4*) pp 169–200

[4] Watson D and Tellegen A 1985 Toward a consensual structure of mood *Psychol. Bull* (vol 98 no 2) p 219

[5] Russell J A 1980 A circumplex model of affect *J. Personality Social Psychol* (vol. 39 no 6) p 1161

[6] Mehrabian A 1995 Framework for a comprehensive description and measurement of emotional states *Genetic, Social, and General Psychology Monographs* (121(3)) pp 339–361

[7] Plutchik R 1980 A general psychoevolutionary theory of emotion *Theories Emotion* (Amsterdam, The Netherlands: Elsevier) pp 3–33

[8] Cambria E 2016 Affective Computing and Sentiment Analysis *IEEE Intelligent Systems* (vol 31 no 2) pp. 102-107

[9] Yadollahi A and Ameneh Gholipour Shahraki and. Zaiane O R 2017 Current State of Text Sentiment Analysis from Opinion to Emotion Mining *ACM Computing Surveys* (Vol 50 No 2)

[10] Poria S and Chaturvedi I and Cambria E and Hussain A Convolutional 2016 MKL Based Multimodal Emotion Recognition and Sentiment Analysis *IEEE 16th International Conference on Data Mining (ICDM)* (Barcelona) pp. 439-448

[11] Sailunaz K and Alhajj R 2019 Emotion and sentiment analysis from Twitter text *Journal of Computational Science* (Volume 36)

[12] Akhtar M S and Chauhan D S and Ghosal D and Poria S and Ekbal A and Bhattacharyya P 2019 Multi-task Learning for Multi-modal Emotion Recognition and Sentiment Analysis *Computation and Language*

[13] S. Imran, S. M. Daudpota, Z. Kastrati and R. Batra, "Cross-Cultural Polarity and Emotion Detection Using Sentiment Analysis and Deep Learning on COVID-19 Related Tweets," in IEEE Access, vol. 8, pp. 181074-181090, 2020, doi: 10.1109/ACCESS.2020.3027350

[14] Pengnate S and Riggins F J 2020 The role of emotion in P2P microfinance funding: A sentiment analysis approach *International Journal of Information Management* (Volume 54)

[15] Kovács G and Bogdanov D and Yussupova N and Boyko M 2015 Informatics Tools, AI Models and Methods Used for Automatic Analysis of Customer Satisfaction *Studies in Informatics and Control.*

IOP Conf. Series: Materials Science and Engineering 1069 (2021) 012014 doi:10.1088/1757-899X/1069/1/012014

On the availability of information exchange in radio notification transmission systems

A V Telny

Vladimir State University, city of Vladimir, Gorky Street, 6 str., 87, 600000, Russia

E-mail andre.izi@mail.ru

Abstract. This article presents the results of studies on assessing the availability of information exchange in radio transmission notifications system (RTNS) of various types. The author proposes to use indicators showing the probability of non-reception of alarm notifications at the scheduled time, the daily average probability of non-reception of data from RTNS, and the average daily occupancy of the RTNS radio channel as parameters for assessing the availability. The article introduces the expansion of the existing assessments of the information exchange availability not only in terms of test messages passing the radio channel control, but also taking into account possible internal time collisions associated with the transmission of service and alarm notifications to RTNS. The author proposes to expand the assessments of availability indicators not only for asynchronous RTNS, but also for combined asynchronous-synchronous systems with working frequency occupancy control, satisfying the conditions stating that the time window for the transmission of alarm and service notifications does not intersect with the time window of test messages for the radio channel control. The calculations show that the obtained estimates of availability are consistent with the established practice of RTNS operation.

Keywords: Centralized security point; Radio system for transmitting notifications; Object transceivers; Synchronous-asynchronous operation mode of the RTNS; Internal time collisions

1. Introduction

Radio systems for transmitting notifications are used for centralized protection and fire monitoring of facilities. Object complexes of security, alarm and fire alarms transmit service and alarm notifications about the state of the object to the centralized protection point (CSP) by means of RTNS. The number of transmitted messages is determined by the information content of RTNS, the tactics of protecting objects and the structure of RTNS. To control the integrity of the radio channel, the object blocks constantly send test messages. For different RTNS, test bursts can have a different set of durations, and the minimum burst period is usually defined as 12-15 seconds. The maximum polling time is limited, and at the same time, the fact of communication failure with each of the protected objects should be determined in a time of no more than 120 seconds. Depending on the hardware and circuitry implementation of RTNS, the duration of the test messages can be from 6 to 15 ms. The duration of a service notifications transmission from the object to the central (base) station for RTNS basically corresponds to the duration of alarm notifications and ranges from 30 to 50 ms, but alarms have a higher priority during transmission. The most significant factor determining availability in RTNS is the number of protected sites. Since all the object transmitters in RTNS operate at the same frequency, the filling of the operating frequency is determined by the number of the objects protected at the same. This factor

IOP Conf. Series: Materials Science and Engineering 1069 (2021) 012014 doi:10.1088/1757-899X/1069/1/012014

also affects time collisions in the operation of RTNS transmitters at other sites. Temporary collisions occur when service and alarm notifications from different objects overlap. As a result, the transmitted notifications disappear.

RTNS types are categorized by design into one-way RTNS (one-way notification, only a transmitter is installed at the site) and two-way RTNS, when the site has a receiver and a transmitter. In addition, in asynchronous RTNS, the transmission of all types of notifications is carried out in a randomly determined time interval and is not synchronized with the operation of other object transmitters. For this reason, all announcements in an asynchronous RTNS are repeated many times. Basically, an asynchronous RTNS transmits notifications either in bursts, with a constant time interval between messages in a burst, or random access to the data transmission channel is used, or mathematical laws are used to ensure the non-repeatability of the number series (for example, a number of primes are used) and to increase the delivery probability notifications. In synchronous notification transmission systems, the operation of object blocks is synchronized from the side of the centralized monitoring panel (each transceiver communicates after the synchronization signal, in its own time window, usually associated with the object number of the block). The transceiver can listen to the operating frequency for its busyness and get in touch after the end of the transmission from another unit. The test messages repetition rate determines the amount of information transmitted in the communication channel and affects the availability of information exchange and the possible number of protected objects, like the information content of RTNS.

The electromagnetic environment in the RTNS coverage area has a strong impact on the state of radio communications. This parameter can be both random (in case of natural interference), and targeted in case of deliberate interference or special purpose sabotage of RTNS. We do not take into account accessibility analysis sabotage performed by an intruder and the influence of external electromagnetic interference. Many publications are devoted to the functioning of various types of RTNS, for example, [1-13] and others. [1,2,3,14,16] deal with ensuring confidentiality of transmitted information, imitation security of RTNS transceivers and secrecy of information exchange. Work [15] is devoted to information exchange algorithms in RTNS and optimization of time windows of pauses in an asynchronous system. In this work, it is proposed to use a relative prime number generator that forms a time sequence with the smallest common multiple exceeding the channel control time. Items [4,8] consider the issues of using radio-technical methods of transmitting notifications with pseudo-random operational frequency readjustment with spreading of the spectrum in asynchronous RTNS, including the use of PROFR only for the choice of the carrier frequency, but also for the time intervals for sending messages. In [16], it is proposed to use the cellular principle of RTNS construction, while the objects transmit all information only to the central nodes of the "cell", which control the communication channels to the objects in their "cells" independently. Only "cell" nodes transmit information to the CSP. The issues of the availability of information exchange for RTNS were directly considered in [17,18,19,20].

2. Problem statement

The availability of information exchange in radio systems for the transmission of notifications can be assessed as the probability of the system fulfilling its main functional purpose of transmitting alarm messages from any protected object to a centralized security point. Determination of availability for RTNS can be interpreted as the possibility of transmitting an alarm notification for a specified time. It is proposed to assess the availability of information exchange in radio systems for transmitting notifications according to the following generalized formula

$$A_D = \int_{t_{AL}}^{t_{AL}+T_D} A_F(\alpha(t), t)dt \quad (1)$$

Here A_D is the directive accessibility of the RTNS; t_{AL} stands for triggering time of technical means of security and alarm on the object device RTNS of a particular object; T_D means directive (maximum possible) time during which it is necessary to transmit the alarm notification to the CSP; A_F stands for actual availability of RTNS for data exchange at time t; $\alpha(t) = \varphi(F(t))$ shows a function of dynamically changing factors; $F(t) = (f_1(t); f_2(t); \dots f_k(t))$ is a vector of factors that can be dynamic and change over time, be constants or random variables; $f_1(t); f_2(t); \dots f_k(t)$ are factors affecting the availability of RTNS (the number of protected facilities, time collisions from the operation of RTNS transmitters at other facilities, the electromagnetic environment in the RTNS coverage area, RTNS type, nature of information exchange, RTNS architecture, duration of service and alarm notifications, RTNS information content, spectrum width, carrier frequency instability, etc.).

According to the Russian standards, the delivery time of an alarm notification for RTNS should not exceed 60 seconds, which is the directive time T_D. The most promising direction in the development of RTNS is the creation of combined asynchronous-synchronous systems in which all service notifications are transmitted in their time windows, and alarm notifications are transmitted in a randomly determined time interval of a special time window. Alarms are repeated (duplicated) until an acknowledgment (notification of notification) is received from the central station to ensure the guaranteed transmission of alarm notifications from the facility to the CSP for bi-directional RTNS. To assess the availability of information exchange in radio systems for transmitting notifications, it is necessary to take some restrictions and assumptions [18]: the vast majority of notifications from object blocks are test signals (more than 95%), the share of alarm notifications (requiring a response from a security squad) is less than 2% of all notifications, while more than 50% of alarms are false (not directly related to the facts of penetration into an object); most of the alarms occur during guarded time outside the "peak hours", when there is no massive process of arming / disarming objects in accordance with the security schedule; propagation time of electromagnetic waves and temporary fading caused by multipathing are not taken into account and do not affect the time synchronization of the RTNS; the probability of errors caused by incorrect transmission or reception of the object number is extremely small. To model the probability of passing notifications from an object to a CSP, it is proposed to use the Poisson distribution for the following reasons: the probability of overlapping notifications from object blocks (time collisions) depends only on the duration of notifications and does not depend on the time of their occurrence, and the notifications themselves are distributed on the time axis with the same average density; the probability of collisions occurring in the allocated time interval does not depend on the probability of overlapping notifications in another time interval; the probability of overlapping three or more notifications in a time interval equal to twice the duration of notifications is negligible compared to the probability of overlapping two notifications in this time interval.

3. Methodology for assessing the availability of information exchange in radio systems for transmitting notifications

We propose to use two indicators to analyze the availability in practical use. These are the messages non-reception probability (the probability of temporary collisions, overlapping messages) P_{not} and the maximum occupancy of the radio channel G_{max} at given levels P_{not} and the number of message repetitions b. The number of message repetitions for different types of RTNS can be either the same or different for service b_{serv} and alarms b_{AL}. Let $b_{serv} = b_{AL} = 6 \div 16$.

According to [18,19,20] for asynchronous RTNS:

$$P_{not} = 1 - \exp(\frac{-2\tau N}{\Delta t_{test}}) \quad (2)$$

Here τ is the duration of test messages ($\tau \approx 6 \div 15$ ms), N means the number of protected objects, Δt_{test} stands for repetition period between test messages. Typically, for RTNS, there are several sets of test transmission periods with a minimum of about 12-15 seconds. T_{contr} maximum time of polls, according to which the fact of communication with the object is established $T_{contr} = 120$ s. It is worth noting that formula (2) gives the probability of non-reception of only test messages and only two messages (one collision), it is possible to refine P_{not} both for test messages and for service and alarm notifications

$$P_{\Sigma not} = 1 - \exp\{-\left(\frac{2\tau N}{\Delta t_{test}} + \frac{2t_{mes}(2N+k_{serv})b_{serv}}{T_{day}} + \frac{2t_{mes}k_{AL}b_{AL}}{T_{day}}\right)\} \quad (3)$$

Here t_{mes} is the duration of alarm and service notifications ($t_{mes} \approx 30 \div 50$ ms), T_{day} is a day ($T_{day} = 86400$ s), k_{serv} and k_{AL} are the average daily number of alarm and service (excluding commands to arm/disarm an object) notifications from RTNS respectively (this data can be obtained from the statistics of RTNS operation on the CSP). It is assumed that each object is armed and disarmed on a daily basis. For practical purposes of protecting objects [18], it is advisable to take the values $P_{\Sigma not} = 10^{-4} \div 10^{-5}$. The maximum occupancy of the radio channel G_{max} for asynchronous systems according to [18] is defined as $G_{max} = \frac{-\ln(1-P_{\Sigma not}{}^{\frac{1}{b}})}{2}$ (4)

Formula (4) is valid only for test messages. The average value of the average daily occupancy of a radio channel (for the time $T_{day} = 86400$ s) for asynchronous systems can be determined as

$$G_{mean} = \frac{1}{T_{day}}\{\left(\tau\left(\frac{T_{day}}{T_{contr}}\right)b\right)N + t_{mes}(2N + k_{serv})b_{serv} + t_{mes}k_{AL}b_{AL}\} \tag{5}$$

The first term in formula (5) means the share of the radio channel occupation with test messages, the second term determines the share of occupation by the arming / disarming commands for each object and the average daily number of other service notices, and the third term determines the share of the radio channel occupation by the average daily number of alarm notifications. An important indicator is to determine the likelihood of non-receipt of exactly alarming notifications at the scheduled time. The number of repetitions of the transmission of alarm notifications for the directive control time $m = \frac{T_D}{\Delta T_{AL}}$.

Here $T_D = 60$ c according to [1], and ΔT_{AL} is the repetition period between alarm messages. At the same time, ΔT_{AL} is a random variable generated by special random number generators for many asynchronous RTNS [15]. In practice, for various types of RTNS, $\Delta T_{AL} = 3 \div 30$ s. Since the overlapping of notifications on the time interval ΔT_{AL} is assumed to be an independent event, the probability of non-reception of alarm notifications at the scheduled time will be

$$P_{\Sigma not}|_{T_D} = (P_{\Sigma not})^{\left(\frac{T_D}{\Delta T_{AL}}\right)}\left(\frac{T_D}{T_{day}}\right) = (P_{\Sigma not})^m\left(\frac{T_D}{T_{day}}\right) \tag{6}$$

In practice, for RTNS, the value $P_{\Sigma not}|_{T_D}$ should be$P_{\Sigma not}|_{T_D} \leq 10^{-6} \div 10^{-7}$ over the time period T_D. For combined asynchronous-synchronous systems with controlled working frequency occupancy, the availability parameters $P_{\Sigma not}$; G_{mean}; $P_{\Sigma not}|_{T_D}$ depend greatly on architecture, operation algorithm and data format in RTNS data exchange protocols. In such systems, the transmission of test messages is synchronized and does not cause time collisions (if the equipment is functioning properly), and the transmission of alarm and service notifications from the facility is a random event. If we assume that the system has the following properties: all test messages are repeated once in the time window set for the object and no temporary collisions occur; the time window for the transmission of alarm and service notices does not overlap with the time window for test messages; only service and alarm notifications are acknowledged from the central station, and after receiving the acknowledgment signal, the object block does not perform further repetitions of notifications.

When these conditions are met for combined asynchronous-synchronous systems, $P_{\Sigma not}$ can be determined as follows

$$P_{\Sigma not} = 1 - \exp\{-\left(\frac{(2(t_{mes}+\tau_p)b_{serv}+\tau_{ack})(2N+k_{serv})}{T_{day}P} + \frac{(2(t_{mes}+\tau_p)b_{AL}+\tau_{ack})\,k_{AL}}{T_{day}P}\right)\} \tag{7}$$

Here τ_{ack} is the duration of acknowledgment notifications (confirmation of information receipt from the object unit from the central station to the central station) (usually$\tau_{ack} = \tau \approx 6 \div 15$ ms), τ_p is the pause time for air control before transmitting each notification ($\tau_p \approx (0{,}2 \div 0{,}3)\, t_{mes}$), P stands for time fraction for service and alarm notifications transmission in the time cycle of information exchange between the object and the CSP. $P = \frac{T_{win\,AL}}{T_{win\,AL}+T_{win\,test}}$, where $T_{win\,AL}$ is the duration of the time window for the transmission of alarm and service notices, $T_{win\,test}$ – is the duration of the time window for the transmission of test messages. $T_{win\,AL} + T_{win\,test}$ – full period of the time cycle of information exchange with the CSP. $T_{win\,test} = (t_{mes} + \tau_p)N$. The minimum value stating the probability of non-receipt of alarms for combined asynchronous-synchronous systems at the scheduled time will be

$$P_{\Sigma not}|_{T_D} = (P_{\Sigma not})^{\left(\frac{T_D - T_{win\ test}}{\Delta T_{AL}}\right)}(T_D/T_{day}P) \quad (8)$$

The average value of the average daily occupancy of a radio channel (for the time$T_{day} = 86400$ s) for combined asynchronous-synchronous systems with control of the working frequency occupancy can be determined similarly to (5) as follows

$$G_{mean} = \frac{1}{T_{day}}\left\{\left((\tau + \tau_p)\left(\frac{T_{day}}{T_{contr}}\right)\right)N + (t_{mes} + \tau_p)(2N + k_{serv})b_{serv} + (t_{mes} + +\tau_p)k_{AL}b_{AL} + \tau_{ack}(k_{serv} + k_{AL})\right\} \quad (9)$$

Taking into account the fact that after receiving an acknowledgment signal for alarms and service notifications, the object block does not perform further repetitions of the notifications transmission, the number of repetitions for service and alarm notifications will be lower for combined asynchronous-synchronous systems with control of the working frequency than for asynchronous systems.

4. Availability assessment comparison for asynchronous and combined asynchronous-synchronous RTNS

For comparison, let us assess the availability of information exchange by the indicators G_{max}; $P_{\Sigma not}|_{T_D}$ for asynchronous RTNS (5), (6) and combined asynchronous-synchronous RTNS with control of the working frequency busy (8), (9). We will take a day (T_{day}=86 400 s) as comparison time. The initial conditions for the systems are assumed to be the same. Let the combined system comply with the above conditions, and for each notification from the object, an acknowledgment follows. Let us take the following initial data for asynchronous RTNS: $\tau = 10$ ms; $T_D = 60$ s; $b = 10$; $b_{serv} = b_{AL} = 16$; $T_{contr} = 120$ s; $\Delta t_{test} = 12$ s; $\Delta T_{AL} = 12$ c; $t_{mes} = 50$ ms; $k_{serv} = 0{,}1N$; $k_{AL} = 0{,}05N$; modified parameter $N = 10 \ldots 410$ in increments of 20. Let us take the following initial data for the combined asynchronous-synchronous RTNS with control of the working frequency occupancy: $\tau = 10$ ms; $T_D = 60$ s; $b_{serv} = b_{AL} = 2$; $T_{contr} = 120$ s; $\Delta t_{test} = 12$ s; $\Delta T_{AL} = 12$ s; $t_{mes} = 50$ ms; $\tau_{ack} = 15$ ms; $\tau_p = 10$ ms; $k_{serv} = 0{,}1N$; $k_{AL} = 0{,}05N$; $P = 0{,}1$; variable parameter $N = 10 \ldots 410$ in increments of 20. The calculation results are shown in Figure 1 and Figure 2.

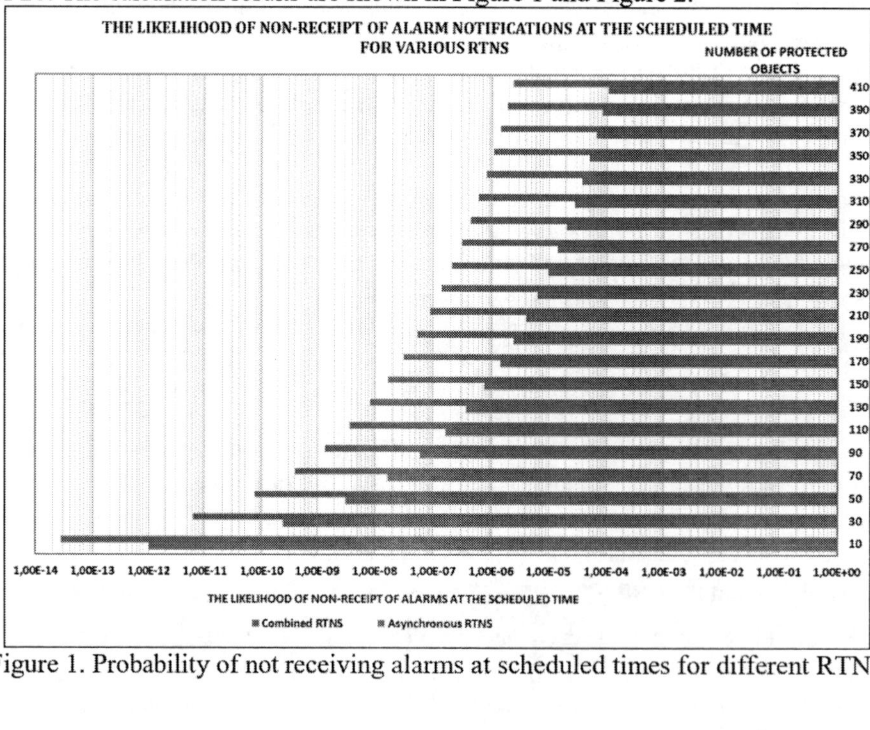

Figure 1. Probability of not receiving alarms at scheduled times for different RTNS

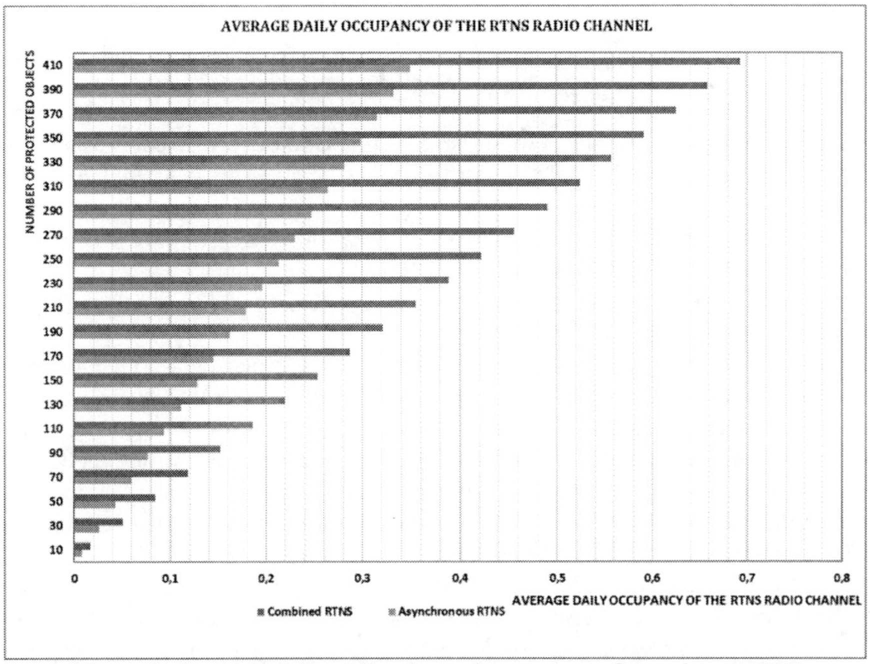

Figure 2. Average daily radio channel occupancy for various RTNS

5. Conclusion

The paper proposes an assessment methodology for the availability of information exchange in radio systems for transmitting notifications of various types, including asynchronous systems and combined asynchronous-synchronous ones with control of the working frequency occupancy. The approaches used can be partially applicable to notification transmission systems of a more complex architecture. Such parameters as the probability of non-reception of alarm notifications at the scheduled time, the average daily probability of non-reception of data from RTNS and the average daily occupancy of the RTNS radio channel are proposed to be used for assessing availability. The novelty in this work is the proposal to calculate the accessibility parameters not only by passing the test messages of the radio channel control, but also taking into account possible collisions associated with the transmission of service and alarm notifications.

The proposed methodology can be useful when comparing RTNS of various types and principles of operation, as well as to justify the maximum possible number of protected objects on one RTNS frequency. Based on the results of the calculations, we can conclude that the combined asynchronous-synchronous RTNS with control of the working frequency occupancy can be more effective than the asynchronous one in terms of the probability of non-receipt of alarm notifications at the scheduled time with a large number of protected objects.

6. Acknowledgment

The study was conducted as part of the implementation of the state task of Vladimir State University FZUN-2020-0013.

References

[1] Fedyaev S L, Maksimov V S and Fedyaev Yu S 2011 *Patent for utility model No. 111938 RU G08B 21/00 27.12.2011. Radio channel alarm complex of protection*

[2] Zavyalov S A, Isupov A V, Kosyih A V, Lepetaev A N, Tsyibenko L V and Zavyalov N S 2008 *Patent for utility model No. 74232 RU G08B 26/00 20.06.2008. Notification Radio System*

[3] Pervuninskikh V A 2016 *Patent for invention No. 2620239 RU G08B 13/00 23.05.2017. Wireless self-organizing network system of protected territory monitoring*

[4] Ji S and Cai Z 2012 Distributed data collection and its capacity in asynchronous wireless sensor networks *Proceedings - IEEE INFOCOM* 2113–21
[5] Ji S and Cai Z 2013 Distributed data collection in large-scale asynchronous wireless sensor networks under the generalized physical interference model *IEEE/ACM Trans. Netw.* **21**(4) 1270–83
[6] Zhang Y, Shum K W and Wong W S 2011 Completely irrepressible sequences for the asynchronous collision channel without feedback *IEEE Trans. Veh. Technol.* **60**(4) 1859–66
[7] Shum K W, Wong W S, Sung C W and Chen C S 2009 Design and construction of protocol sequences: Shift invariance and user irrepressibility *IEEE International Symposium on Information Theory – Proceedings* 1368–72
[8] Wong W S 2007 New protocol sequences for random-access channels without feedback *IEEE Trans. Inf. Theory* **53**(6) 2060–71
[9] Chen S, Wang Y, Li X Y and Shi X 2009 Order-optimal data collection in wireless sensor networks: Delay and capacity *2009 6th Annual IEEE Communications Society Conference on Sensor, Mesh and Ad Hoc Communications and Networks* pp 1–9
[10] Marler R T and Arora J S 2004 Survey of multi-objective optimization methods for engineering *Struct. Multidiscip. Optim*, **26**(6) 369–96
[11] Chamberland J F and Veeravalli V V 2007 Wireless sensors in distributed detection applications *IEEE Signal Process. Mag.* **24**(3) 16–25
[12] Rani B S and Shyamala K 2019 Energy Efficient Load Balancing Approach for Multipath Routing Protocol in Ad Hoc Networks *2019 Second International Conference on Advanced Computational and Communication Paradigms (ICACCP)*, pp. 1–5
[13] Bishop A N, Fidan B, Anderson B D O, Doğançay K and Pathirana P N 2010 Optimality analysis of sensor-target localization geometries *Automatica* **46**(3) 479–92
[14] Osipov D L, Zhuk A P and Gavrishev A A 2015 *Patent for invention No. 2560824 RU H04L 9/20 20.08.2015. Apparatus for protection against imitation of controlled objects with high structural security of carrier signals*
[15] Tsyibenko L V, Zavyalov S A and Tihonov S V 2008 Improving the algorithm of operation of security radios *The Journal Omsk Scientific Bulletin* (*Omsk: Publishing House of Omsk University*) **70**(3) 126–29 (in Russian)
[16] Partyka A 2005 *Patent for invention No. 6870875 US H04B 1/69 22.03.2005. Transmission of urgent messages in frequency hopping system for intermittent transmission*
[17] Spiridonova N E and Zarubin V S 2017 The cellular principle of creation of perspective RSPI for the organization of the protection of stationary objects *Okhrana, bezopasnost, sviaz* (*Voronezh: Publishing House of Voronezh Institute of the Ministry of Interior of Russia*) **1-1** 175–78 (in Russian)
[18] Zuev P I 2011 Funktsionalnyie vomozhnosti asinhronno adresnyih radioohrannyih sistem *Algoritm Bezopasnosti magazine* **4** 78–82 (in Russian)
[19] Zuev P I 2004 Veroyatnostnyiy analiz funktsionalnyih vozmozhnostey asinhronno-adresnyih radioohrannyih sistem *Algoritm Bezopasnosti magazine* **2** 52–55 (in Russian)
[20] Belkin V. 2006 Veroyatnost dostavki soobscheniy v radiosistemah peredachi izvescheniy asinhronno-adresnogo tipa *Algoritm Bezopasnosti magazine* **2** 60–62 (in Russian)

Artificial intelligence technologies to identify similar objects and formalization of inference rules

N I Yusupova[1], O N Smetanina[1], E Yu Sazonova[1], A I Agadullina[1]

[1] Ufa State Aviation Technical University, Ufa, K. str. K. Marxa 12, 450008, Russia

E-mail yussupova@ugatu.ac.ru

Abstract. Analysis of data accumulated during the activities of the enterprise may be applied for creating novel business models and more efficient use of information resources. Specific requirements for analysis of data in socio-economic areas entail the need to develop or select the required tools and methods. This article considers identification of implicit knowledge on a complex of historical and operational data. The task statement and the methodology of solution are given. Data mining technologies are applied for identifying hidden common factors in data. Interpretation of the results, formalization of knowledge, formation of the knowledge base and use of the inference mechanism allow us to organize decision support. As an example, the task of personnel selection is considered. The choice of certain factors that are significant for each enterprise, which are the basis for the selection of personnel, will make it possible to spread the solution method for enterprises where strict requirements for information security are imposed. An analysis of decision-making support research in this area, is conducted with the use of artificial intelligence technology. Neural network technologies for data analysis are proposed to be used by the authors.

Keywords: Neural network technology; Data mining; Decision support; Human resource management

1. Introduction

Effectiveness of the enterprise depends on the quality of the decisions taken (including in the field of organization and information security) at the human resource management. Large amounts of information are accumulated during the company's activities in this area. Therefore, we attempted to use this data to identify implicit knowledge and use it to support decision making.

Issues of decision making in the field of human resource management are considered by many Russian and foreign researchers: Azhmukhametov A.I. [1], A. Buzuev [2], Dergacheva M.Yu. [4], Evstratova E. I. [3], Kuts Yu.I. [6], M.A. Navrotsky [4], Prokushev Ya.E. [9], Protalinskiy O.M., Fedyakina OP [10], Shagieva Yu.R. [11], Farrés Rocabert A. [12], Hamidah Jantan [13], Hossein Hakimpoor [14], Steven Walczak [16], Loo-See Beh [17], Hao G. [18], Raphael Olufemi [19], Stefan Strohmeier [20], Ernst Biesalski [22]. Analysis of research has revealed that the decision support for various staff management tasks using neural network technology.

Here we propose and implement a methodology for organizing decision support for the selection of staff based on an analysis of collected historical data (for example, when filling out a questionnaire) and operational data (personnel assessment) and its implementation. The proposed generalized methodology can be in the complex developed applications for filling out questionnaire questions and collecting data in the form of completed questionnaires and forming a knowledge base can be used for other tasks in the field of human management resource, and many other areas. Part of the steps of the methodology is carried out using the analytical platform Deductor.

2. State of the problem

Authors [1,2,6,9] address issues of mathematical models development of the staff selection problem and methods for solution in order to create software.

There are many researches on decision support using artificial intelligence techniques. So, the authors [4] presented the concept of organizing intelligent decision support in the field of staff management, based case reasoning and ontology. It allows increasing the effectiveness of decisions in this subject area. Fedaykina I.P. [10] addresses decision support issues with using fuzzy knowledge base.

Evstratova E.I. [3] considers the possibility of using artificial neural networks in order to simplify and improve the efficiency of staff management processes. It should be noted that neural network technologies are widely used for organizing decision support in the field of staff management.

Farrés Rocabert A. [12] describes a Matlab toolbox for career-oriented social network that allows recommending a candidate at the request of an employer for a specific task. Authors [13]. explore the possibility of using machine learning to solve staff problems, and Daniel Faggella consider machine learning to select new team members. Hossein Hakimpoor [14] describes the use of neural networks for business intelligence in general. Steven Walczak [16] proposes to ensure an increase in the efficiency of the manager's work and, in some cases, management actions using the neural network apparatus.

Loo-See Beh et al. [17] propose the use of hybrid intelligent methods. Example methods of this kind are machine learning and knowledge-based approach for extracting new knowledge. Hao G.[18] describes neural network model for solving typical task of staff planning — the task of drawing up a list of airport ground staff.

Author [19] describes the system that collects data from candidates via web-interface and compares them with the relevant posts. The basis is neural network web model of human resource management system.

Stefan Strohmeier [20] offers the potential of artificial intelligence in human resource management for using in six selected scenarios (prediction of turnover with artificial neural networks, search of candidates with intelligent search engines, staff rostering with genetic algorithms, analysis of personnel tonality with text mining, resume data collection with information extraction and self-service of employees with inter-active voice response). Robert Charlier [21] uses artificial intelligence technology to speed up the search for talent.

Ernst Biesalski [22] notes possibility of using knowledge-based technologies to support decision-making in staff management.

The issues of decision support in staff management are discussed in this article. The proposed method allows implementing a number of tasks (functions), in particular, staff selection functions.

The authors of this article have experience in conducting research using artificial intelligence technologies [2, 12-13], in particular, using neural network technologies and technologies based on knowledge.

The general formulation of the task of processing large-scale data and identifying unobvious, but useful knowledge to use in decision-making will be formulated below. The range of tasks for which such an approach is characteristic, and this method of solution are much wider than it presented in the article

3. Identifying non-obvious knowledge: task statement and methodology

The general task of identifying non-obvious knowledge based on data is as follows. There are historical and current data. It is necessary to reveal knowledge based on these data that can be interpreted and used to support decisions in staff management. A feature of the statement is the separation of data into historical and current (operational). The mathematical formulation of the problem can be stated as follows:

Given $X = \{x_1, x_2, \ldots, x_l, x_{l+1}, \ldots, x_n\}$ - historical data with $i = 1, \ldots, l$ and current data with $i = l + 1, \ldots, n$.

Find: $F: X \rightarrow Z$, function that maps data to knowledge.

The method for solving the problem (Fig. 1) includes the following stages: data collection, accumulation and preliminary processing (for example, transformation of non-numeric data into nominal data); checking the correctness of data sets and their cleaning (for example, based on filtering (allows working with data that meets specified conditions), evaluating data quality (filling in missing data, editing anomalous values, etc.); conducting data mining; interpretation of results (for example, clustering carried out at the previous stage, analysis of objects of one cluster and interpretation of results taking into account subject semantics), formalization of knowledge and formation of a knowledge base (choice of knowledge representation model and formation of a knowledge base), organization of support for management decisions based on the identified knowledge (application of inference engine, an adequate knowledge representation model).

Regarding data mining, among the methods for identifying and analyzing knowledge, we can distinguish predictive methods (classification, regression, time sequence prediction (series)) and descriptive (to describe patterns in data) clustering, association, sequence.

Here we will rely on descriptive methods. Neural networks can be used as tools and methods for analysis.

4. Solution for staff selection

To solve the task of staff selection, the following steps are implemented.

Stage 1. Description of the collection, accumulation of data and preliminary processing. It is assumed that for this task historical data (collected earlier in the selection of applicants), and operational data (as an assessment of the "active" staff), are used.

The stage is implemented using an application developed in framework of the research.

During the selection process information about the applicant was collected: age, education (secondary, secondary vocational, higher, humanitarian, technical, name of the university, specialized or non-core education), additional education, retraining, obtaining additional skills (advanced computer use, knowledge of a foreign language, driving skills and driving license), work experience (work experience in a similar position, work experience in a managerial position, length of service, terms of work one place, reasons for dismissal), level of wages, marital status and children, bad habits, reasons for choosing a company, etc.

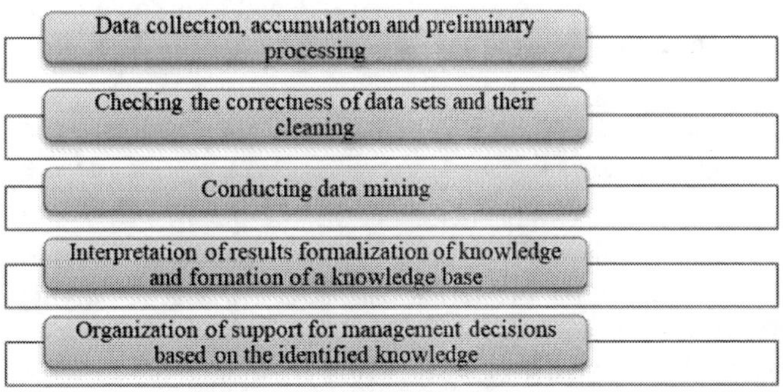

Figure 1. Method for solving the problem

Professional testing has revealed some information about the level of communication, ability to work in a team, ability to work on a computer at the level of an advanced user. All information is stored in the database.

For staff assessment the "360 degrees" method was used. The assessment is conducted according to the list of competencies presented to the position of applicant. The method can be adapted for other company positions.

E.g. the requirements for the seller-consultant are the following: customer-oriented (ability in its activity to proceed from the needs of the client; in the process of concluding a transaction shows what this proposal is beneficial for the client), knowledge of the whole range of equipment sold and its technical characteristics, ability to demonstrate the operation of equipment; ability to work on a computer (software product 1C: Enterprise 8), teamwork (ability to listen; ability to persuade; influence colleagues; ability to transmit information without loss; ability cooperate; cooperate; constructively overcome differences; use the potential of the group and achieve collective results; focus on team achievement of a common goal); communication skills (conflict-free, communication skills, diplomacy, ability to express thoughts verbally and in writing), etc.

When using the scoring method, it should be noted that the staff assessment is carried out by those who indirectly or directly interact with applicants. The "grouping system" is based on the fact that in each applicant gains one or another number of points. Depending on this mumber each employee can be assigned to one of the groups. Comprehensive assessment is used as an output for clustering. The higher the assessment, the more the employee meets the criteria.

After the data for analysis has been collected, they are pre-processed, in particular, non-numeric data is replaced with nominal data.

Stage 2. Checking the correctness of data sets and their cleaning. Verification is carried out using the analytical platform Deductor. The presence of special nodes allows for cleaning. In particular, filtering was carried out, which allows selecting the most important data and improving the quality of analysis. The data quality assessment allowed filling in the missing data, editing anomalous values. The factor analysis allows reducing the data dimension.

One of the methods of factor analysis is the principal component method, which is based on determining the minimum number of factors that make the greatest contribution to data variance.

The general model of factor analysis is: $X_j = \sum_{i=1}^{k} b_{ij} * F_i + \sum_{i=k+1}^{m} b_{ij} * U_i + e_j, i = 1..m, k < m$; where X_j- are original variables in space of m measurements; F_i - are common factors in k measurements; U_i - are specific factors in the space of m - k measurements; $b_{ij}$are factor loads; $e_j$are random errors.

The principal component model is: $X_j = \sum_{i=1}^{m} a_{ij} * Z_i; i = 1..m; j = 1..m$, where X_j- are original variables; Z_i are principal components; a_{ij} are transition factors from a system of variables X to the system of components Z, constituting the matrix of eigenvectors A, which is a solution of the matrix equation: $R * A = \lambda * A$, where $R -$ a correlation matrix between the original variables; $\lambda -$ matrix eigenvalue vector R.

Moreover, since the matrix A is orthogonal, its inverse matrix is equal to the transpose: $A^{-1} = A'$.

Therefore, the matrix equation can be rewritten as: $R = \lambda * A * A'$ or $R = B * B'$, where B is the factor load matrix.

Estimates of factor loads are the values $l_{ij} = a_{ij} * \sqrt{V(Z_i)}$, and estimates of specific factors are given by equalities: $e_i = \sum_{i=k+1}^{m} a_{ij} * Z_i$.

Thus, an estimate of the factor model is obtained.: $X_j = \sum_{i=1}^{k} l_{ij} * Z_i + e_j, j = 1..m$.

Stage 3. Conducting data mining. To identify implicit knowledge, clustering, with Kohonen self-organizing maps is used. To solve the clustering task, it is necessary to divide into clusters applicants with similar characteristics.

The set of applicants X is given, from which the training sample $X^m = \{x_1, x_2, \dots, x_m\} \subset X$ is selected, as the initial information for the considered task. The output is represented by a set of clusters $Y = \{y_1, y_2, \dots, y_l\}$, including objects with similar characteristics and estimates. The function of the distance between objects $\rho(x, x')$ is given. It is required to split the learning sample into clusters based on the similarity of features of objects for one cluster and the difference of features of objects of different clusters. Moreover, each object $x_i \in X^m$ is assigned a cluster number y_j. Further, each object under consideration belongs to one of the classes.

To solve such tasks, a number of algorithms and methods have been developed, in particular, on the basis of artificial intelligence technologies: K-means fuzzy clustering method, Kohonen's genetic algorithm and neural network.

The solution of the problem with the help of Kohonen's self-learning network is based on the fact that the neural network, focusing on the structure of the input vectors supplied, relates the object to a specific class.

The number of neurons in input and output layers, the learning rate and the stopping criterion are determined in the process of Kohonen network development. The number of layers is two; the number of neurons in the second layer is determined by dimension of feature vector; the number of neurons in output layer is determined by the number of classes; the learning rate is 0.3; stop criterion is error less than 0.05

The neurons in the output layer compete with each other in the process of inputting features of an object for the right to be a "winner" whose incoming weights ω are closest to the input image. The relation $|\omega_j - x| \leq |\omega - x|$ is for all j for the winning cell j'. The winner has the right to regulate their weight.

The Kohonen network learning algorithm is defined by the following steps.

Step 1. Initiate the weight, set the parameters of neighborhood function and learning speed.

Step 2. If the stop condition is not satisfied, go to steps 3-9.

Step 3. Perform steps 4-6 for each input vector x.

Step 4. Calculate $D(j) = \sum_i (\omega_{ij} - x_i)^2$ for each j, where ω_{ij} – synaptic weights between neurons i and j.

Step 5. Find an index j 'such that $D(j') = min$.

Step 6. Calculate for all cells j within a certain neighborhood from j', find for all i $\omega_{ij}(new) = \omega_{ij}(old) + \eta\Lambda(j,j')[x_i - \omega_{ij}(old)]$, where $\omega_{ij}(new)$ and $\omega_{ij}(old)$ – new and old weights, η – learning speed, $\eta\Lambda(j,j')$ – neighborhood function.

Step 7. Change the learning speed.

Step 8. Reduce the radius of the neighborhood function.

Step 9. Check the stop condition (reduce the learning speed to zero).

Kohonen's self-organizing map, as one of the varieties of neural networks, is used for clustering. The analytical platform Deductor is used for implementation. The algorithm determines the location of clusters in a multidimensional feature space, and the construction of a map of Kohonen allows to display multidimensional space in more convenient two-dimensional space for visualization.

The clustering scenario is ensured by the following steps:

- importing source data from a file and specifying input columns.
- settings of dividing method for the original data set into train (95% of sample) and test (5% of sample);
- settings of the map parameters (the number of cells along axes X – 16 and Y – 12, the shape of cells - hexagonal),
- setting of the training parameters (speed at the beginning of training – 0.3 and at the end - 0.005; radius at the beginning of training - 4 and at the end - 0.1; neighborhood function is step; a fixed number of clusters — 4) and the learning stops (authors consider an example recognized if the error is less than 0.05);
- implementation of the learning process (the number of recognized examples and the current error values are determined);
- settings of mapping for Kohonen network.

Kohonen maps (Fig. 2) allow to demonstrate visually the obtained results.

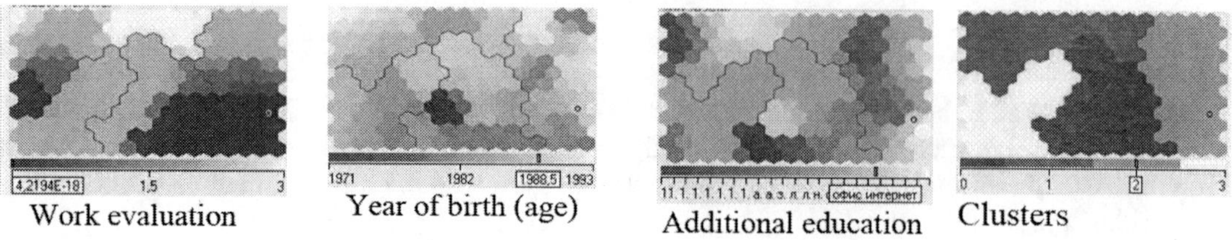

Figure 2. Examples of Kohonen maps

Stage 4. Interpretation of results, formalization of knowledge and formation of a knowledge base.

Interpretation of the results is made with the participation of an expert and taking into account the subject area. It allows us to reveal new knowledge. Knowledge formalization as a way of representing the knowledge of subject area using sign systems with the participation of a knowledge engineer, allows deciding that the identified knowledge can be represented by production rules.

Let's consider the cluster with the highest employee scores. This cluster is characterized by the presence of family and children; age from 25 to 43 (the absence of older staff in the position, more than 43); the presence of technical education (better higher education); the presence of additional education or additional knowledge (for example, increasing knowledge in the field of information security); length of experience at the previous job from 25 to 50 months (in positions different from the considered); the number of jobs over the past 2 years is no more than one; the reason for leaving a previous job is the lack of professional growth or low wages; high level of stress resistance; having the ability to learn; the true answer to the question about the presence of friends in the company at the time of arrival, etc.

All clusters are described in the same way. Results allow us to formulate the production rules. As part of the study, an application was developed to form a knowledge base and recommendations for decision-making.

The main operations of the developed software solution are discovery, creation, editing and saving of the knowledge base.

Inference engine allows us to receive the appropriate recommendations.

The quality evaluation of the applicant's selection is made using the formula: $S = K / L * 100\%$, where K is the number of correctly recommended, L is the total number of applicants.

100 employees who have worked in the organization for more than a year, and in-formation about which was not included in the initial sample for factor and cluster analysis, were selected to assess the correct functioning of the software solution developed using the proposed approach.

Thus, the following information were used: information from the questionnaires, additionally collected data in the selection process, data collected in the process of evaluating their activities. Also, production rules from the knowledge base. As a result, for 96 employees out of 100, the software solution gave the correct recommendation.

The rules of logical inference are based on the interpretation of Kohonen maps obtained during the clustering process.

5. Conclusion

Research analysis by foreign and Russian authors have shown that neural network technologies demonstrate good results in solving personnel management problems. A number of publications devoted, among other things, to the identification of implicit knowledge for decision support in this area. As a rule, operational knowledge is used in these cases.

We propose the common method for identifying implicit knowledge and organizing decision support on the example of personnel selection using both operational knowledge and historical knowledge. The neural network technologies are used, in particular, Kohonen maps of analytical platform Deductor to identify non-obvious knowledge through clustering.

The data collected about candidates in the process of their selection is used as initial data for clustering, including personal data, and evaluation data in real-time of those candidates who have been selected.

The results of clustering allow us to reveal implicit knowledge based on the similarity of characteristics (features) of those workers, who in the process of their professional activities can be assessed as effective or not.

Interpretation of the results of cluster analysis with the participation of an expert and a knowledge engineer allows us to formulate production rules.

The performed experiments using the developed software solutions for formation of the knowledge base and its use showed a good result in 96% of cases. Validation of the decision made it possible to confirm the possibility of using the proposed approach for new applicants.

6. Acknowledgements

The results of the research presented in this article were supported by Grants RFBR 19-07-00709 and 18-07-00193. The part of study which associated with use of methods for identifying similar objects and formalizing inference rules is conducted with financial support from the Ministry of Education and Science of the Russian Federation as part of the basic part of the state assignment to higher education educational institutions # FEUE-2020-0007.

References

[1] Azhmukhamedov A.I., Protalinskiy O.M. 2016 Mathematical model of motivational personnel management *Vestnik of Astrakhan State Technical University. Series: Management, Computer science and Informatics* vol 2 pp 50-59

[2] Buzuev A.I. 2014 Matematicheskoye i metodicheskoye obespecheniye podderzhki prinyatiya resheniy v sisteme upravleniya personalom krupnoy organizatsii *Phd Diss* [Mathematical and methodological support of decision support in human resource management of large organization. *Phd Diss.*]. Penza State University Penza p 146

[3] Evstratova E.I. 2018 The use of neural network analysis in personnel management processes [Primenenie nejrosetevogo analiza v processah upravleniya personalom] *Young Scientist* vol 35 pp 31-33 URL:https: //moluch.ru/archive/221/52515/

[4] Navrotskiy M. A., Zhukova I. G., Dergacheva M. Y. 2014 Intelligent support of decision making in human resource management using case-based reasoning and ontology *Izvestia VSTU* vol 12(139) pp 91-98

[5] Danilenko A.N. 2015 Developing methods and algorithms for a decision-making intellectual support in personnel management systems *Proceedings of International Conference Information Technology and Nanotechnology (ITNT-2015)* pp 381-388 URL: http://ceur-ws.org/Vol-1490/paper45.pdf

[6] Kuts Yu.I. 2008 On development of automated support system for development and decision-making on personnel management [O razrabotke avtomatizirovannoj sistemy podderzhki vyrabotki i prinyatiya reshenij po upravleniyu personalom]. *Programmnye Produkty i Sistemy* vol 1 URL: http://www.swsys.ru/index.php?page=article&id=105

[7] Guzairov M.B., Yusupova N.I., Kudryashev A.F., Elkhova O.I., Popov D.V., Rizvanov D.A., Enikeeva K.R., Dolomatov M.Yu., Yartsev R.A., Smetanina O.N., Verkhoturov M.A., Verkhoturova G.N., Makeev G.A., Gayanova M.M., Boyko M.V., Bogdanova D.R., Klimova A.V., Galyamov A.F., Orekhov Yu.V., Orekhov E.Yu., Verkhoturova O.M., Minigulova I.R., Malakhov N.E., Malakhov E.N., Furman K.G., Vinogradova N.V., Sultanova L.B., Salakhova N.R., Varygin D.V. 2014 Methodical aspects of artificial intelligence (in Russian) (Moscow) *Mechanical Engineering*

[8] Yusupova N.I., Smetanina O.N., Rizvanov D. A., Enikeeva K.R. 2016 Knowledge representation models for decision support in the management of complex systems taking into account

uncertainty and resource constraints [Modeli predstavleniya znanij dlya podderzhki prinyatiya reshenij pri upravlenii slozhnymi sistemami v usloviyah neopredelennosti i resursnyh ogranichenij] *Proceedings of the 4th International Conference Information Technologies for Intelligent Decision Making Support (ITIDS'2016)* (USATU, Ufa) vol 1 pp 24-27

[9] Prokushev Y.E. 2015 Support of decision-making for the procedure of selection of the personal for the training and development (in Russian) *Statistics and Economics* vol 4 pp 144-148 URL: https://doi.org/10.21686/2500-3925-2015-4-144-148

[10] Fedyakina O.P. 2004 Intellektualizaciya podderzhki reshenij po upravleniyu personalom organizacii na osnove nechetkih modelej znanij. *Phd Diss* [Intellectualization of support of organization's personnel management solutions based on fuzzy knowledge models. *Phd Diss.*] Voronezh State Technical University Voronezh p 194

[11] Shagieva Yu.R. 2013 Informacionnaya podderzhka prinyatiya reshenij pri upravlenii processom otbora tekhnicheskogo personala na osnove kompleksnoj diagnostiki. *Phd Diss* [Information support for decision-making in management of process of selection of technical personnel based on integrated diagnostics. *Phd Diss.*] Ufa State Aviation Technical University Ufa p 200

[12] Farrés Rocabert A. 2017 Artificial Neural Network System Applied to Human Resource Management *ETSEIB* Barcelona Spain

[13] Jantan H., Hamdan A. R., Othman Z. A. 2010 Intelligent Techniques for Decision Support System in Human Resource Management Decision Support Systems *Advances Book edited by: Ger Devlin* (INTECH, Croatia) ISBN 978-953-307-069-8 pp 261-276

[14] Hakimpoor H., Arshad Kh. A. B., Tat H. H., Naser Khani, Rahmandoust M. 2011 Artificial Neural Networks' Applications in Management *World Applied Sciences Journal* 14 (7) pp 1008-1019

[15] Faggella D. *Machine Learning in Human Resources – Applications and Trends* https://www.techemergence.com/machine-learning-in-human-resources last accessed 2020/10/20

[16] Walczak S. 2016 Artificial Neural Networks and other AI Applications for Business Management Decision Support *International Journal of Socio technology and Knowledge Development archive* 8 (4) pp 1-20

[17] Masum A. K., Beh L.-S., Azad A.-K., Hoque K. 2018 Intelligent Human Resource Information System (iHRIS): A Holistic Decision Support Framework for HR Excellence *The International Arab Journal of Information Technology* vol 15(1) pp 121-130

[18] Hao G & Lai, Kin Keung & Tan, Manzhi. A. 2004 Neural Network Application *Personnel Scheduling. In: Annals of Operations Research* vol 128 pp 65-90 DOI: 10.1023/B:ANOR.0000019099.29005.17

[19] Akinyede R. O., Daramola O. A. 2013 Neural Network Web-Based Human Resource Management System Model (NNWBHRMSM) *International Journal of Computer Networks and Communications Security* vol 1 No 3 pp 75–87

[20] Strohmeier S., Piazza F. 2015 Artificial Intelligence Techniques Human Resource Management—A Conceptual Exploration *Kahraman C., Çevik Onar S. (eds) Intelligent Techniques in Engineering Management. Intelligent Systems Reference Library* (Springer, Cham) vol 87 pp 149-172

[21] *Artificial Intelligence in HR: a No-brainer*, https://www.pwc.nl/nl/assets/documents/artificial-intelligence-in-hr-a-no-brainer.pdf, last accessed 2020/10/19

[22] *Knowledge Management and e-Human Resource Management*, https://km.aifb.kit.edu/ws/LLWA/fgwm/Resources/FGWM03_08_Ernst_Biesalski.pdf, last accessed 2020/10/22

A Review of Smart Off-Grid Power Systems Optimization Models for the Oil and Gas Industry

V V Fedorenko[1], V V Samoylenko[1], I V Samoylenko[2], Yu K Dimitriadi[1]

[1] North-Caucasus Federal University, Department of Oil and Gas Wells Construction, 1 Pushkin str., Stavropol, 355000, Russian Federation

[2] Stavropol State Agrarian University, Department of Information Systems, 12 Zootechnicheskiy Ln, Stavropol, Russian Federation

E-mail vvsamoylenko@ncfu.ru

Abstract. The article discusses current issues of using alternative energy sources for electricity supply of consumers in the oil and gas industry. Examples of off-grid power systems used for the electrification of different facilities of oil and gas production are shown. This paper introduces a review of mathematical optimization models of modes in micro-energy systems with different distributed generation resources. Optimization methods based on mathematical programming in game theory are studied in details. It is concluded that the use of adaptive-game modelling is perspective in solving optimization problems.

Keywords: Off-Grid Power Systems, Mathematical Programming, Game Theory, Micro-Energy System, Oil and Gas Industry, SCADA

1. Introduction

Intense development of informational systems, power electronics led to wide use of new technologies of generation and energy distribution. Important structure element of such technologies is Smart grid and micro-grid [1]. Smart grid conception is government policy in technological development of energy future in such countries as USA, Canada, China. This concept is considered as "intelligent power system with active-adaptive net".

The implementation of power supply of remote objects based on renewable energy sources, including Smart Grid technology, is widely used in various areas of industry. The oil and gas industry is targeted to achieve maximum efficiency in oil and gas production in numerous fields. This industrial sector is considered as one of the dirtiest and most energy-consuming. In particular, in the USA mining and processing companies consume about 26% of all electricity used in industry [2]. Technological operations cover exploring, mining and transportation of these natural resources, using a wide range of equipment that requires a reliable electricity supply. The territorial remoteness of oil and gas industry facilities from centralized power supply systems make the use of alternative energy sources attractive. Renewable sources replace traditional methods of generating electricity for the needs of oil and gas wells, such as diesel generators, pneumatic systems, and the use of natural resources, such as natural gas from a well. The disadvantage of these methods is a low economic efficiency and a number of environmental problems that force oil companies to more active application of renewable energy sources, such as solar energy and wind energy with backup storage.

2. Examples of renewable energy sources application in oil and gas industry

Reliable operation of remote wells, gas and oil pipelines implies wireless telemetry elements of the technical condition of the objects. The supply issues of this equipment are a complex technological problem. In particular, data acquisition is realized for pipelines monitoring, regulatory equipment and emergency response. Alternative power supplies are also used for dispatching of these objects, including SCADA and RTU. In addition, system integrators often offer projects for the complex use of solar or wind energy [3,4] for the following energy-consuming processes:

- Well Site Monitoring & Control;
- Multi-well Pad Sites;
- Emergency Shutdown Valves;
- Wireless Monitoring;
- Chemical Injection;
- Natural Gas Processing;
- Tank Level Monitory & Control;
- Remote & Process Automation;
- Cathodic Protection;
- Valve Automation & Control;
- Burner Management;
- Fixed Flame & Gas Detection;
- Flow Computers.

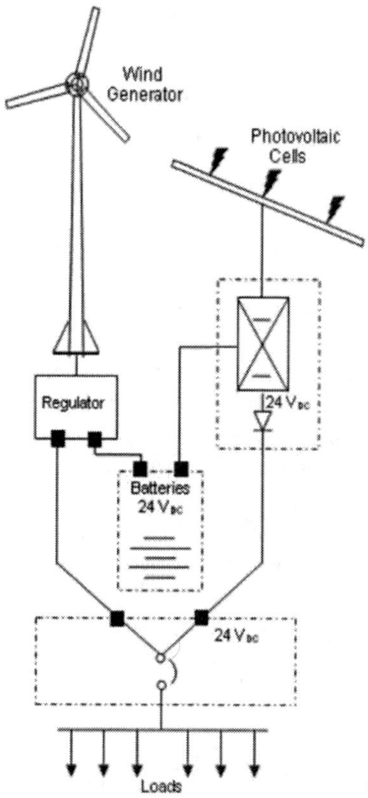

Figure 1. Hybrid power system on the offshore oil platform in Brazil

IOP Conf. Series: Materials Science and Engineering 1069 (2021) 012016 doi:10.1088/1757-899X/1069/1/012016

A wide range of the represented operations has the main task: replacement of brown energy in mining and oil and gas processing with renewable power plants installed near industrial sites. Therefore, the green energy provides the main power for pumps, heaters and compressors.

The scientific literature presents implementation examples of this concept. In the mining industry a wind power plant is used to feed the cement plant [5], in the oil and gas industry the wind power plant is also used at surface oil production field.

The first offshore oil pumping unit operating at the onshore wind farm was registered in the UNFCCC as the Clean Development Mechanism (CDM). Other specific applications in the oil and gas industry can be used as solar heating to supply thick oil and produce steam for injecting oil wells [6]. Another possible application is the use of solar radiation for photovoltaic panels that supply critical electrical loads.

One of the most effective options for the use of renewable energy sources is a hybrid system, which includes both wind and solar stations. In the Brazilian Northeast, there is an application of green technology at the offshore oil platform [7].

One of the main initial problems is argued choice of primary power sources and optimal ratio of rated power of all generating sources, included in created micro-grid structure. The correct solution to this problem is possible by optimization function choice based on adequate modeling of micro-grid control processes.

3. Optimization models of the Off-Grid Power systems

Nowadays a lot of scientists suggest different methods of mathematical modeling of micro-grids with several sources of distributed generation. Optimized micro-grid indices are output power parameters, such as: the load curve of consumers; the quality of electric energy; the reliability of the power system and its operational conditions. Optimization models and methods of mathematical programming, mostly used in studies of micro-grids are:

1. Linear programming (LP) methods are simple optimization methods in which objective function and the constraints are defined by linearity. These methods are used for searching the parameters of economic power distribution, designing and power systems operation, the consistency of its protection, insurance of the load curve, estimation of the system state [8–10]. A special case of the PL is a method of integer programming in which some or all variables are defined as discrete integer values. This method is used for security assessment of energy systems, transmission lines optimization and design, reliability analysis, design of distributed systems and load management [11,12].

2. Methods of nonlinear programming are used for optimization problems of dynamic security of power systems, control of reactive power flows, design and operation of power systems, optimal power flow in the system, optimal placement of reactive power compensation and optimal location of energy sources [13].

3. Methods of dynamic programming are used for optimization of sources power of electricity generation and the optimal functioning of the electrical system at the level of a distributed system [14].

Particle Swarm Optimization (PSO) takes a special place among publications about micro-grid optimization using simulation algorithms. The main property of PSO is that the complexity and nonlinearity of the task do not affect the finding of the optimal value of the objective function [15,16].

4. Differential evolution approach method of proposed in [17]. Micro-grid is optimized with the aim of cost estimation of harmful emissions into the environment and ongoing costs of operation and maintenance of the system. A problem decision is an algorithm that takes change in the load curve in time with a corresponding variation of the batteries charge level into account. The practical implementation of the algorithm is based on central controllers for adequate response to variable conditions, such as: the weather, the intensity of solar radiation, wind speed, operating costs.

4. Application of game theory in smart grid

The solution of the optimization tasks is often carried out in conditions of uncertainty in its various forms. Correct and appropriate account of uncertainty (primary information, objectives, conditions) in appropriate mathematical modeling is often a fundamental requirement. It ensures objectivity and factual efficiency of decisions based on model analysis. Therefore, game theory is the key analytical tool in Smart grid design in order to explore complex interactions among independent rational players.

In a systematic description of game theory use in smart grid is presented. The work gives a clear idea about the basic benefits and challenges of using classic and new game-theoretic methods in the context of intelligent networks and consists of three logically related parts:

1) complex description of the existing game-theoretic applications in smart grid;

2) determination of the main open issues in smart grid that are required to be solved using game theory;

3) definition of the basic game-theoretic tools that can be adopted for the smart grid design.

Game theory is a mathematical basis that can be divided into two main branches: the theory of non-cooperative games and the theory of cooperative games. The theory of noncooperative games can be used to analyze strategic decisions of players who have fully or partially opposing interests. In fact, noncooperative games can be considered as a distributed decision-making process. The players optimize decisions without any coordination or communication between them.

In classical power grid solutions of system optimization problems are based on a centralized objective function. However, if micro-grid is included in power grid, its optimization is possible only in case of definition a specific objective function for each micro-grid. This is mainly due to heterogeneous nature of micro-grids, which often consist of different sources of energy, such as energy storage (e.g. batteries), diesel generators, wind turbines and solar panels. The values of the objective functions are elements of payment matrices (matrix games) for various strategies of selecting the energy source in variable operating conditions of the power system [18,19].

Saad et al. showed a mechanism of cooperative energy exchange in micro-grid based on the model of coalitional game in excess of domestic consumption [20]. It is necessary to create joint groups, i.e., coalitions that represent local currency energy market. Inside the coalition micro-grids can exchange power with each other locally, thus reducing power loss and increasing the independence of the micro-grid.

Another paper proposes the use of noncooperative game for the adequate modeling of interactions between sources and loads in small-scale power systems [21]. For example, advanced analytical methods and algorithms for finding Nash equilibrium and efficiency increase of power transmission are used in the problem of power control.

Pan used the theory of evolutionary game to study the cooperation between the micro-grid and the common system, focusing on the factors that affect the choice of strategy players [22]. The results of the analysis based on the evolutionary game model suppose that both parties hold a joint strategy. Unanimous choice of the players positively correlates with the benefits of their joint cooperation. Pan proposed ideas about cooperation deepening between the micro-grid and the common system.

The main problem of various control loops from the level of decision-making in the micro-grid to the level of predicting the evolutionary development of power systems with dispersed generation is the uncertainty of the upcoming operating conditions. Two fundamentally different approaches can be singled out among known approaches to solve this problem:

1) optimistic approach is able to predict the parameters of these terms and conditions accurately;

2) pessimistic one recommends to rely on the worst case.

The basis for the legitimacy of the first approach is the existence of destabilizing factors, whose parameters vary slowly (e.g., seasonal demand for electricity). It allows accurate estimating and managing the choice of energy source on the basis of monitoring of the performance indicator implementation of the micro-grid. Mathematical description of the interaction of power grid components is possible on the basis of coalitional games, which allow energy exchange in the networks [20].

The basis for eligibility of the second approach is the existence of destabilizing factors caused by the competition for energy resources, which occurs between the load and the competition for the energy supply between the sources. The model of noncooperative games for optimal control of elements is proposed [21].

The intermediate case should be presented as models of account of destabilizing factors that change faster than you can manage to predict their exact implementation, but have slowly changing limits of the possible variations in the form of some statistical averages.

5. Conclusion

The analysis in the review article shows the high energy intensity of the oil and gas industry. The given examples indicate the relevance of replacing brown energy sources with environmentally friendly renewable green energy sources. Using only one type of renewable energy source is incorrect solution. Therefore, hybrid systems including both solar energy and wind energy gain wide popularity. The management of these systems is most often carried out using Smart Grid technology and the remoteness of objects positions this class of systems as Off-Grid Power Systems.

The mathematical optimization of micro-grid data plays an important role at the design stage. The article presents a detailed analysis of the optimization models of these systems. Apparently, a reasonable compromise should be searched in integrating adaptive and game models of the grid control. Elements of adaptation are more or less evident included in many game models. This is especially noticeable in dynamic games, such as differential and multistage.

These games are composed of game situations with the changing constraints of the strategies at each step. Elements of adaptation are less obvious and less connected presented in task of strategies constraints in one-step games. Game situations are a special case of involving more than two players, and not all of them pursue antagonistic goals. In this case, we can expect a positive effect off adaptation and a joint management optimization and implementation of part of the destabilizing factors, that express the common interests of producers and consumers of electricity.

Reference:

[1] Qiao L 2013 A Summary of Optimal Methods for the Planning of Stand-alone Microgrid System *Energy Power Eng.* **05** 992–8
[2] Malmedal K, Sen P K and Candelaria J 2011 Electrical energy and the petro-chemical industry: Where are we going? *Record of Conference Papers - Annual Petroleum and Chemical Industry Conference*
[3] Anon Oil & Gas - SunWize | Power Independence
[4] Anon Solar Power Solutions - Oil and Gas Industries | Ameresco Solar
[5] Shafer B 2009 Electrical wind generation for cement manufacturing, a case study: Calportland Mojave Plant's 24 MW wind project *IEEE Cement Industry Technical Conference (Paper)*
[6] Anon Oil From The Sun
[7] Oliveira M F, Saidel M A, Queiroz A R S and Filho E N 2012 Renewable sources at offshore petroleum and gas production platforms *Record of Conference Papers - Annual Petroleum and Chemical Industry Conference*
[8] Khodr H M, Gómez J F, Barnique L, Vivas J H, Paiva P, Yusta J M and Urdaneta A J 2002 A linear programming methodology for the optimization of electric power-generation schemes *IEEE Trans. Power Syst.* **17** 864–9
[9] Kurucz C N and Brandt D 1996 A linear programming model for reducing system peak through customer load control programs *IEEE Trans. Power Syst.* **11** 1817–24
[10] Delson J K and Shahidehpous S M 1992 Linear Programming Applications To Power System Economics, Planning and Operations *IEEE Trans. Power Syst.* **7** 1155–63
[11] Lima F G M, Galiana F D, Kockar I and Munoz J 2003 Phase shifter placement in large-scale systems via mixed integer linear programming *IEEE Trans. Power Syst.* **18** 1029–34

[12] Alguacil N, Motto A L and Conejo A J 2003 Transmission expansion planning: A mixed-integer LP approach *IEEE Trans. Power Syst.* **18** 1070–7
[13] Pudjianto D, Ahmed S and Strbac G 2002 Allocation of VAr support using LP and NLP based optimal power flows *IEE Proc. - Gener. Transm. Distrib.* **149** 377
[14] Kanchev H, Francois B and Lazarov V 2011 Unit commitment by dynamic programming for microgrid operational planning optimization and emission reduction *International Aegean Conference on Electrical Machines and Power Electronics, ACEMP 2011 and Electromotion 2011 Joint Conference* (IEEE Computer Society) pp 502–7
[15] Arya L D, Titare L S and Kothari D P 2010 Improved particle swarm optimization applied to reactive power reserve maximization *Int. J. Electr. Power Energy Syst.* **32** 368–74
[16] Laskari E C, Parsopoulos K E and Vrahatis M N 2002 Particle swarm optimization for integer programming *Proceedings of the 2002 Congress on Evolutionary Computation, CEC 2002* vol 2 (IEEE Computer Society) pp 1582–7
[17] Vahedi H, Noroozian R and Hosseini S H 2010 Optimal management of microgrid using differential evolution approach *2010 7th International Conference on the European Energy Market, EEM 2010*
[18] Mohamed F A and Koivo H N 2011 Multiobjective optimization using modified game theory for online management of microgrid *Eur. Trans. Electr. Power* **21** 839–54
[19] Maity I and Rao S 2010 Simulation and pricing mechanism analysis of a solar-powered electrical microgrid *IEEE Syst. J.* **4** 275–84
[20] Saad W, Han Z and Poor H V 2011 Coalitional game theory for cooperative micro-grid distribution networks *IEEE International Conference on Communications*
[21] Weaver W W and Krein P T 2009 Game-theoretic control of small-scale power systems *IEEE Trans. Power Deliv.* **24** 1560–7
[22] Pan C and Long Y 2015 Evolutionary Game Analysis of Cooperation between Microgrid and Conventional Grid *Math. Probl. Eng.* **2015**

A method for transmitting information based on the stochastic use of binary quasi orthogonal code sequences

A P Zhuk[1], D V Orel[1], V P Pashintsev[1], I A Kalmykov[1]

[1]North-Caucasus Federal University, Stavropol, Pushkin str., 1, 355017, Russia

E-mail azhuk@ncfu.ru

Abstract. The objective of the article is to develop the structure of a system for transmitting information with code division of channels and stochastic usage of systems of binary quasi-orthogonal code sequences obtained on the basis of functional transformations of pseudo-random arguments, as well as a description of the principle of its functioning. The aim of the article is to develop a method for transmitting information in satellite communication systems and global navigation satellite systems with code division of channels and stochastic usage of binary quasi-orthogonal code sequences obtained on the basis of functional transformations of pseudo-random arguments, providing an increase in structural secrecy. The elements of novelty are obtaining a new structure of an information transmission system with code division of channels and increased structural secrecy, as well as a description of the main stages of the information exchange method based on this system. The method differs from the known ones by stochastic use of binary quasi-orthogonal code sequences obtained on the basis of functional transformations of pseudo-random arguments. The main result of the article is that the presented method of information transmission makes it possible to reduce the impact of deliberate imitating interference and cases of substitution of the signals used by increasing the structural (signal) secrecy of the information transmission system with code division multiplexing and stochastic usage of systems of binary quasi-orthogonal code sequences.

Keywords: binary quasi-orthogonal code sequences; method of information transfer; functional transformations of pseudo-random arguments; code division of channels; structural secrecy

1. Introduction

The technology of code division of channels (CDC) allows you to get a number of advantages when organizing information exchange in satellite communication systems (CCS), which determines the feasibility of using CDC in the design of new CCS and global navigation satellite systems (GNSS) [1-6]. In a number of works [7-9] it is noted that at present, due to the increase in the volume of transmitted data in the CCS and GNSS, there is a high probability of realization of threats to disrupt their stable functioning due to the impact of deliberate imitating interference and substitution of the signals used. At the same time, the methods used to counter this interference do not provide effective protection for CCS and GNSS due to their limited capabilities.

In works [7,8], it is shown that increasing the noise immunity of CCS and GNSS with CDC when organizing imitating interference and substitution of NS is quite effective due to an increase in the structural secrecy of NS based on the stochastic usage of systems of code sequences. The work proved that the required number of non-repeating binary code sequences (BCS), which can provide an increase in the structural secrecy of the NS GNSS when they are stochastic use, is $A_r = 2{,}3652 \cdot 10^{13}$.

To assess the structural (signal) secrecy in [8-10], is used an indicator of the complexity of solving the structure of code sequences (CS) in the form of a coefficient H. The coefficient H serves as a measure how the CS is estimated as random when analyzed by the radio monitoring subsystem of the radio-electronic interference station (REI). The most common variant for assessing the complexity of solving the structure of a code sequence is the Berlekamp-Messi algorithm. In this case, the coefficient H is determined based on the following expression:

$$H = \frac{M}{N}, \qquad (1)$$

where M - is the number of consecutive symbols of the code sequence that must be received by the monitoring subsystem of the REI station to build the structure of an equivalent shift register with linear feedback (SRLFB); N - code sequence length.

Since the normalized linear complexity of each individual BCS from the system of binary quasi-orthogonal code sequences (SBQCT) may differ, the averaged H values were found. Table 1 shows the values of the averaged length-normalized code sequence complexity of solving the structure for the known types of code sequences Gold, Kasami, Kamaletdinov, Kerdock, Weil and Bent sequences, considered in [8] and for code sequences obtained on the basis of functional transformations of pseudo-random arguments [11,12].

Comparative analysis shows that the average complexity of solving the structure H for BCS obtained on the basis of functional transformations of pseudo-random arguments for all three considered lengths 4095, 8191 and 10230 has the highest value, which is approximately equal to the length of the BCS itself $M \approx N$ and is comparable with the same indicator for Kamaletdinov's BCS, Weil and Bent-sequences. However, the advantage of BCS obtained on the basis of functional transformations of pseudo-random arguments is that the number of nonrepeating BCSs is significantly larger than all known BCSs, and also closest of all analyzed to the required value $A_r = 2{,}3652 \cdot 10^{13}$. For example, for BCS length equal to 4095 the advantage over Kerdock sequences is more than 10,000 times, for BCS length equal to 8191 the advantage over Weil sequences is about 10,000,000 times, for BCS length 10230 the advantage over Kerdock sequences is more than 1000 times. This circumstance indicates that with the stochastic usage of binary quasi-orthogonal code sequences obtained on the basis of functional transformations of pseudo-random arguments, they provide a higher structural (signal) secrecy in a code division multiplexing system in comparison with the case of usage known discrete quasi-orthogonal code sequences.

In works [1,4,8,9,11-19] approaches to obtaining various types of SBQCT are considered and their properties and the possibility of practical application in information transmission systems are evaluated. However, they are limited either using previously known SBQCTs, or by describing only the methods of their formation.

Information exchange methods based on the stochastic usage of binary quasi-orthogonal code sequences obtained on the basis of functional transformations of pseudo-random arguments were not considered in these works.

Proceeding from this, the development of the structure of the information transmission system with code division of channels and the stochastic usage of systems of binary quasi-orthogonal code sequences obtained on the basis of functional transformations of pseudo-random arguments, and the description of the principle of its functioning is an urgent task.

The aim of the article is to develop a method for transmitting information in CCS and GNSS with code division of channels and stochastic usage of binary quasi-orthogonal code sequences obtained on the basis of functional transformations of pseudo-random arguments, providing an increase in structural secrecy.

2. Method of information transmission with code division of channels and stochastic usage of binary quasi-orthogonal code sequences

The developing method of information transmission in CCS and GNSS with code division multiplexing and stochastic usage of binary quasi-orthogonal code sequences obtained on the basis of functional transformations of pseudo-random arguments develops a method of information transmission based on chaotically forming ensembles discrete multilevel orthogonal signals proposed in [15].

In the considered method of information transmission for message transmission, it is proposed to use binary quasi-orthogonal code sequences obtained by functional transformations of random functions [8, 14], based on the invariance property of the probability differential

$$g(\tau)d\tau = f(x)dx, \tag{2}$$

where x - is the initial random variable;; $f(x)$ - distribution density of the initial random variable; τ - a random time interval between instantaneous values converted by random variables; $g(\tau)$ - the distribution density of the transformed function, and the random variables x and τ are connected by an unambiguous deterministic functional dependence $\tau = \varphi(x)$, which is an unambiguous differentiable function that admits an inverse transformation $x = \psi(\tau)$.

It follows from this property that the distribution function of the random variable is

$$G(\tau) = \int_0^{\psi(\tau)} f(x)dx \tag{3}$$

and allows us to determine the value of the random variable τ:

$$\tau = \left\{F^{-1}\left[rnd(1) + F(a)\right]\right\}, \tag{4}$$

where F^{-1} - is the function inverse to the distribution function of the random function x, and $F'(x) = f(x)$; $rnd(1)$ - numerical values generated by a random number generator, evenly distributed over the interval [0,1]; a - is the parameter determined from the normalization condition

$$\int_0^{\infty} f(x)dx = 1. \tag{5}$$

To obtain a random variable with a given distribution function, it is necessary to construct a deterministic function $\tau = G^{-1}(x)$ and obtain the desired random values of this function from the argument determined by the number, which is a random variable with a uniform distribution law on the interval [0,1]. Taking this remark into account, we rewrite the formula (3) for τ

$$\tau = \varphi\left\{F^{-1}\left[rnd(1) + F(a)\right]\right\}. \tag{6}$$

The analysis of the above shows that it is quite possible to obtain a binary code sequence in which both the pulse durations and the intervals between them would be a random value of the unit amplitude, it is quite possible if the sequence is fed to the counting C input of the RST-trigger, provided that it provides the distribution of trigger overturning according to the law, defined by expression (6).

This method for constructing SBQCT is based on the usage of pseudo-random numbers obtained as a result of functional transformation as a parameter that determines the length of series of consecutive elements of the same character in a discrete code sequence and is described in detail in [8, 14]. Taking into account the fact that each time the set of random numbers, in the amount determined by the length of the generated code sequence intended for information exchange in the developed method, will differ from each other on the interval [0,1], then the code sequences themselves will be different. In order for the code sequences obtained in this way to satisfy the quasi-orthogonality condition, in the process of their formation, their correlation properties can be checked for compliance with this condition.

We will describe the proposed method of information transmission using the example of an information transmission system with code division of channels and stochastic usage of systems of binary quasi-orthogonal code sequences, the structure of which is shown in Fig. 1.

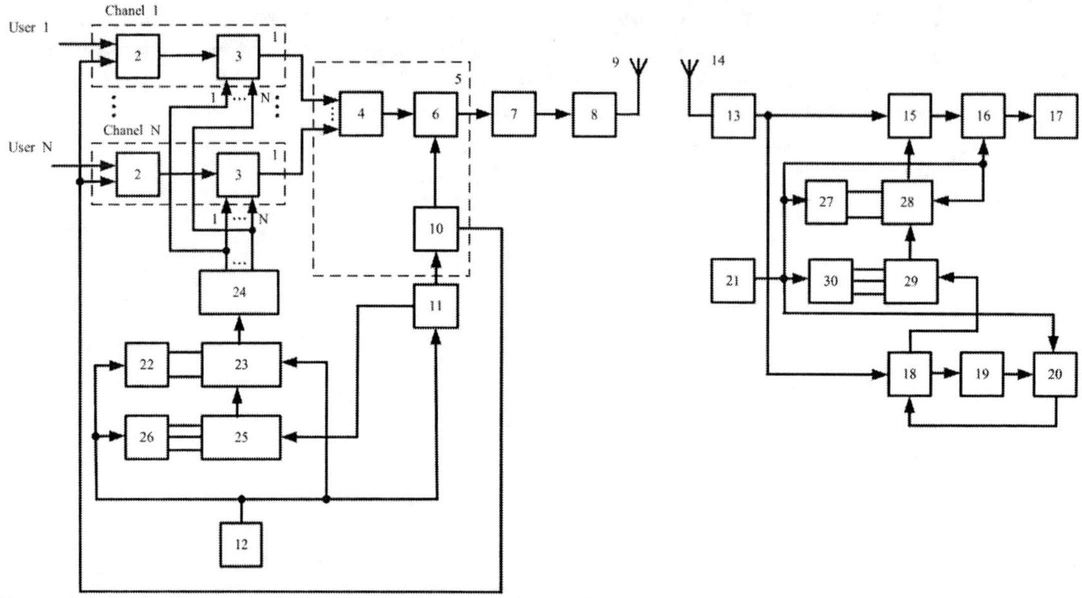

Figure 1. Block diagram of a system for transmitting information with code division of channels and stochastic usage of systems of binary quasi-orthogonal code sequences

In fig. 1 the following notation is used: 1 – information channels, 2 – digital information block, 3 – modulator, 4 – input combiner, 5 – group signal forming unit, 6 – group signal forming unit modulator, 7 – phase modulation unit, 8 – power amplifier, 9 – transmitting antenna, 10 – synchronization signal generator, 11 – service information block, 12 – clock pulses generator, 13 – high-frequency selection unit, 14 – receiving antenna, 15 – correlation processing unit, 16 – information extraction unit, 17 – information recipient block, 18 – sync signal detection unit, 19 – search block, 20 – sync signal copy block, 21 – clock pulses generator unit, 22 – block for setting initial functions, 23 – functional transformation block, 24 – converter, 25 – random number generator, 26 – block for setting initial values, 27 – block for setting copies of original function, 28 – block of copies of functional transformations, 29 – random number copies generator, 30 – block for specifying copies of initial values.

The above system for transmitting information with code division multiplexing and stochastic usage of binary quasi-orthogonal code sequences includes transmitting and receiving equipment. The transmitting equipment contains N information channels 1, each of which consists of a block of digital information 2 connected by its output to the first input of the modulator 3, the output of which is the output of each information channel. Information transmitted from an individual user enters the information channel 1 of the information transmission system shown in Fig. 1.

In each information channel, information symbols of an individual $User\ N$ are entered into digital information blocks 2 and modulated using modulator 3.

For the formation and stochastic usage of SBQCT as spreading sequences obtained on the basis of functional transformations of random functions in accordance with the procedures described by expressions (2) - (6), in the transmitting equipment it is proposed to use a block for setting initial functions 22, a block of functional transformations 23, a generator random numbers 25 and block for setting initial values 26. In the receiving equipment, it is similarly proposed to use a block for setting

IOP Conf. Series: Materials Science and Engineering 1069 (2021) 012017 doi:10.1088/1757-899X/1069/1/012017

copies of the original functions 27, a block for copies of functional transformations 28, a generator of copies of random numbers 29 and a block for setting copies of initial values 30.

Let us consider the operation of the information transmission system shown in Figure 1. The signals generated at the output of the modulator 3 of each information channel 1 are simultaneously fed to the group signal generation unit 5. Further, the information flows of each information channel 1 are combined using a combiner 4 of the group signal forming unit 5 and fed to the first input of the modulator of the group signal forming unit 6, on which the combined group signal is modulated. The second input of the modulator of the group signal generating unit 6 receives a signal from the output of the synchronization signal generator 10, which is controlled by the service information unit 11 and the clock pulses generator 12. In addition, the signal from the output of the synchronization signal generator 10 is fed to the inputs of the digital information units 2 of each N information channels 1. Under the control of the signals of the synchronization generator 10, clocked synchronous input of information symbols into the blocks of digital information 2 is carried out. After combining the information streams of each information channel 1 with the help of a combiner 4, in the block for forming the group signal 5 and superposition in the modulator 6 of the synchronization signal generated by the synchronization signal generator 10, a group signal is formed, the spectrum of which, after being transferred to the carrier frequency region in the phase modulation unit 7 and the power amplifier 8, is emitted into the air through the antenna 9.

The radio signal, after propagation in space, is fed to the input of the receiving antenna 14 of the receiving equipment. On the receiving side, the signal received by the receiving antenna is subjected to preprocessing in the high-frequency selection unit 13. From the output of this unit, the signal is simultaneously fed to the correlation processing unit 15 and the synchronization signal detection unit 18. In this case, the synchronization signal detection unit 18 together with the search unit 19 are brought into synchronism synchronization signal copy generator 20. The signal from the output of the correlation processing unit 15 is fed to the first input of the information extraction unit 16, which is connected by its output to the information recipient unit 17. Correlation processing in the correlation processing unit 15 is carried out using the binary quasi-orthogonal code sequences used by this receiving equipment, obtained on the basis of functional transformations random functions implemented using the block for setting copies of the original functions 27, the block for copies of functional transformations 28, the generator of copies of random numbers 29 and the block for setting copies of the initial values 30.

Since the block of functional transformations 23 on the transmitting side and the block of copies of functional transformations 28 on the receiving side are important elements in the proposed information transmission system, we will consider it in more detail. Taking into account that the block of functional transformations 23 on the transmitting side and the block of copies of functional transformations 28 on the receiving side are identical in their structure and functions, we will consider the composition and operation of the functional transformations block 23 on the transmitting side, taking into account that the composition of the elements and the operation of the block copies of functional transformations 28 will be similar.

In fig. 2 shows a block diagram of the block of functional transformations 23. The designations of elements for the case when they are included in the block of copies of functional transformations 28 on the receiving side are indicated in brackets.

In fig. 2 the following designations are used: 12 (21) - clock pulse generator; 22 (27) - block for setting initial functions; 23 (28) - block of functional transformations; 24 - converter; 25 (29) - random number generator; 31 - functional converter; 32 - block for calculating the inverse function; 33 - trigger; 34 - normalization block, 35 - segment selection block; 36 - frequency divider.

The procedure for generating a block of functional transformations of 23 binary quasi-orthogonal code sequences intended for use in the process of transmitting and receiving information in a code division multiplexing system is as follows:

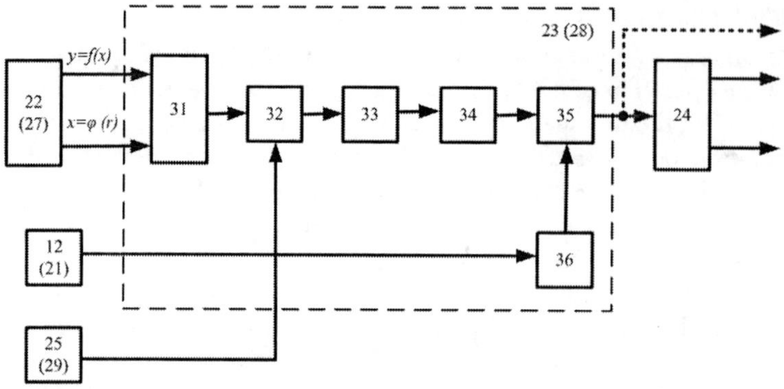

Figure 2. Block diagram of the block of functional transformations

1. Using the block for setting the initial functions 22, the initial functions are set to be used in the functional transformations. As a result, tables of initial functions are formed, which will be used to form working quasi-orthogonal code sequences. As the initial functions, analytical expressions are considered, which can be considered as distribution functions taking into account the formal properties and the field of assignment. As an example, the following function can be used:

$$f(x) = \begin{cases} 0, \text{with } x < a, \\ \frac{x-a}{\beta-a}, \text{with } \beta < x < a, \\ 1 \text{ with } x > \beta. \end{cases} \quad (7)$$

$$x = \varphi(\tau) = \left(\frac{k}{\tau}\right)^a. \quad (8)$$

2. The setting of the initial values is carried out by the unit for setting the initial values 26 on the transmitting side and by the unit for setting copies of the initial values 30 on the receiving side. For instance, $m_k = 1, q = 5^{17}, M = 2^{40}$.

3. Functional converter 31 of the block of functional transformations 23 carries out the functional transformation of the initial functions used in accordance with the expression (2)

$$G(\tau) = 1 - \left(\frac{k}{\tau}\right)^a. \quad (9)$$

4. A sequence of random numbers $rnd(1)$ is generated using a random number generator 25, and the numerical values $rnd(1)$, generated by the random number generator are uniformly distributed over the interval [0, 1]. For instance

$$m_{k+1} = m_k q - \left]\frac{m_k q}{M}\right[\cdot M. \quad (10)$$

5.Using the block for calculating the inverse function 32, the inverse function F^{-1} in accordance with expression (3), where F^{-1} - is the function inverse of the distribution function of the random function x, moreover $F'(x) = f(x)$. For the function under consideration with $a = 2$ we have

$$\tau_n = \frac{k}{(1-rnd)^{\frac{1}{a}}}. \quad (11)$$

6. In accordance with the obtained inverse distribution function of the random function x at the output of the trigger 33, a sequence $U_T(t)$ is formed, shown in Fig. 3b.

7. Using normalization block 34, the generated sequence $U_T(t)$ standardized.

8. By means of the segment separator 35, operated under the control of the frequency divider 36, the normalized sequence is segmented. Figure 3 shows timing diagrams of the operation of the main elements of the block of functional transformations 23.

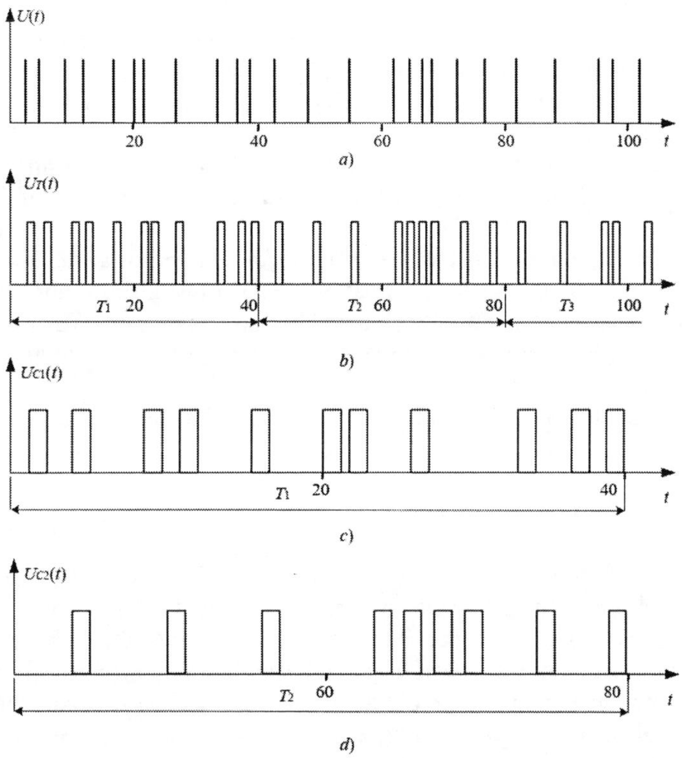

Figure 3. Timing diagrams of the operation of the main elements of the block of functional transformations

On the receiving side, a similar procedure for generating copies of code sequences used in the process of receiving information in a code division multiplexing information transmission system is carried out using a block of copies of functional transformations 28. Figure 3 shows the following timing diagrams: signals generated at the output of the random number generator 25 (Fig. 3a); signals received at the output of trigger 33 (Fig. 3b); signals at the output of the segment allocation block 35 for $T_1 = [0-40]$ (Fig. 3c); signals at the output of the segment allocator 35 for $T_2 = [40-80]$ (Fig. 3d).

Received signal systems $U_{C1}(t)$, $U_{C2}(t)$, etc. at the output of the segment allocation unit are segment systems [4] and are formed from sequence segments, the lengths of which are determined by the period of operation of the random number generator 25. The segment length determines the working length of the code sequence and is specified as a characteristic of the transmission system and determines the signal base ($B = N$).This length is formed in the segmentation unit 35 by dividing the frequency of the clock generator 12 using a frequency divider 36, the division factor of which determines the segment length.

Both on the transmitting and on the receiving side, tables of initial functions are preliminarily formed, which at the time of information transmission will be used to form working code combinations. These tables are stored in the memory of the transmission system on the transmitting side. Copies of these tables are stored on the receiving side. The order of using and changing the functions presented in these tables at each given moment of time is set by the service information block 11 on the transmitting side and its copy 18 on the receiving side.

In view of the above, the proposed method for transmitting information based on the stochastic use of binary quasi-orthogonal code sequences obtained on the basis of functional transformations of pseudo-random arguments is carried out in the following sequence:

Stage 1. Before the start of the information transfer process, the formation, manipulation and transmission of an auxiliary synchronization signal by the transmitting equipment is carried out. As a result, service information is transmitted to each subscriber station (a single initial cyclic phasing marker for all subscriber stations).

Stage 2. After the receiving equipment receives the service information (a single initial cyclic phasing marker), the receiving equipment of each subscriber station enters a state of synchronism with the transmitting equipment.

Stage 3. Using a random number generator, a block for setting initial functions, a block for setting initial values and a block of functional transformations in the transmitting equipment, a system of binary quasi-orthogonal code sequences is formed, obtained by functional transformation of random functions based on the invariance property of the probability differential (2), which is used as carriers of information.

Stage 4. Using a generator of copies of random numbers, a block for setting copies of initial functions, a block for setting copies of initial values and a block of copies of functional transformations in the receiving equipment, a set of copies of quasi-orthogonal code sequences is formed, obtained by functional transformations of random functions similar to those that were carried out in the transmitting equipment.

Stage 5. On the transmitting and on the receiving side, before the start of the information transfer process, tables of initial functions are formed, which at the time of information transfer are used to form working quasi-orthogonal code sequences. These tables are stored in the memory of the transmission system on the transmitting side. Copies of these tables are stored on the receiving side. The order of using and changing the functions presented in these tables, at each given moment in time, is set by the service information block in the transmitting equipment and, in a similar way, by its copy in the receiving equipment.

Stage 6. A modulated signal of an individual information channel is formed using individual quasi-orthogonal code sequences obtained by functional transformations of random functions obtained at the previous stage.

Stage 7. With the help of the formed set of quasi-orthogonal code sequences, digital information is simultaneously transmitted by all active subscribers of the transmitting equipment via a wireless communication channel. In this case, in each information path, a complex signal is assigned to an individual information bit, the structure of which is determined by the signal number of the original system of binary quasi-orthogonal code sequences and the value of the transmitted bit. If the information bit is zero, then during its duration one period of the quasi-orthogonal code sequence of the direct structure is transmitted, and if the information bit is equal to one, one period of the quasi-orthogonal code sequence of the inverse structure is transmitted.

Stage 8. A combined baseband signal is formed using a phase modulation unit and a power amplifier, which is emitted by the transmitting antenna of the transmitting part of the information transmission system.

Stage 9. With the help of the generated set of copies of the system of binary quasi-orthogonal code sequences in the receiving equipment, high-frequency selection, correlation processing and extraction of the information bit are carried out, which is issued to the recipient of information.

Stage 10. After the transmission of the next information bit in each information path on the transmitting side, the system of binary quasi-orthogonal code sequences used as information carriers is changed on the basis of a new functional transformation, carried out as described in the third step of the method under consideration. This procedure is performed by a command issued by the service information block of the transmitting side.

Stage 11. After high-frequency selection, correlation processing and extraction of the information bit issued to the recipient, at the command issued by the service information unit of the transmitting side, on the receiving side, synchronously with the transmitting side, the system of binary quasi-orthogonal code sequences is changed, which is used as information carriers based on a new functional transformation, carried out as described in the third step of the considered method. In this case, a copy of the system of binary quasi-orthogonal code sequences generated and used in the receiving equipment for correlation processing will have a structure that coincides with the system of binary quasi-orthogonal code sequences generated and used in the transmitting equipment.

Stage 12. The information transfer process is carried out similarly to steps 6-11 of the considered method using a new system of binary quasi-orthogonal code sequences. Since the structure of quasi-orthogonal code sequences after the transmission of the next information bit changes synchronously in the transmitting and receiving equipment, and is also not reused due to their increased number in comparison with the known sequences, this provides an increase in the structural secrecy of the code division multiplex information transmission system. The process of transmitting information is carried out until a command is received from the service information block of the transmitting side to suspend information exchange.

3. Conclusion

The advantage of BCSs obtained on the basis of functional transformations of pseudo-random arguments is that the number of non-repeating BCSs is significantly greater than all known ones. For a BCS length of 4095, the advantage over the most numerous Kerdock sequences is more than 10,000 times, for a BCS length of 8191, an advantage over the most numerous Weil sequences is about 10,000,000 times, for a BCS length of 10230, an advantage over the most numerous Kerdock sequences is more than 1000 times.

With the stochastic use of binary quasi-orthogonal code sequences in a code division multiplexing information transmission system, obtained on the basis of functional transformations of pseudo-random arguments, they provide a higher structural (signal) secrecy in comparison with the case of using the known discrete quasi-orthogonal code sequences of Gold, Kasami, Kamaletdinov, Weil and Bent sequences.

The developed information transmission system with code division of channels and stochastic use of systems of binary quasi-orthogonal code sequences has the structure shown in Fig. 1, a distinctive feature of which is the presence of a block of functional transformations on the transmitting and receiving sides.

For the formation and stochastic use of SBQCT as spreading sequences obtained on the basis of functional transformations of random functions in accordance with the procedures described by expressions (2) - (6), it is proposed to use a block for setting initial functions, a block of functional transformations, a random number generator in the transmitting equipment and a block for setting the initial values. In the receiving equipment, it is similarly proposed to use a block for specifying copies of the original functions, a block for copies of functional transformations, a generator of copies of random numbers and a block for specifying copies of initial values.

The proposed method for transmitting information based on the stochastic use of binary quasi-orthogonal code sequences is carried out through the implementation of 12 stages.

A distinctive feature of the developed method is the formation using a random number generator, a block for setting initial functions, a block for setting initial values and a block of functional transformations in the transmitting equipment of systems of binary quasi-orthogonal code sequences obtained by functional transformations of random functions based on the invariance property of the probability differential (2), which are stochastically used as carriers of information.

4. Acknowledgment

This work was supported by the Russian Foundation for Basic Research, project No. 18-07-01020.

References

[1] Golomb S W and Gong G 2005 *Signal design for good correlation for wireless communication, cryptography, and radar* (Cambridge: University press)

[2] Viterbi A J 1995 *CDMA: Principles of spread spectrum communication* (Reading, MA: Addison-Wesley)

[3] Ipatov V P 2005 *Spread Spectrum and CDMA: Principles and Applications* (Portland, OR: Book News, Inc.)

[4] Varakin L E 1978 *Complex signals theory* (Moscow, Soviet radio)

[5] Petrovich N T, Razmahin M K 1969 *Telecommunication systems with noise-like signals* (Moscow, Soviet radio)

[6] Kosjakin S I, Moskvitin I A, Smirnov A A 2004 *Method and device for data transfer in code-division systems* (Patent of Russian Federation for invention RU 2234191 C2. 10.08.2004)

[7] Dyatlov A P, Dyatlov P A, Kulbikayan B H 2004 *Electronic warfare with satellite radio navigation systems* (Moscow, Radio and communications)

[8] Orel D V 2013 *Modeling of stochastic systems of binary quasi-orthogonal code sequences based on the method of functional transformations* (Ph.D. thesis on 05.13.18 Mathematical modeling, numerical methods and software packages. Stavropol, North Caucasus Federal University)

[9] Pashintsev V P, Kalmykov I A, Zhuk A P, Orel D V, Zhuk E P 2018 Formation Algorithms and Properties of Binary Quasi-Orthogonal Code Sequence of Modern Satellite Systems *International Journal of Engineering & Technology* vol 7 (4.38) pp 1205-1209 DOI: 10.1117/12.2324249

[10] Zhuk A P, Ivanov A S 2011 Increasing the structural secrecy of the information transmission system with code separation of channels *High-tech technologies in Earth space research* No1 pp 26-28

[11] Zhuk A P, Orel D V, Kalmykov I A, Studenikin A V 2019 Improved Method of Formation of an Increased Number of Binary Quasi-Orthogonal Code Sequence Systems with the Required Statistical and Correlation Characteristics *Atlantis Highlights in Computer Sciences* vol 3 pp 209-214 DOI: 10.2991/csit-19.2019.36

[12] Borisov V I, Zinchuk V M 2008 *Noise immunity of radio communication systems. Probabilistic-time approach* (Moscow, RadioSoft)

[13] Borisov V I, Zinchuk V M, Limarev A E, Muhin N P, Nahmanson G S 2003 *Noise immunity of radio communication systems with signals range expansion of modulated carrier pseudorandom sequence* (Moscow, Radio and communications)

[14] Fomin L A, Budko P A, Zhuk A P, Shlaev D V 2007 Modeling of self-similar processes in information and communication systems *Telecommunications Journal* No3 pp 34-37

[15] Zhuk A P, Ivanov A S, Golub U S, Orel D V 2011 *Method to transfer information based on chaotically generated ensembles of discrete multi-level orthogonal signals* (Patent of Russian Federation for invention RU 2428795 C1. 10.09.2011)

Software-hardware complex for modeling secure IIoT distributed ledger

A R Makhmutov[1], S V Trishin[1], K V Mironov[2,1], A M Vulfin[1]

[1]Ufa State Aviation Technical University, Ufa, Karla-Marxa str., 12, 450000, Russia

[2]Ural Federal University, Ekaterinburg,

E-mail Mironovconst@gmail.com

Abstract. This paper describes the modeling of the industrial Internet-of-Things network for further exploration of its security properties. The goal is to increase the security of process control systems for industrial facilities based on the development of conceptual foundations, methodology, mathematical models, methods and software tools for comprehensive analysis and cyber security risk management of process control systems for industrial facilities using cognitive modeling technology and data mining. One specific task is modeling DLT implementation for industrial network. Common structure of the proposed modeling network is described. Example of smart contracts for the distributed ledger are presented. Performance of lightweight implementations for ciphers and hash functions are given.

Keywords: Data acquisition system; Queueing theory; Markov model; Analogue digital converter

1. Introduction

In recent years, the use of distributed ledger technology along with Internet of Things systems has been actively developing [1]. Distributed ledger technologies can ensure the security of data storage and transmission between IoT devices, trust between system users, and the system's resilience to technical failures and malicious attacks.

Comparing distributed ledgers with centralized storage systems, the following advantages can be distinguished [2]:

- distribution - data is stored on many independent devices, which make the system resistant to unauthorized data changes;
- stability - data in the ledger is almost impossible to change or delete, which can be used to store records of financial transactions or any other data that requires an audit trail. Every change is tracked and continuously recorded in a distributed and public ledger;
- transparency - when using distributed ledgers, it becomes possible to almost completely get rid of intermediaries in financial and other transactions, since the correctness of the data entered, and the transactions carried out can be carried out by users themselves using mining. This allows you to establish trust between network participants and reduce commission costs.

Among the disadvantages of using technology are:

- lack of legislative regulation and technology standardization;

- complexity of management in the case of using open distributed ledgers;
- the need for nodes with high computing power to maintain the network;
- the presence of vulnerabilities specific only to this technology, and, accordingly, the need to develop new methods to eliminate these vulnerabilities;
- vulnerabilities of client applications, with the help of which access to the distributed registry is carried out.

One example of the convergence of distributed ledger technologies and the Internet of Things is decentralized generation and distribution of electricity, in which small renewable energy producers are able to sell the surplus to nearby consumers. The development of such systems becomes relevant with the increasing demand for clean energy [3]. Such systems are developing in the United States and the European Union countries, where renewable energy sources are actively used and there is a need to create a new model of a decentralized energy market [4]. In Russia, this industry is poorly developed, energy is generated by a limited number of large producers who have little interest in the decentralization of the system. However, the creation of such systems is relevant for settlements in remote regions of the North of Russia, Siberia and the Far East, as well as villages and villages located on the territory of reserves.

2. A developing solution

In the course of our project, a model of a distributed energy generation system was proposed, which includes four main components that ensure its operation, a smart contract, a local aggregator of a distributed electricity system, smart meters, and a mobile application. The users of such a system are divided into consumers and producers. Consumers send requests to buy energy to a local aggregator, which provides a list of current offers from producers who own various renewable energy generation facilities. Energy producers can also act as consumers. As a rule, distributed generation systems are divided into communities. The local aggregator acts as an intermediary, registers new users in the system, maintains the network and exchanges real currency for virtual currency. Smart meters record energy production and consumption, calculate hash functions of the received data, electronically sign transactions, and periodically transmit them to the registry. The mobile application provides users with an interface for accessing a smart contract, receiving information about current tariffs and supply offers.

The proposed system solves the problems of unauthorized access to data and their privacy since data on power consumption and transactions are transferred to the register in the form of cryptographically protected transactions. The data entered in the registry cannot be changed or destroyed, and any attempts to change it will be automatically detected, since the transactions in the registry are interconnected and each complete node stores a complete copy of the registry. It also improves device availability by using a distributed storage structure that does not have a single point of failure. The logic for executing transactions is fixed in a smart contract, the code of which cannot be changed after being loaded into the registry.

After the local aggregator uploads the contract to the registry, users can send their account addresses to the aggregator to register in the contract. The aggregator creates a database of user addresses within the contract and assigns a balance of funds to each user. Users can choose their type - consumer or producer of electricity. In the case of the manufacturer, the value of the electricity produced is recorded by a smart meter and transferred to the contract. After consumers know the current supply offers from the aggregator, they submit a request for a deal. It is checked if there are enough funds on the user's balance. If not, the user is notified of this. If there are enough funds, the funds are transferred to the balance of the selected manufacturer and the amount of available electricity is updated.

3. Smart-contracts

For the first time, the concept of smart contracts was proposed by the scientist in the field of computer science and cryptography Nick Szabo in 1996. He described a smart contract as a set of certain promises written in digital form, including protocols according to which the parties fulfill these promises [5]. When concluding a transaction between several parties, it becomes necessary to create a contract in which all the conditions, rights and obligations of the parties will be recorded. The conclusion of a paper contract requires the involvement of intermediaries who will control the process of completing the transaction. In the case of using a smart contract, the computing system automatically analyzes the fulfillment of the terms of the contract and makes settlements between the parties to the transaction. Nick Szabo described three characteristics of smart contracts:

- the ability to monitor the execution of the contract by the parties;
- verifiability, which is understood as the ability of a party to prove that the contract has been performed or violated;
- secrecy, the principle according to which control over the content and execution of the contract should be provided to the parties to the extent that it is necessary for its execution.

With the advent of distributed ledger technology, in particular blockchain, it became possible to implement this concept. In 2015, the Ethereum platform appears, the blockchain of which was able to store not only information about all transactions that have ever been committed, but also smart contracts - programs stored in blocks and called by transferring transactions to them, the response transaction will occur according to a set of conditions fixed in the code contract. The advantage of smart contracts is that they are encrypted in the registry and cannot be changed or destroyed. The costs of intermediaries are also reduced.

However, the use of smart contracts has several weaknesses, including:

- smart contract is a program code that is written by a person, and accordingly is prone to errors made at the development stage. The situation is aggravated by the fact that the smart contract uploaded to the distributed registry cannot be changed, and, accordingly, there is no way to correct the errors of the used smart contract;
- based on the previous point, the development of a smart contract requires the involvement of an experienced developer, which will entail additional implementation costs;
- at the moment, smart contracts do not have a specific legal status.

The smart contract was implemented on the Ethereum platform, since at the moment it is the most widely used platform that allows you to create custom smart contracts, has its own Turing-complete Solidty programming language, similar in syntax to JavaScript. Ethereum transactions are data packets signed by the account owner. An Ethereum transaction contains the amount of Ether (the currency of the Ethereum platform), the address of the recipient of the STARTGAS message, GASPRICE and optional data fields. The STARTGAS value is the maximum number of computational steps that are allowed to be performed while executing a transaction. The GASPRICE value is the fee that the sender is willing to pay for gas, one unit of gas corresponds to one computational operation.

Below is a smart contract that serves to register users and form their virtual balance, obtain information about registered users by the aggregator of the distributed system.

```
contract smart_cl{
// event NewCustomer occurs when a new customer is registered
event NewCustomer(address indexed client, address client_contract);
// Create an array of registered customers address / address of the client
mapping (address => address) contract_registered;
// Create an array of user address / balance value
    mapping (address => uint256) public balances;
    address dso;
//Конструктор для загрузки контракта в сеть Ehthereum
    constructor(uint8 tokens) public {
// assign the dso variable the address of the sender of the transaction
dso = msg.sender;
// Set the sender's balances to tokens
    balances[msg.sender] = tokens;
    }
// register_client function is called to register a new client
function register_client(address client, uint256 tokens )public returns (address
contract_address)
{
                  require(contract_registered[client] == 0x0);
address new_client = new client_contract(tokens, client);
```

IOP Conf. Series: Materials Science and Engineering 1069 (2021) 012018 doi:10.1088/1757-899X/1069/1/012018

```
// Call event NewCustomer
                emit NewCustomer(client, new_client);
// Recording a new client in the client array
contract_registered[client] = new_client;
                // The function returns the address of the new client
return new_client;
}
// The function is called when checking if the client is registered in the contract base
function getClient(address client_address) public returns (address client_contract)
{
   // Returns the address of the registered client
return contract_registered[client_address];
}
}
```

Next, a smart contract allows you to determine the type of client (prosumer or consumer), pay for the consumed electricity and receive information about the user's balance.

```
contract contract_cli{
// Variable threshold for notification of insufficient funds
int256 threshold;
//event Lowcredit occurs when funds are insufficient
event Lowcredit(address indexed client, uint256 _value);
// event Transfer occurs when funds are transferred
event Transfer(address indexed _from, address indexed _to, uint256 _value);
address owner;
    // Constructor for uploading a contract to the network
constructor(uint256 tokens, address new_client) public {
// assign the variable owner the address of the sender of the transaction    owner =
new_client;
// Set the sender's balances to a value tokens
balances[msg.sender] = tokens;
}
      // The user structure containing the type
customer and electricity volume
struct user{
    uint8 client_type;
uint256 total_electricity;
    }
// Create an array of user balance values user address / balance value
    mapping (address => uint256) public balances;
// Payment function for payment of received electricity
function payment(address receiver, uint amount) public returns(bool sufficient) {
if (balances[msg.sender] < amount) return false;
    balances[msg.sender] -= amount;
    balances[receiver] += amount;
    emit Transfer(msg.sender, receiver, amount);
    return true;
    }
// Payment getBalance to get information about the client's balance
function getBalance() public view returns(uint) {
    return balances[msg.sender];
    }
}
```

4. Network architecture

During the work on our project an architecture of data transmission system for distributed networks was developed. Schematic view of this architecture with used hardware is shown on the Figure 3.

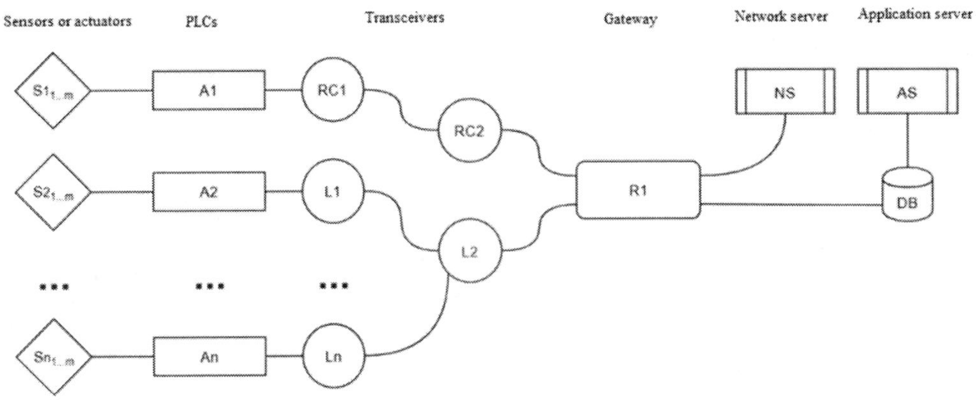

Figure 1. Network architecture

Following shortcuts are used:
S1…n1…m – Various end devices, for example, sensors or actuators.
A1…n – Arduino Uno microcontrollers.
RC1,2 – Radio transceivers HC12.
L1, 2,…n – Radio transceivers LoRa.
R1 – Raspberry Pi computer.
NS – Network hardware management server (Network server).
DB – End devices information database.
AS – Application server that working with information from the database.
These gears have following tasks:

- End devices (sensors or actuators) generate data stream automatically according to the pre settled schedule;
- Arduino Uno PLCs receive such a generated data, encrypt it, create appropriate data frames and send them to the gateway with the help of radio transceivers;
- Gateways receive data that has been sent from PLCs via radio channel and forward it to the database server using standard TCP/IP stack;
- Database server receives and stores received data. сервер баз данных осуществляет прием отправленной со шлюза информации и ее хранение;
- Network server manages all network equipment from the end device level (if possible) to the gateway level. Also establishes data transmission schedule;
- Application server operates with data stored in database.

5. Tests of used hardware

Considering the planned use of the developing system and its operational features, we can say that the most important aspect of information security is the integrity of data received from end devices.

Due to the features of used hardware and development goals, key requirements to the information security in our project was set. These features mean potentially low computational power, firstly, of Arduino Uno microcontrollers. They can be the weakest link of the whole hardware set since lower (in accordance with the network architecture) devices are sensors and actuators that have no computing tasks, and upper devices have much stronger hardware stuffing. An important task is to test Arduino Uno in working with cryptography: encryption, hashing, message authentication. Also, it is important to find out the actual working distance between radio transceivers.

Arduino Uno specifications are in the table 1 [6].

Table 1. Characteristics of a single-board microcontroller Arduino Uno

Microcontroller	ATmega 328
Clock frequency	16 MHz
RAM	2 KB
Flash-memory	32 KB
Working voltage	5 V
Digital I/O	14 (6 can be used with pulse-width modulation)
Analog inputs	6
DC through I/O	40 mA
Width	53 mm
Length	69 mm

LoRa SX1278 transceiver specifications are in the table 2 [7]. This radio module will be tested for communication range.

Table 2. LoRa SX1278 transceiver specifications

Transceiver	LoRa SX1278
Supported modulation types	FSK, GFSK, LoRa
Transmitter output power	≤20 dbm
Receiver sensitivity	-139 dbm
Working voltage	1,8 – 3,6 B (recommended 3,3 B)
Rated current	Receiver mode: 11 мA Transmitter mode: up to 120 мA
Data transfer rate	FSK mode: 300 kb/s LoRa mode: 0.018 – 40 kb/s
Weight	3 g
Dimensions	18mm × 17mm × 4mm
Interface	SPI

Next crypto algorithms will be used in Arduino Uno benchmarking:

1) AES (ECB mode)
2) AES (CTR mode)
3) SHA256
4) SHA3 256

Testing was carried out for the communication range of the Arduino Uno + LoRa SX1278 bundles. The first bundle was a transmitter, the second one – a receiver. The transmitting bundle was sending a symbol array from different spots. The maximum distance between two beams with a stable signal in a dense urban development was 250m. At each spot transmitter sent a symbol array trice. Power supply was provided with power bank via USB interface. Received by the second bundle messages can contain some part of collisions because of weak signal. Similar experiment was carried out in the countryside.

Range test results:
- Stable communication in dense urban areas is provided at distances up to 250m;
- Stable communication in rural areas is provided at distances up to 600m.
- For better result a more powerful transmitter is required. Also, for the best result, it is necessary to place the receiving device at a dominant height.

In our computational test the Arduino Uno board was used. It is significant task to check this element because Arduino Uno can be the weakest link in our hardware chain. While choosing cryptographic algorithms we need to consider its computing power.

The results of the performance testing that was carried out are shown in tables 3 and 4.

Table 3. Cipher validation.

Algorithm	Encryption time, microseconds per byte	Decryption time, microseconds per byte
AES128 ECB	33.30	72.15
AES192 ECB	39.97	87.44
AES256 ECB	46.65	102.74
AES128 CTR	36.03	36.03

Table 4. Hash validation.

Algorithm	Hashing time, microseconds per byte
SHA3 256	60.12
SHA256	44.75

If uploaded source code takes up a lot of memory space Arduino IDE shows a notification that this code can work unstable because of a lack of memory.

Information about required memory space for each algorithm is in the table 5.

Table 5. Required memory space for tested algorithms.

Algorithm	Used device memory, bytes (% of total memory)	Used dynamic memory of the device, bytes (% of the total memory)
AES-ECB	8318 (25)	1388 (67)
AES128 CTR	9078 (28)	1316 (64)
SHA3 256	11610 (35)	1708 (83)
SHA 256	11064 (34)	1146 (55)
Transmitter mode	3846 (11)	312 (15)
Receiver mode	3698 (11)	345 (16)

Performance test results:

- Single-board Arduino Uno microcontrollers are able to work with cryptographic algorithms;
- To encrypt a payload, it is recommended to use algorithms that need no more memory space than AES128 CTR. In other case a lack of memory will occur;
- To provide an integrity control of messages SHA256 algorithm can be used;
- It is impossible to use AES and SHA256 at the same Arduino Uno board. In order to be able to implement both payload encryption and integrity control, it is necessary to use more "lightweight" algorithms.

Also, the performance test showed that Arduino Uno board should be tested for work in DLT systems since such board is able to work with popular in DLT systems SHA256 hash function and enough memory remains to provide other functions except hashing.

6. Conclusion

The analysis of the possibilities of using the distributed ledger technology in distributed power generation systems is carried out. Distributed ledger technologies allow solving the problems of unauthorized access to data and their changes. Smart contracts allow you to reliably record the terms of the transaction.

A model of a distributed power generation system operating on the basis of distributed ledger technologies and the Internet of Things, and a smart contract algorithm for this system are presented.

A smart-contract algorithm for such a system is presented.

The analysis of the applicability of cryptographic algorithms in the implementation of information security in such systems is carried out.

A test of the radio communication equipment used in the project was carried out.

Acknowledgements

The reported study was funded by RFBR according to the research projects № 19-09-00972 and 20-08-00668. The work of Konstantin Mironov is supported by the scholarship of Russian President for young scientists.

References

[1] Mironov K. et al. 2019 On Application of Distributed Ledgers for Internet of Things in Russia *VIth International Workshop «Critical Infrastructures: Contingency Management, Intelligent, Agent-Based, Cloud Computing and Cyber Security (IWCI 2019)* Atlantis Press,.

[2] Fernández-Caramés T M, Fraga-Lamas P A 2018 Review on the Use of Blockchain for the Internet of Things *Ieee Access* 6 pp 32979-33001.

[3] Liu W, Zhan J, Chung C Y 2018 A novel transactive energy control mech-anism for collaborative networked microgrids *IEEE Transactions on Power Systems* **34** 3 pp 2048-2060.

[4] Ahl A et al 2019 Review of blockchain-based distributed energy: Implications for institutional development *Renewable and Sustainable Energy Reviews* 107 pp 200-211.

[5] Szabo N 1996 Smart contracts: building blocks for digital markets *EXTROPY: The Journal of Transhumanist Though* **18** 2.

[6] Arduino Uno [Website, viewed on 10.11.2020] URL: http://arduino.ru/Hardware/ArduinoBoardUno.

[7] SX1276-7-8.book [Website, viewed on 10.11.2020] URL: https://cdn-shop.adafruit.com/product-files/3179/sx1276_77_78_79.pdf

[8] Vysotskiy N, Makhmutov A, Mironov K, Meisel M, Sauter T 2018 Secure Communication Technology for Devices with Limited Resources *IWCI 2018 (*Atlantis Press) 158 pp 207-210.

Possible variant of the confidential data leak detection model

I S Chernousov[1], A E Borshevnikov[1], D V Zakharchenko[1], Y V Dobrzhinskii[1]

[1]Far Eastern Federal University, Vladivostok, Russky Island, Ajax Bay, 10, 690922, Russia

E-mail chernousov.is@students.dvfu.ru

Abstract. The paper discusses a model for warning about leaks of electronic documents, describes a possible structure of a system for monitoring document leaks. Within the framework of the paper, the model is verified by evidence and formal methods.

Keywords: Data leaks; Leak detection; Corporate data protection; Electronic document tagging

1. Introduction

As the number of corporate data leaks is growing more and more, the question of the need for measures to combat data leaks arises. One of the options for increasing the efficiency of information protection systems is to warn owners about leaks that have already occurred, which allows them to take timely actions to minimize damage [1].

This mechanism cannot be considered as an independent element of the electronic data protection system, but only as an additional component to ensure awareness of data theft [2].

2. Sensitive data leak detection model

During the developing process of the mathematical model, the next main principles were taken [3]:

1. Control of the host;
2. Reference database;
3. Use of hash functions;
4. Control of the traffic.

Let us go directly to the description of the developed mathematical model. In the development process, a set-theoretic method will be used. Let's define the subjects of interaction:

Consider the set U is the set of all possible users (access subjects):

$$U = \{U_1, U_2, \dots, U_s, E_1, \dots, E_k, T\},\ n = s + k + 1, \tag{1}$$

where, n is the number of all users; s is the number of all valid users; k is the number of all intruders; T is the trusted subject of a computer network, which is a computer network administrator and a trusted server.

Let $U_l, E_g = \{UID_{U_l(E_g)}\}$ be a user (attacker) identifier.

Continuing with the introduction of the concept of access objects:

$$O = PC_i \times M_j, \tag{2}$$

where × - cartesian product of sets; PC_i - i-th object of a computer network.

Consider j-th file:

$$M_j = (m_j, mark_j, rand_j), \quad (3)$$

where m_j is a message; $mark_j$ is a mark itself; $rand_j$ is a random number.

Based on the above entities, the following sets will be formed:

1. Set of all files $M = \{M_j : j = \overline{1,f}\}$;
2. Set of all devices $PC = \{PC_i : i = \overline{1,k}\}$.

Let $|O| = r$.

Continuing with the introduction of i-th object of a computer network:

$$PC_i = (IDT_i, IP_i, AM_i), \quad (4)$$

where, IDT_i is the identifier of i-th object of a computer network; IP_i – IP address of i-th object of a computer network; AM_i - matrix of access of subjects of access to objects of access.

Consider the trusted subject of a computer network T:

$$T = \{(t_{mark}, UID_l, IDT_i, h(M_j), rand_j, IP_i)\}, \quad (5)$$

where, t_{mark} is the time the mark was created; $h(M_j)$ - hash value from j-th file; $rand_j$ is a random number[4].

For a better understanding of the interaction between users and the object, we introduce the access function α:

$$\alpha : U \times O \to \{0,1\} \quad (6)$$

$$\alpha(U_p, O_q) = \begin{cases} 0, O_q \notin U_p \\ 1, O_q \in U_p \end{cases}, p = \overline{1,n}, q = \overline{1,r}. \quad (7)$$

We will assume that

$$O_q \in U_p \leftrightarrow (PC_q \in U_p) \wedge (M_q \in U_p) \quad (8)$$

In the same connection, through the access function, it is possible to describe the violator of the access control rules who applied to documents to which he does not have access:

1. $UID_{U_l} \neq 0$ for $\alpha(U_p, O_q) = 0$ - internal violator;
2. $UID_{U_l} = 0$ for $\alpha(U_p, O_q) = 0$ - external violator

The access matrix can be described as follows:

$$AM = \begin{pmatrix} AC_{1,1} & \cdots & AC_{1,n} \\ \vdots & \ddots & \vdots \\ AC_{r,1} & \cdots & AC_{r,n} \end{pmatrix}, \quad (9)$$

where $AC_{p,q} = \alpha(U_p, O_q), p = \overline{1,n}, q = \overline{1,r}$.

Unauthorized access to an object should be considered the case when $(AC_{p,q} = 0) \wedge (M_q \in U_p)$.

For a mathematical description of the interaction of access objects O, which in accordance with this scheme are user devices and files of critical importance to be protected, we introduce a membership function β such that:

$$\beta : O \to \{0,1\} \quad (10)$$

$$\beta(O_q) = \begin{cases} 0, M_j \notin PC_i \\ 1, M_j \in PC_i \end{cases} \quad (11)$$

The belong matrix can be described as follows:

$$BM = \begin{pmatrix} B_{1,1} & \cdots & B_{1,f} \\ \vdots & \ddots & \vdots \\ B_{k,1} & \cdots & B_{k,f} \end{pmatrix}, \quad (12)$$

where $B_{i,j} = \beta(O_q), q = \overline{1,r}, i = \overline{1,k}, j = \overline{1,f}$.

By leakage we mean the situation:

$$(B_{i,j} = 0) \wedge (M_j \in PC_i) \quad (13)$$

Let's define the following operations:

1. Access to a file

$$U_l \xrightarrow{give} PC_i: UID_l, ID(M_j) \quad (14)$$

The user contacts the device, to which he transmits his personal unique identifier and the identifier of the document to which he requests access [5].

1.1. $PC_i: AC_{l,i} == 1$, checking for access rights;
1.2. $M_j \xrightarrow{give} PC_i: M_j$;
1.3. $PC_i \xrightarrow{give} U_l: M_j$.

2. Copy file

2.1 $U_l \xrightarrow{give} PC_i: UID_l, ID(M_j)$;
2.2 $PC_i: AC_{l,i} == 1$;
2.3 $M_j \xrightarrow{give} PC_i: M_j$;
2.4 $PC_i \xrightarrow{give} PC_k: M_j$;
2.5 $PC_k \xrightarrow{give} M_j: M_j$.

3. File marking

3.1 $T \xrightarrow{give} PC_i: mark_{j,}rand_j$;
3.2 $PC_i: AC_{l,i} == 1$;
3.3 $PC_i \xrightarrow{give} M_j: mark_{j,}rand_j$;
3.4 $M_j: m_j, mark_j, rand_j$;
3.5 $M_j \xrightarrow{give} PC_i: h(M_j)$;
3.6 $PC_i \xrightarrow{give} T: h(M_j), IDT_i, IP_i$.

4. User access to the marked file

4.1 $U_l \xrightarrow{give} PC_i: UID_l, ID(M_j)$;
4.2 $PC_i: AC_{l,i} == 1$;
4.3 $M_j \xrightarrow{give} PC_i: mark_j, rand_j, h(M_j)$;
4.4 $PC_i \xrightarrow{give} T: mark_j, rand_j, h(M_j), IDT_i, IP_i$;
4.5 $T \xrightarrow{give} PC_i: mark_j$;
4.6 $PC_i \xrightarrow{give} M_j: mark_j$.

Let's proceed to the verification of the mathematical model of leak detection.

3. Evidence-based verification

Let us verify the constructed model with respect to the Bell-LaPadula model. The main provisions of the Bell-LaPadula (BLP) model are as follows[7].

The system model $\sum(\vartheta_0, Q, \mathcal{F}_T)$ is represented by the set:

1. Set of the access objects O;
2. Set of the access subjects S;
3. Sets of access rights R (in the classical model "read", "write");
4. Access matrix A[s,o];
5. Lattices Λ_L of security levels of L subjects and objects of the system;
6. Function $\mathcal{F}_L: S \cup O \rightarrow L$, that maps elements of the set S and O to the set L;
7. The set of states of the system V, which is determined by the set of ordered pairs $(\mathcal{F}_L, A)$;

8. Initial state $\vartheta_0 \in V$;

9. A set of requests of Q subjects for access (performing an operation on objects, the execution of which transfers the system to a new state);

10. The transition function $\mathcal{F}_T: (V \times Q) \rightarrow V^*$, which transfers the system from one state to another V^* when executing queries from Q.

The states of the system are divided into dangerous and safe. To analyze and formulate conditions that ensure the safety of system states, the following definitions are introduced:

1. A state is called read-safe $\leftrightarrow$ when, for each subject performing read access to an object in this state, the security level of that subject dominates over the security level of that object:

$$\forall s \in S, \forall o \in O, read \in A[s,o] \rightarrow \mathcal{F}_L(s) \geq \mathcal{F}_L(o) \quad (15)$$

2. A state is called write-safe $\leftrightarrow$ when, for each subject accessing an object in this state, the security level of that object dominates the security level of that subject:

$$\forall s \in S, \forall o \in O, write \in A[s,o] \rightarrow \mathcal{F}_L(s) \geq \mathcal{F}_L(o) \quad (16)$$

3. A system state is safe when it is both read and write safe.

Let's correlate the elements of the developed and basic models:

1. $O \sim O = PC_i \times M_j$;
2. $S \sim U$;
3. $A[s,o] \sim AM$;
4. $\Lambda_L \sim O_i = \alpha(U_1, O_i) \cdot U_1 + \alpha(U_2, O_i) \cdot U_2 + \cdots + \alpha(T, O_i) \cdot T$;
5. $\mathcal{F}_L : S \cup O \rightarrow L \sim \alpha(U_p, O_q)$;
6. $V = \{(\mathcal{F}_L, A)\} \sim AC_{p,q} \wedge (M_q \in U_p)$;
7. $\vartheta_0 \in V$ – initial state of the system;
8. Q – operations;
9. $\mathcal{F}_T: (V \times Q) \rightarrow V^* \sim B_{i,j} \wedge (M_j \in PC_i)$;
10. $L = \{0,1\}$.

Let us assume that read and write operations are indistinguishable within the framework of this model.

Consider two cases of leak $(B_{i,j} \wedge (M_j \in PC_i) = 0)$.

Consider the first case of a leak:

1 $E_l \xrightarrow{give} PC_i: UID_l, ID(M_j)$;

2 $PC_i: AC_{l,i} == 1;$ ↯

3 $M_j \xrightarrow{give} PC_i: M_j$;

4 $PC_i \xrightarrow{give} PC_k: M_j$;

5 $PC_k \xrightarrow{give} M_j: M_j$.

The controversy arises in step 2, but the system cannot be brought to a safe state since no leak can be detected.

Let the file be marked, then consider the following situation:

1. $E_l \xrightarrow{give} PC_i: UID_l, ID(M_j)$;

2. $PC_i: AC_{l,i} == 1;$ ↯

3. $M_j \xrightarrow{give} PC_i: mark_j, rand_j, h(M_j)$;

4. $PC_i \xrightarrow{give} T: mark_j, rand_j, h(M_j), IDT_i, IP_i$;

5. $T \xrightarrow{give} PC_i: mark_j$;

6. $PC_i \xrightarrow{give} M_j: mark_j$;

7. $M_j \xrightarrow{give} PC_i: M_j;$

8. $PC_i \xrightarrow{give} U_l: M_j.$

The controversy also arises at the step 2; however, unauthorized access cannot be detected either. You can get out of this situation by excluding illegitimate users from the system at the stage of preparing to launch the system.

Consider the second case of a leak:

1. $U_l(E_l) \xrightarrow{give} PC_i: UID_l, ID(M_j);$
2. $PC_i: AC_{l,i} == 1;$ ↯
3. $M_j \xrightarrow{give} PC_i: M_j;$
4. $PC_i \xrightarrow{give} PC_k: M_j;$
5. $PC_k \xrightarrow{give} M_j: M_j;$
6. $U_l(E_l) \xrightarrow{give} PC_i: UID_l, ID(M_j);$
7. $PC_k: AC_{l,k} == 1;$ ↯
8. $M_j \xrightarrow{give} PC_k: M_j;$
9. $PC_k \xrightarrow{give} U_l(E_l): M_j.$

A contradiction can occur in steps 2 and 7, but the system cannot be brought to a safe state as no leak can be detected.

1. $U_l(E_l) \xrightarrow{give} PC_i: UID_l, ID(M_j);$
2. $PC_i: AC_{l,i} == 1;$ ↯
3. $M_j \xrightarrow{give} PC_i: M_j;$
4. $PC_i \xrightarrow{give} PC_k: M_j;$
5. $PC_k \xrightarrow{give} M_j: M_j;$
6. $U_l(E_l) \xrightarrow{give} PC_i: UID_l, ID(M_j);$
7. $PC_k: AC_{l,k} == 1;$ ↯
8. $M_j \xrightarrow{give} PC_k: mark_j, rand_j, h(M_j);$
9. $PC_k \xrightarrow{give} T: mark_j, rand_j, h(M_j), IDT_i, IP_i;$
10. $T \xrightarrow{give} PC_k: mark_j;$
11. $PC_i \xrightarrow{give} M_j: mark_j;$
12. $M_j \xrightarrow{give} PC_k: M_j;$

13. $PC_k \xrightarrow{give} U_l(E_l): M_j$.

The contradiction arises at the steps 2 and 7, but the detection of the contradiction occurs at the 9 step. For each of the detected cases, the system is brought to a safe state in a countable number of steps by removing the user from the system. Let us bring the described procedures into a safe state due to the methods described above → ϑ_0- safe (it is this state that should be considered the initial one).

Due to the condition adopted within this model that the read and write operations are indistinguishable, it is necessary and enough to understand, that $\mathcal{F}_L^*(s) = \mathcal{F}_L^*(o)$ and $read, write \in A^*[s,o]$, if $\mathcal{F}_L^*(s) \neq \mathcal{F}_L^*(o)$.

Let for the i-th user:

$$AC_{i,l} = 0 \to \alpha(U_i, O_l) = 0 \to O_l \notin U_i \to \mathcal{F}_L(s) = \mathcal{F}_L(o) \quad (17)$$
$$AC_{i,l} = 1 \to \alpha(U_i, O_l) = 1 \to O_l \in U_i \to \mathcal{F}_L(s) = \mathcal{F}_L(o) \quad (18)$$
$$AC_{i,l} = 0 \to \alpha(U_i, O_l) = 1 \to \lightning \to read, write \notin A^*[s,o] \quad (19)$$
$$AC_{i,l} = 1 \to \alpha(U_i, O_l) = 0 \to \lightning \to read, write \notin A^*[s,o] \quad (20)$$

Then, by the main security theorem, the system is safe.

4. Conclusion

Here are the main results of the study:

1. A confidential data leak detection model was developed and verified by evidence-based method;
2. Among the main aspects characterizing the developed mathematical model, one can single out the ability to detect leaks outside the information system, which has a positive effect on the information security of the organization as a whole.

Thus, this work presented a new look at the issue of detecting leaks of confidential data.

References

[1] Papadimitriou P and Garcia-Molina H 2011 *Data Leakage Detection* (*Knowledge and Data Engineering, IEEE Transactions*) pp 51–63

[2] Riya N and Manisha G 2019 *Data Leakage Detection in cloud using Watermarking Technique (ICCCI)* pp 10–15

[3] Suryawanshi M and Patil S 2016 *Avoiding the data leakage and providing privacy to data in networking (ICCUBEA)* pp 1–4

[4] Salvia R, Ferrara R, Spoto F and Cortesi A 2018 *SDLI: Static Detection of Leaks across Intents* (*IEEE TrustCom-18)* pp 33–52

[5] Huang X, Lu Y, Li D and Ma M *A Novel Mechanism for Fast Detection of Transformed Data Leakage* (*IEEE Access* vol 6) pp 35926–35936

[6] Bidmeshki M, Antonopoulos A and Makris Y 2017 *Information Flow Tracking in Analog/Mixed-Signal Designs through Proof-Carrying Hardware IP* (*Design, Automation and Test in Europe*) pp 1703–1708

[7] Bell D *Looking Back at the Bell-LaPadula Model* (*Proceedings of the 21st Annual Computer Security Applications Conference*) pp 337–351

[8] Raja Koti B, Raj Kumar G and Naveen Kumar K 2019 *Identification of Information Leakage and Guilt Agent by using MAC-IP Binding and Recursive Partitioning Algorithm to Modulate the Uncertainty in the Organization's Network* (*International Journal of Innovative Technology and Exploring Engineering* vol 9) pp 4091–4096

[9] Cinque M, Cotroneo D and Pecchia A 2018 *Challenges and Directions in Security Information and Event Management (SIEM)* (*ISSREW* vol 1) pp 95–99

A method for detecting false positives in procedure of malware analysis

M I Starodubov[1], A E Borshevnikov[1], S A Bystrevskii[1]

[1]Far Eastern Federal University, Vladivostok, Russky Island, 10 Ajax Bay, 10, 690922, Russia

E-mail maxhays@live.ru

Abstract. The paper discusses the possibility of determining false positive by the clustering method. A model of malicious software based on the "behavioral" attribute of malicious objects is described in the paper. Described is a method for studying a sample of malicious objects using a clustering algorithm based on spatial density for applications with noises - DBScan.

Keywords: Clustering; Malware; Mathematical model; Malware behavior; False positive

1. Introduction

Modern operating systems and software products are grandiose and complex products. Every day, researchers and cybercriminals discover new and new vulnerabilities that are exploited by malicious software, and many users knowingly refuse security updates, which puts their systems at risk.

Malicious software often disguise themselves as regular software. As a result, modern anti-viruses, along with malicious components, give out the fact of detection to legal software.

In connection with the above, research into new methods of detecting malware, detecting false positives of malware detection, as well as related research will never lose its relevance.

2. Clustering as a method of detecting malware

Malicious software is a broad term that refers to a part of software code, software tool, or special module that is deliberately created to damage a computer system. Often, many researchers also refer to this concept as errors of software developers, which can lead to damage. However, this statement is erroneous. By themselves, these errors cannot be called malicious, since it is impossible to prove the malicious intent of the developers who made them. However, many companies try to prevent them, as this directly affects the security of their products and opens up new opportunities for attackers.

Malware always seeks to infiltrate a computer system, inflict damage, take over control of processes, and sometimes disable the system. The purpose of the malware is quite obvious - to make a profit at the expense of the system owner. Malicious software is capable of stealing, deleting, altering data, changing the functions of a computer, and taking control of it.

There are many ways to distribute malware. Malicious software most often reaches the user's device via the Internet. It can be installed when a user browses infected websites, opens malicious email attachments, downloads hacked versions of paid programs, downloads data from unconfirmed sources, etc. Thus, system infection can occur anytime the device is connected to the Internet.

Very often, malicious software is hidden in popular applications or embedded in their updates. As a result, the likelihood of deceiving even the most attentive user increases significantly.

Clustering is one of the methods for solving the problem of detecting malware.

Another concept is closely related to clustering, classification is the process of assigning an object to a certain class. Clustering differs from classification as follows:

1. The number of object groups is unknown in advance;
2. The content of the groups is unknown in advance;
3. Groups are formed based on a certain proximity of objects.

Cluster analysis can be represented as the following algorithm:

1. Formation of a set of sample objects under study;
2. Determination of the features of objects to solve the problem of drawing up vectors of characteristics;
3. Normalization of vectors of characteristics;
4. Determination of similarity for a given metric;
5. Application of the method of cluster analysis in the framework of solving the problem of partitioning into clusters;
6. Presentation of results in a form convenient for the researcher.

There are a lot of clustering algorithms. Almost all algorithms were studied in the study of the set of malicious objects. Here are the main types:

1. Hierarchical cluster analysis [1];
2. Square-error schemes [2];
3. Fuzzy algorithms [3];
4. Graph clustering algorithms [4]:

4.1. Connected components algorithm [5];

4.2. Minimum Spanning Tree Algorithm;

4.3. Layer-by-layer algorithms [9].

These algorithms are currently used in a large number of problems: in detection problems, an ensemble of decision trees, locally stable convolutions are used. However, in this case, there is a possibility of a large number of false positive detection of malware.

Machine learning clustering is often more efficient than other methods [6].

3. Malware model

The problem of clustering malware has many solutions. Let's describe the model of malicious software.

Suppose that the source code of the program, obfuscation methods are not important to us, but only the behavior is important. In reality, the choice of a class of malware depends only on behavior, just like the choice of a particular family. This paper considers the problem of creating a method for studying anomalous points of the considered data sample.

Consider the set U is the set of all possible programs:

$$U = \{PR_1, \dots, PR_s, R_1, \dots, R_k, T_1, \dots, T_m\}, n = s + k + m, \quad (1)$$

where, n is the number of all programs; s is the number of all programs, probably not RansomWare; k is the number of all RansomWare programs; m is the number of programs that are not trusted by RansomWare.

Let the set $Api = \{Api_1, \dots, Api_{84}\}$ is the set of Boolean functions.

Each element of the set Api has the following meaning: the function Api_i returns the value 1 if and only if during the execution of the program the i-th API-function of the system from the set Api was called.

Let the set $Dll = \{Dll_1, \dots, Dll_{35}\}$ is the set of Boolean functions.

Each element of the Dll set has the following meaning: the Dll_i function returns the value 1 if and only if during the execution of the program the i-th dynamic link library of the system from the set Dll was called.

Let the set $File = \{File_1, \dots, File_{20}\}$ is the set of Boolean functions.

Each element of the $File$ set has the following meaning: the $File_i$ function returns the value 1 if and only if the i-th operation with a file from the set $File$ took place during the program execution.

Let the set $Reg = \{Reg_1, \dots, Reg_{261}\}$ is the set of Boolean functions.

Each element of the Reg set has the following meaning: the Reg_i function returns the value 1 if and only if during the execution of the program there has been a change in the i-th registry key from the set Reg.

Let us define any element of the set U as a set of the following vectors.

Each element U_i from the Set U has the form presented in formula:

$$U_i = \left\{ \begin{bmatrix} Api_{1,i} \\ \dots \\ Api_{84,i} \end{bmatrix}, \begin{bmatrix} Dll_{1,i} \\ \dots \\ Dll_{35,i} \end{bmatrix}, \begin{bmatrix} File_{1,i} \\ \dots \\ File_{20,i} \end{bmatrix}, \begin{bmatrix} Reg_{1,i} \\ \dots \\ Reg_{261,i} \end{bmatrix} \right\}, i = \overline{1,n} \quad (2)$$

Let $\varphi(U_i): U_i \to \{0,1\}$ be a functional denoting the execution of the program U_i and leading either to a safe state of systems (0) or an unsafe state (1). Let us define the area of probabilities of malicious software behavior as $V = \{U_i : U_i \in U, P(\varphi(U_i) = 1)\}$.

Let us define the sets $G = \{PR_1, \dots, PR_s, T_1, \dots, T_m\}$ and $R = \{R_1, \dots, R_k\}$.

The selection of the sets $\{G_1, \dots, G_s\}$ (a set of programs not verified by analysts) and $\{T_1, \dots, T_m\}$ (a set of programs tested by analysts) from the set G is conditional and is set only for the convenience of virus analysts.

4. Experiment

As the set under study, a sample of 994 files with the following properties was formed:

1. All files are executable objects in the operating system Windows, i.e. have a PE format;
2. The files must have a detection verdict of the type Trojan-Ransom.Win32.Agent;
3. Files must be emulated by means of Cuckoo Sandbox.

Popular resources for storing and analyzing malicious objects were selected as a source of objects: VirusTotal, VirusShare.com, Malware.lu, etc.

The study of the executable code of each object in modern realities often takes a lot of analyst's time and effort. In this work, a different approach was used, a behavioral model of malicious objects was investigated and developed.

The following configuration was chosen as an environment for investigating the behavior of objects: Cuckoo Sandbox, a virtualization product of Oracle VM VirtualBox with a virtual machine on which operating system Windows 7 is installed and popular libraries (vc ++, .net and others) are preinstalled.

This configuration was chosen for the following reasons:

1. Ability to connect additional analysis systems to Cuckoo Sandbox, including traffic analysis systems;
2. Cuckoo Sandbox is an open source project, which is why the source code is available for study and it is possible to change it for a specific one;
3. Operating system Windows 7 was chosen as a guest operating system, as it is the second most Internet-users system of the Windows family, but at the same time it is much less secure than operating system Windows 10.

During the analysis, the following actions take place:

1. The executable object is sent for analysis;
2. A virtual machine with known parameters is launched on the physical computer and the object being analyzed is passed to it for execution;
3. The parameters changed after the execution of the object are transferred back to the physical computer;
4. A behavioral report of the object is generated.

A behavioral report is one of the most important parts of an analyst's analysis of malicious objects, along with the source code, and can give a complete picture of the hidden capabilities of the object under investigation.

The formation of a vector of characteristics is the starting point for solving the clustering problem. As a result, from the available data of the Cuckoo Sandbox report and the typical behavior of malicious objects, the most important characteristics for analysis were identified, namely, important API functions [7], important registry keys, calls to important dynamic link libraries, changes to important system files and directories [8].

Each characteristic of a vector is a boolean value:

- 0 – the object did not call the API function, did not change the registry key, etc.
- 1 – performed one of the above actions.

Each vector of characteristics is saved to the database of all analyzed objects. As a result, a base of vectors of characteristics is formed. After adding another vector, a new object is sent for analysis. The local csv table is temporarily used as the base used.

Cluster analysis works best on a set of normalized vectors. This problem was solved at the stage of forming the vector of characteristics. Since the set of values for each of the characteristics is clearly defined and discrete, it is easy to check that each value falls into the segment [0;1].

As can be seen in figures (figure 1, figure 2), the reduction of vectors to the same length is not only meaningless, but also harmful [9] within the framework of solving our problem. After bringing the vectors to the same length, the possibility of finding anomalous points is lost.

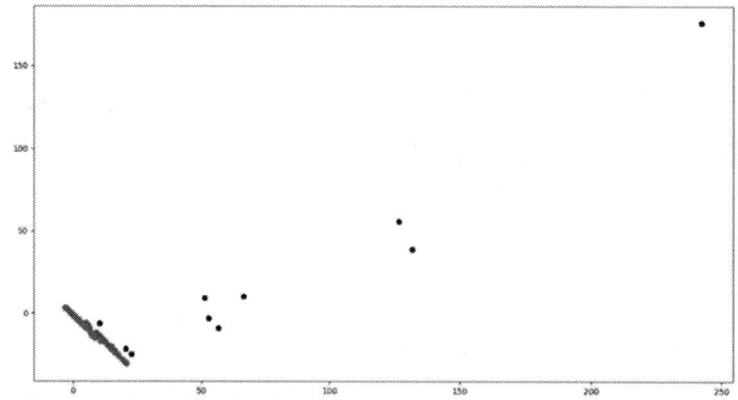

Figure 1. Data distribution, before converting vectors to the same length

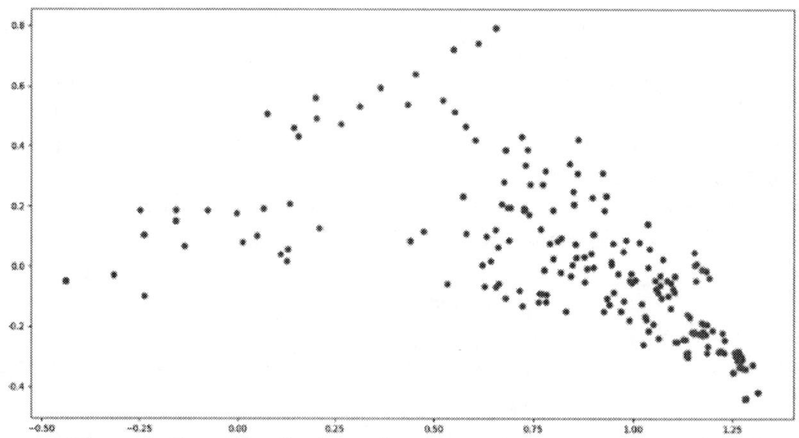

Figure 2. Data distribution, after converting vectors to one length

The DBSCAN method was used to perform the clustering task. The implementation of this method within the sklearn library takes the following parameters as input:

1. eps - maximum distance to the nearest object of one cluster;

2. min_samples - the minimum number of objects in the cluster.

Experimentally, it was found that the optimal maximum distance to the nearest object of one cluster in the frame of work with the current sample is 25.4.

From the analysis of anomalous points, it became clear that the minimum number of objects in the cluster should be set to 3.

After execution, the DBSCAN function returns the original set of data under investigation with the added characteristic labels, each i-th element of which indicates the belonging of the i-th object to a specific cluster. If labels has a value of -1, then such a point is considered noise, and the object will be called anomalous.

One of the simplest and most convenient methods for visualizing clustering results is a projection onto a two-dimensional plane. As part of the study, it was decided to use the principal component analysis to reduce the dimension of the parameters under study.

As can be seen in figures (figure 3, figure 4), the shapes of the clusters are quite clearly visible, and also, what is more important in the framework of our work, anomalous points are visible (marked in black).

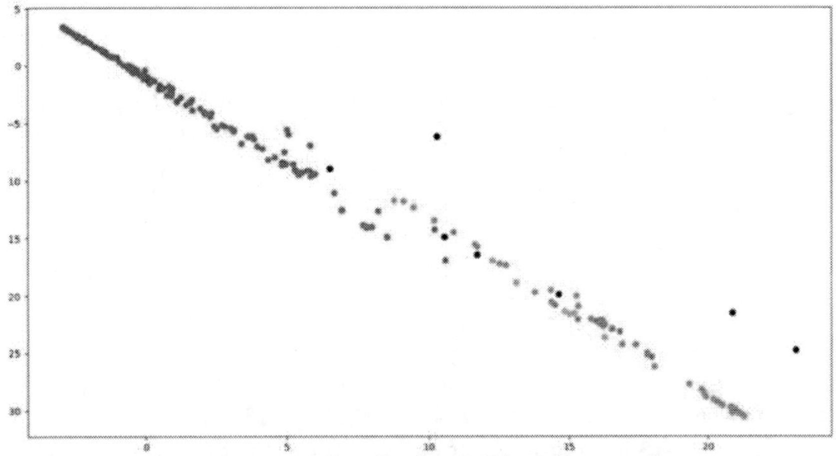

Figure 3. Visualization of clustering results

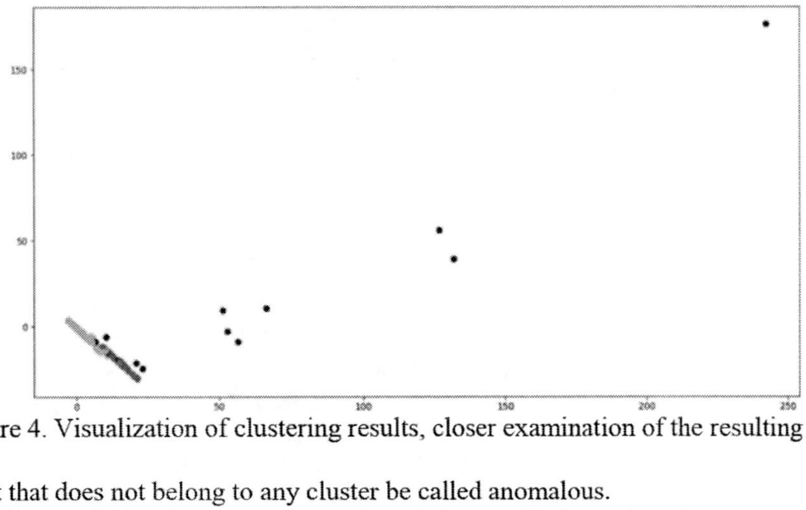

Figure 4. Visualization of clustering results, closer examination of the resulting clusters

Let a point that does not belong to any cluster be called anomalous.

Analysis of anomalous points is the key point of this work. The points that get out of the general picture of the family's behavior are obviously not in their own class, family, or are very specific representatives of this family. In our sample of data for anomalous points, the following options are possible:

1. A specific representative of his family;
2. False positive detection of malware;
3. Incorrect class, family of the fact that malware was detected.

Analyzing the data sample, 14 anomalous points were identified.

5. Conclusion

Here are the main results of the study:

1. The proposed method shows its productivity;
2. All anomalous points found are objects either mistakenly assigned to the Trojan-Ransom.Win32.Agent class, or specific representatives of this class;
3. This method depends on the professionalism of the analyst who performs the manual analysis.

The results obtained within the framework of the study can be applied to modify algorithms for the behavioral analysis of malware.

References

[1] Sisaat K, Kittitornkun S, Kikuchi H, Yukonhiatou C, Terada M and Ishii H 2016 A Spatio-Temporal malware and country clustering algorithm: 2012 IIJ MITF case study *Int. J. Inf. Secur.* **16** pp 459-473

[2] Cordeiro de Amorim R and Komisarczuk P 2014 Partitional Clustering of Malware Using K-Means *Cyberpatterns Unifying Design Patterns with Security and Attack Patterns* pp 223-233

[3] Pai S, Di Troia F, Corrado V A, Austin T H, Stamp M 2016 Clustering for malware classification *J. Comput. Virol. Hack. Tech.* **13** pp 95-107

[4] Chen Y, Liu F, Shan Z and Liang G. 2018 MalCommunity: A Graph-Based Evaluation Model for Malware Family Clustering *4th Int. Conf. of Pioneering Computer Scientists, Engineers and Educators (Zhengzhou)* part 1 pp 279-297

[5] Kinable J and Kostakis O 2011 Malware classification based on call graph clusterin *J. in Computer Virology* **7** pp 233–245

[6] Narra U, Di Troia F, Corrado V A, Austin T H and Stamp M 2015 Clustering versus SVM for malware detection. *J. Comput. Virol. Hack. Tech.* **12** pp 213–224

[7] Szeles G, Cole A 2019 Malware Clustering Based on Called API During Runtime *Information and Operational Technology Security Systems* pp 110–121

[8] Ashraf A, Aziz A, Zahoora U, Khan A 2019 Ransomware Analysis using Feature Engineering and Deep Neural Networks *arXiv: Cryptography and Security*

[9] Milosevic N and Huang J 2019 Deep learning guided Android malware and anomaly detection *arXiv: Cryptography and Security*

Detection of network anomalies in log files using machine learning methods

V A Skazin, A V Pavlychev, S S Zotov

Polytechnic Institute, Far Eastern Federal University, Vladivostok, 10 Ajax Bay, Russky Island, Russia

E-mail skazin_va@dvfu.ru

Abstract. Detection of network anomalies plays an important role in ensuring information security and countering unauthorized access to information infrastructure, including critical facilities. Detecting of abnormal events in log files is complicated by the fact that individual events without any context may be uninformative. The growing importance of log file analysis in large computer systems requires the development of automated methods for processing unstructured data that retrieves information from large log files without human intervention. This article discusses K-means data clustering method and Isolation forest and OSVM machine learning algorithms in terms of searching network anomalies in network log files in order to detect malicious domains.

Keywords: Machine learning, Isolation Forest, One class SVM, K-means, Text data processing, GrayLog, Network anomaly search

1. Introduction

Nowadays, there is a great interest in detection of "abnormal" states in data sets using machine learning methods. The process of detecting "abnormal" states is known as anomaly or outlier detection. The initial definition of this process was given by Grubbs in 1969 [1]: “an outlying observation, or outlier, is one that appears to deviate markedly from other members of the sample in which it occurs”. Although this definition is still relevant today, the goals for detecting outliers are different now. Initially, the main task of detecting anomalies was to remove outliers from the training data, since the pattern recognition algorithms were quite sensitive to them. This procedure is also called data cleansing. After the development of more reliable classifiers, attention to detecting anomalies has diminished significantly. However, critical point was around 2000, when researchers paid more attention to the anomalies themselves, as they are often associated with specific events or suspicious data records. Since then, new algorithms for detecting anomalies have been actively studied. Some algorithms will be studied in this article. Since the purpose of using anomaly detection algorithms has changed, the Grubbs’ definition can be expanded by two important characteristics:

- Anomalies differ from the norm in terms of their characteristics
- Anomalies are rare in the dataset compared to normal instances.

Currently, anomaly detection algorithms are widely used in different areas and often improve traditional rule-based detection systems.

The prime example of using an anomaly detection algorithm is modern intrusion detection tools. A first stage presents tracking of network traffic and server application activity by pattern matching according to pre-set rules (heuristic analysis). In the second stage, the anomaly detection module attempts to identify unknown suspicious activity (behavioral analysis), which identifies potential intrusion attempts and exploits.

Anomaly detection algorithms cover different requirements. In some cases, the simple algorithms are used, which must be very fast to work in near-real-time mode. In other cases, complex algorithms are used when detection efficiency is important due to the high criticality of anomaly missing. Intrusion detection tools may be classified in relation to the time when the anomaly should be detected. In addition to post-incident analysis and real-time detection, there is predictive motivation, known as early warning. Obviously, this is the most difficult task of detecting anomalies. However, major incidents are often preceded by minor signs that can be detected.

In this article, we present a comparative assessment of some algorithms for detecting anomalies. We describe the strengths and weaknesses of the algorithms in terms of their usefulness for specific application scenarios. As a result, this work is aimed to choose a suitable algorithm for detecting uncontrolled anomalies for the intrusion detection problem.

2. Methodology for finding anomalies in system logs

The algorithm for finding anomalies using log file analysis consists of four main stages: collecting of log files, pre-processing data, clustering data and searching for anomalies.

Collecting of log files. Security systems in modern corporate networks generate logs with a large amount of data about the state of the network and its processes. The first stage includes selection of the system for collecting log files and the main parameters which will be included in the final upload. Data can be uploaded to a csv file for more convenient analysis.

Preliminary data processing. As a rule, uploaded CSV files of system logs contain poorly structured data and are often presented in an incorrect format. The purpose of preprocessing is to remove events or features that are not used in the algorithm in the future, as well as to present the necessary data of the algorithm to a unified form for better clustering in the future.

Clustering. Clustering is the task of grouping a set of objects in **clusters.** Each cluster must have "similar" objects, and objects must be different from each other. The most popular clustering algorithm used in anomaly search methods is K-means.

K-means. This clustering algorithm divides data into appropriate groups based on the information available in the algorithm. Data is divided into K remote clusters in order to produce effective data mining results. Each cluster has a center called a centroid. The data point is grouped into a specific cluster based on how close the objects are to the centroid.

The K-means algorithm iteratively minimizes the distance between each data point and its centroid to find the most optimal solution for all data points. The stages of the algorithm are:

1. K-random points of the data set are selected as centroids;
2. Distances between each data point and the centroids K are calculated and stored;
3. Each point is assigned to the nearest cluster according to the calculation of the distance;
4. New positions of the cluster centroids are updated similar to finding the average value in locations;
5. If the location of the centroids has changed, the process is repeated from step 2 until the calculated new center remains the same, which indicates that the cluster members and centroids are now installed.

Minimum distances between all points are found if the data points were separated in such way as to form the most compact clusters with the least variance. In other words, no other iteration can have a smaller average distance between the centroids and the data points inside them.

The defined K-means algorithm aims to minimize the objective function. In this case it is presented as the squared error function:

$$J = \sum_{i=1}^{k} \sum_{j=1}^{n} (||x_i - u_j||)^2 = 1$$

where, $||x_i - u_{j||}$ is the Euclidean distance between the point x_i and the centroid u_j, repeated over all points k in the i cluster for all n clusters.

Basically, the objective function tries to select centroids which minimize the distance to all points in its corresponding cluster, that increases accuracy of centroids for the surrounding cluster of data points. This number is set based on the results of previous research and theoretical considerations.

Search for anomalies. The results of clustering the initial data set can be used in machine learning methods to search for anomalous data. This paper discusses the following methods: Isolation Forest, OSVM.

Isolation forest. This method builds a set of trees. Each branch corresponds to a chosen random attribute and an outlier. The tree is built until each object (or the specified number of objects) is in a separate sheet. In this approach, the average path length from root to leaf is an anomaly measure of the corresponding object, since the leaves have outliers at earlier stages than the typical points.

Figure 1 illustrates that anomalies are more susceptible to isolation under random partitioning. We observe that the normal point x_i generally requires more separations for isolation. The x_0 anomaly point required fewer sections for isolation.

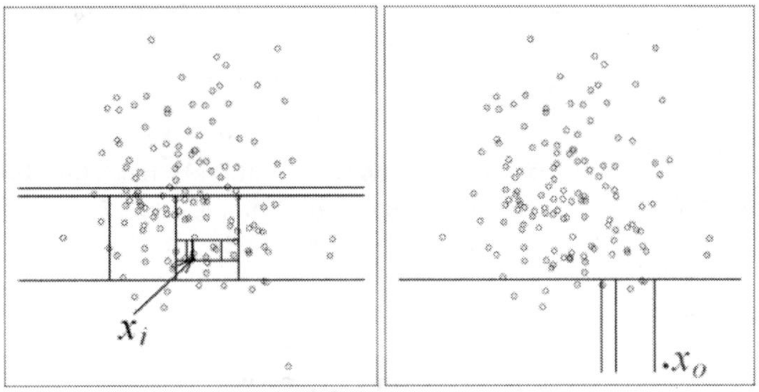

Figure 1. Isolated points $\boldsymbol{x_i}$ and $\boldsymbol{x_0}$

Anomalies are more susceptible to isolation and have a shorter path length. In the example above, x_i requires allocation of twelve random partitions (Figure 1, left); x_0 anomaly requires allocation of only four partitions (Figure 1, right). Since each section is randomly generated, individual trees are generated with different sets of sections. We average the path length across multiple trees to find the expected path length. Figure 2 shows that the average path lengths x_0 and x_i converge as the number of trees increases. When using 1000 trees, the average path lengths x_0 and x_i are 4.02 and 12.82, respectively. This shows that anomalies have a shorter path length than normal instances [2].

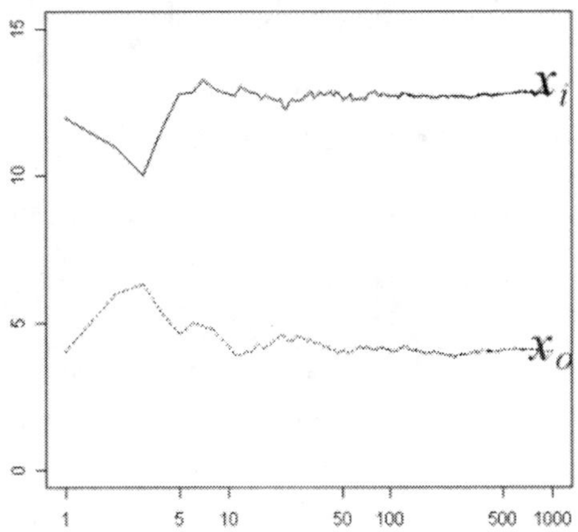

Figure 2. The average path lengths x_i and x_0 converge as the number of trees increases.

OSVM (One Class Support Vector Machine). OSVM model learns on data from a single class (usually a "normal" class). The trained model can be used to classify new data that is both similar and different from the training set. This is useful for detecting outliers, since there are usually few examples of outliers (anomalies), which makes it difficult to train a two-class classifier to distinguish them.

In order to get more universal decision boundaries by OSVM, kernel functions are often used to map the source object space to multidimensional spaces. Let $\varphi:X\rightarrow Z$ be a mapping from the source data space to the multidimensional object space. To train the OSVM model, we solve the following quadratic program:

$$\min_{\omega,\varepsilon_i,b} \frac{1}{2}\|\omega\|_2^2 + \frac{1}{\nu m}\sum_{i=1}^{m} \varepsilon_i - b \tag{1}$$

$$\forall_i = 1,2,\ldots,N$$

$$\nu \in (0,1],\ \varepsilon_i \geq 0,$$

$$\omega \cdot \varphi(z_i) \geq b - \varepsilon_i$$

where ε_i is a variable for the data point i. ν takes a value from 0 to 1; it limits a part of training errors from above and a part of support vectors from below. Consequently, a non-zero ν may create the OSVM model and exclude some of the training data as outliers from the normal class. To define belonging x to a normal class, we use the following function:

$$F(z) = sign\ (\omega \cdot \varphi(x) - b) = \begin{cases} +1\ normal, \\ -1\ outlier. \end{cases} \tag{2}$$

In this work, the kernels of radial basis functions (RBF) are selected. They do not take any parametric form of data distribution and as a result suit to capturing complex data.

3. The result of the methodology application

Let us examine the described methodology for finding anomalies in system logs using a special example. At the first stage we selected the popular GrayLog log file storage system, from which the csv file was uploaded for a specified period of time for further processing. The choice of the system was based on the fact that each event record was shown in raw form with the maximum possible information about the network packet. The studied upload per day from the GrayLog system was about 40 million records.

The second stage presented application of regular expressions during data preprocessing to remove unnecessary characters. Furthermore, empty rows were deleted, since their presence incorrectly affected further clustering. The data type "datetime64" was set for records containing data about the date and time of the event. The original domain names were assigned to second-level domains, because they could fall into different clusters during clustering when working with third-level and higher domains. The final step at this stage was to upload a smaller size to the final file.

At the third stage, data clustering process is aimed to identify atypical objects that cannot be attached to any cluster, or to simplify further data processing and decision-making by applying different analysis methods to each cluster.

Since we used K-means clustering algorithm of unsupervised learning, we had to select the number of clusters that would give us an optimal separation for further machine learning. The number of clusters was determined by iteration. The studied data set included 4 clusters.

At the fourth stage, abnormal records (domains) are detected using the described machine learning methods. The result of the algorithms is a csv file containing information about anomalous domains. The popular VirusTotal service was used for their subsequent evaluation and identification of malicious domains.

Machine learning algorithms revealed 84 Isolation forest and 96 OSVM anomalous domains. The result of domain verification is shown in Tables 1 and 2.

IOP Conf. Series: Materials Science and Engineering 1069 (2021) 012021 doi:10.1088/1757-899X/1069/1/012021

Table 1. Selection of malicious domains from abnormal Isolation forest domains.

№	Isolated Forest	Type
1	au79nt5wic4x.com	Malware
2	fbwat.ch	Phishing
3	rncdn7.com	Malicious
4	nbf9b5aurl.com	Malicious
5	reichelcormier.bid	Malicious
6	pubt.in	Malicious
7	m.me	Malicious
8	adsco.re	Malicious
9	aletorrenty.pl	Malware

Table 2. Selection of malicious domains from abnormal O-SVM domains.

№	OSVM	Type
1	pubt.in	Malicious
2	mgtracker.org	Malicious
3	ceoworldeventsummit.info	Spam
4	m.me	Malicious
5	au79nt5wic4x.com	Malware
6	nbf9b5aurl.com	Malicious
7	emlstart.com	Spam
8	aletorrenty.pl	Malware
9	klike.net	Malicious

This sample includes domains only with a negative rating on the VirusTotal site. Domains classified as "Suspicious" but having a neutral rating on VirusTotal were not entered. Furthermore, the domains that were not classified as malicious but had a negative rating from users of the VirusTotal service were not included. The domains of torrent trackers, sites with prohibited content and *.onion domains were also detected as a result of the algorithm, since they were prohibited on the corporate network.

The algorithm process identified 13 unique domains that were classified as malicious by the VirusTotal service.

4. Conclusion

This article discusses machine learning methods - Isolation forest and OSVM, used for anomalies search in log files. As a result, the following results were obtained:

- Raw log file of 40 million records was uploaded from the GrayLog system in "*.csv " format;
- Preprocessing of the resulting csv file was performed, which resulted the creation and processing of a data set of 21 million records;
- Selection of parameters for optimal data clustering using the K-means method;
- Search for anomalies in the prepared data set using Isolation Forest and OSVM methods;
- A manual check of "abnormal" domains for malicious ones using the VirusTotal service; identification of 13 unique domains found using machine learning methods.

This study shows the effectiveness of using machine learning methods in system log files analyzing. Existing security systems did not identify requests on domains previously, and the presence of these connections may pose a potential threat to the network infrastructure.

References

[1] Grubbs F E 1969 *American Society for Quality Procedures for Detecting Outlying Observations in Samples* vol 11

[2] Liu F T, Ting K M and Zhou Z H 2008 Isolation forest *Proceedings - IEEE International Conference on Data Mining, ICDM* pp 413–22
[3] Callegari C, Giordano S and Pagano M 2017 Entropy-based network anomaly Detection *2017 International Conference on Computing, Networking and Communications, ICNC 2017* (Institute of Electrical and Electronics Engineers Inc.) pp 334–40
[4] Kohout J, Komárek T, Čech P, Bodnár J and Lokoč J 2018 Learning communication patterns for malware discovery in HTTPs data *Expert Syst. Appl.* **101** 129–42
[5] Thomas T, Vijayaraghavan A P and Emmanuel S 2019 *Machine learning approaches in cyber security analytics* (Springer Singapore)
[6] Du M, Li F, Zheng G and Srikumar V 2017 DeepLog: Anomaly detection and diagnosis from system logs through deep learning *Proceedings of the ACM Conference on Computer and Communications Security* (Association for Computing Machinery) pp 1285–98
[7] Bertero C, Roy M, Sauvanaud C, Trédan G and Tredan G 2017 *Experience Report: Log Mining using Natural Language Processing and Application to Anomaly Detection*
[8] Hunt K 2016 *Log Analysis for Failure Diagnosis and Workload Prediction in Cloud Computing* (STOCKHOLM, SWEDEN)

Possible lightweight modification of hashing algorithms GOST R 34.11-94 and GOST R 34.11-2012

S A Bystrevskii [1], A E Borshevnikov[1], M I Starodubov [1]

[1]Far Eastern Federal University, Vladivostok, Russky Island, 10 Ajax Bay, 10, 690922, Russia

E-mail serg.by.and97@mail.ru

Abstract. The paper discusses the development of a lightweight modification of the hashing algorithms GOST R 34.11-2012 and GOST R 34.11-94. The quality of optimization of the obtained algorithms and their safety are considered.

Keywords: Lightweight cryptography; Hash function; Complexity of the algorithm; Algorithm optimization

1. Introduction

Most modern algorithms for protecting information and, in particular, encryption, are designed for use in a PC as part of software systems without taking into account optimization at the hardware level. This fact makes it impossible to apply most of the existing cryptographic algorithms in devices with limited computing power, small volume and low power consumption. Cryptographic data protection techniques in low cost systems have become the backbone of lightweight cryptography.

Lightweight cryptography is gaining special relevance in the light of the development of the idea of the Internet of Things, which is a wireless self-configuring network between objects of various classes, examples of which are household appliances, vehicles, smart sensors and radio frequency identification (RFID) tags [1].

The goal of this work is to develop and analyze lightweight implementations of cryptographic hash functions standardized in the Russian Federation: GOST R 34.11-2012 and GOST R 34.11-94.

2. Development of lightweight hashing algorithms

Let us first consider the optimization of the GOST R 34.11-2012 algorithm.

First of all, it is worth reducing the block size of the hash function. Perhaps this will slow down the algorithm, but it will clearly reduce its size. Let the block size be 64 bits, let's call this algorithm Stribog-64 accordingly. Moreover, this block size should be throughout the entire algorithm, and not just cut at the end, as in Stribog-256. Let's consider the modified transformations for "Stribog-64.

The nonlinear bijective transformation $\pi: V_8 \rightarrow V_8$ will not change. But it is worth replacing the permutation $r: Z_{64} \rightarrow Z_{64}$ with $r: Z_{16} \rightarrow Z_{16}$, we will set it below as an array $r = (r(0), r(1), \ldots, r(16))$:(0, 4, 8, 12, 1, 5, 9, 13, 2, 6, 10, 14, 3, 7, 11, 15).

To convert $l: V_{64} \rightarrow V_{64}$, it is worth changing the constants, or rather, generating them, since they take up a lot of space. 64 constants, each 64 bits in size, totaling 4096 bits only for storing these constants. It makes sense to store only one constant, and generate the rest from it. Since the transformation l is used frequently in the algorithm, this will slow down the speed of the algorithm, so the generation should not be difficult.

Also, in order for the transformation result to be any vector from V_{64}, it is necessary that the newly formed constants form the basis of the V_{64} space. This result can be achieved if we take 0x8e20faa72ba0b470 as the first number, and each next constant is obtained from the previous one by cyclic shift by any prime number less than 64. For definiteness, take the number 5, then the process of generating a constant vector from the standard can be expressed below as:

$$A_i = \begin{cases} 8e20faa72ba0b470, \text{if } i = 0 \\ A_{i-1} << 5, if\ 1 \leq i \leq 63 \end{cases}, \quad (1)$$

where $A_i \in V_{64}$ are elements of the constant vector A, defined in GOST R 34.11-2012.

The transformation $X[k]: V_{512} \to V_{512}$ will be changed only in dimension, that is, it will become $X[k]: V_{64} \to V_{64}$.

The transformation $S: V_{512} \to V_{512}$ will accordingly become $S: V_{64} \to V_{64}$, and will definitely be like:

$$S(a) = S(a_7||...||a_1||a_0) = \pi(a_7)||...||\pi(a_1)||\pi(a_0), \quad (2)$$

where $a = a_7||...||a_0 \in V_{64}, a_7, ..., a_0 \in V_8$.

The permutation conversion $P: V_{512} \to V_{512}$ is redefined as $P: V_{64} \to V_{64}$ as follows:

$$P(a) = P(a_{15}||...||a_1||a_0) = a_{r(15)}||...||a_{r(1)}||a_{r(0)}, \quad (3)$$

where $a = a_{15}||...||a_0 \in V_{64}, a_{15}, ..., a_0 \in V_4$, r is a new permutation.

The linear transformation $L: V_{512} \to V_{512}$ becomes $L: V_{64} \to V_{64}$ and will be equal to the new linear transformation $l: V_{64} \to V_{64}$.

Thus, the basic transformation used in the Stribog algorithm was simplified. Now we need to define a one-step compression function.

It makes no sense to make major changes in it. But it's worth understanding the generation of thirteen round keys using twelve constants for the algorithm. These constants take up a lot of space. Each is 512 bits and thus they take up 6144 bits in total. In key generation, they are used only for modulo two addition, which is not justified memory consumption. It makes sense to leave one constant and use it only when adding modulo two with the first key. So now in the one-step compression function, instead of N, some special vector will be used. But in order to have the effect of changes, this vector must change every run of the hash function. To achieve this goal, you can use some light pseudo-random number generator, for example, created on the basis of the linear shift register proposed in the cipher KATAN [2,3].

To implement the idea described above, we introduce a new transformation $C_{RI_1,RI_2}: V_{64} \to V_{64}$, which we define as follows:

$$C_{RI_1,RI_2}(N) = C_{RI_1,RI_2}(N_{63}||...||N_0) = N_{62}||...||N_{32}||RI_2 \oplus N_{31} \oplus N_{24} \oplus (N_{20}\&N_{15}) \oplus (N_3\&N_1)||N_{30}||...||N_0||RI_1 \oplus N_{63} \oplus N_{56} \oplus (N_{52}\&N_{47}) \oplus (N_{35}\&N_{33}), \quad (4)$$

where $RI_1, RI_2 \in V_1$ are stochastic bits obtained from the hash, $N = N_{63}||...||N_0 \in V_{64}$, $N_{63}, ..., N_0 \in V_1$.

This function consists of two linear shift registers that write new data to each other. The peculiarity is that it uses stochastic bits. The bits RI_1, RI_2 will be the last and first bits of the hash, respectively. And the first value of N will be b1085bda1ecadae9.

Now let's review the transformations of the one-step hash function.

Key generation is now set like this:

$$K_i = \begin{cases} K, \text{for } i = 1 \\ LPS(K_{i-1}), \text{for } i = 2, ..., 13 \end{cases} \quad (5)$$

where $K \in V_{64}$ is the value for the first key.

The transformation $E: V_{512} \times V_{512} \to V_{512}$ will remain the same, except for the change in dimension $E: V_{64} \times V_{64} \to V_{64}$:

$$E(K, m) = X[K_{13}]LPSX[K_{12}] ... LPSX[K_1](m), \quad (6)$$

where $K, m \in V_{64}$, $K_{13}, \dots, K_1$ are generated keys, where the value of the first key is K.

Compression function $g_N: V_{512} \times V_{512} \to V_{512}$, also changes only in dimension $g_N: V_{64} \times V_{64} \to V_{64}$:

$$g_N(h, m) = E(LPS(h \oplus N), m) \oplus h \oplus m \quad (7)$$

where $h, m, N \in V_{64}$.

Finally, by redefining all the functions, you can define the algorithm itself. The algorithm for generating the value of the hash function Stribog-64 is defined as follows.

Input: $M \in V^*$.
1. $h = 0^{64}$.
2. $N = Vec_{64}(0xb1085bda1ecadae9)$, $\Sigma = 0^{64}$.
3. If the condition $|M| \geq 64$ is satisfied, then go to step 4, otherwise go to step 10.
4. Calculate $M', m: M = M' || m$, where $m \in V_{64}$.
5. $h = g_N(h, m)$.
6. $N = C_{h_{63}, h_0}(N)$, where $h = h_{63} || \dots || h_0$, $h_{63}, \dots, h_0 \in V_1$.
7. $\Sigma = Vec_{64}(Int_{64}(\Sigma) \boxplus Int_{64}(m))$.
8. $M = M'$.
9. Go to step 3.
10. $m = 0^{62-|M|} || 1 || M$.
11. $h = g_N(h, m)$.
12. $N = C_{h_{63}, h_0}(N)$, where $h = h_{63} || \dots || h_0$, $h_{63}, \dots, h_0 \in V_1$.
13. $\Sigma = Vec_{63}(Int_{63}(\Sigma) \boxplus Int_{63}(m))$.
14. $h = g_N(h, \Sigma)$.

Output: h – hash value.

Now let's start considering the algorithm for generating the value of the hash function GOST R 34.11-94.

If the block size is reduced, then this algorithm meets the size requirements for lightness. Therefore, it is necessary to increase the performance of this algorithm.

Since the algorithm uses the Magma cipher, the performance of the GOST R 34.11-94 algorithm will directly depend on the performance of the Magma cipher. This problem can be solved by reducing the number of rounds in the encryption algorithm. According to research, the number of rounds can be reduced to sixteen without a strong loss of cryptographic strength.

In what follows, the transformation $E_K: V_{64} \to V_{64}$ will mean the sixteen-round cipher "Magma" with the key $K \in V_{256}$.

In the algorithm GOST R 34.11-94, we will reduce the block length to 64 bits. And we will redefine transformations accordingly.

First, let's define an algorithm for generating a key. Here we have to change something, since we need to collect one key from 64 bits with a length of 256 bits. We will use the same principle as in the original algorithm and generate four 64-bit keys, and then glue them into one.

We redefine the transformation $A: V_{256} \to V_{256}$ as $A: V_{64} \to V_{64}$. This transformation is a linear register shift and can be expressed as:

$$A(X) = A(x_3 || x_2 || x_1 || x_0) = (x_0 \oplus x_1) || x_3 || x_2 || x_1, \quad (8)$$

where $X = x_3 || x_2 || x_1 || x_0 \in V_{64}$, $x_3, x_2, x_1, x_0 \in V_{16}$.

We redefine the supporting permutation $P: V_{256} \to V_{256}$ as $P: V_{64} \to V_{64}$. Let's set it as follows:

$$P(X) = P(x_8 || \dots || x_1) = x_{\varphi(8)} || \dots || x_{\varphi(1)}, \quad (9)$$

where $X = x_8 || \dots || x_1 \in V_{64}$, $x_8, \dots, x_1 \in V_8$; $\varphi(i + 1 + 4(k-1)) = 2i + k$, $i = 0, \dots, 3$, $k = 1,2$.

Now let's start redefining the key generation algorithm.

Key generation is a 256-bit key generation from vectors $M, H \in V_{64}$. We define the algorithm as follows.

Input: $M, H \in V_{256}$.
1. $C_2 = C_4 = 0^{64}$, $C_3 = 1^{16}0^{24}1^{16}0^{8}$.
2. $i = 1, U = H, V = M$.
3. $W = U \oplus V, K_1 = P(W)$.
4. $i = i + 1$.
5. If the condition $i = 5$ is satisfied, then go to step 9, otherwise go to step 6.
6. $U = A(U) \oplus C_i, V = A(A(V))$.
7. $W = U \oplus V, K_i = P(W)$.
8. Go to step 4.
9. $K = K_1||K_2||K_3||K_4$
Output: $K \in V_{256}$ – generated key.

Now consider mixing transformation $\Psi: V_{256} \to V_{256}$, which needs to be redefined as $\Psi: V_{64} \to V_{64}$. Let's set it up as follows:

$$\Psi(X) = \Psi(x_8|| \dots ||x_1) = x_1 \oplus x_3 \oplus x_7 \oplus x_8||x_8|| \dots ||x_2, \quad (10)$$

where $X = x_8|| \dots ||x_1 \in V_{64}, x_8, \dots, x_1 \in V_8$.

Since the algorithm uses the function $\Psi(X)$ 61 times and 12 times in a row, it makes sense to calculate the values $\Psi^{61}(X)$ and $\Psi^{12}(X)$ in advance, rather than using a loop in the one-step compression function.

The composition of these functions can be set as follows:

$$\Psi^{12}(X) = \Psi^{12}(x_8|| \dots ||x_1) = x_3 \oplus x_7 \oplus x_8 \,||\, x_2 \oplus x_6 \oplus x_7 \,||\, x_1 \oplus x_5 \oplus x_6 \,||\, x_2 \oplus x_4 \oplus x_5 \oplus x_6 \oplus x_7 \oplus x_8 \,||\, x_1 \oplus x_3 \oplus x_4 \oplus x_5 \oplus x_6 \oplus x_7 \,||\, x_3 \oplus x_4 \oplus x_5 \oplus x_7 \oplus x_8 \,||\, x_2 \oplus x_3 \oplus x_4 \oplus x_6 \oplus x_7 \,||\, x_1 \oplus x_2 \oplus x_3 \oplus x_5 \oplus x_6, \quad (11)$$

$$\Psi^{61}(X) = \Psi^{61}(x_8|| \dots ||x_1) = x_2 \oplus x_3 \oplus x_5 \oplus x_6 \oplus x_8 \,||\, x_1 \oplus x_2 \oplus x_4 \oplus x_5 \oplus x_7 \,||\, x_1 \oplus x_2 \oplus x_3 \oplus x_4 \oplus x_7 \oplus x_8 \,||\, x_1 \oplus x_3 \oplus x_8 \,||\, x_6 \oplus x_8 \,||\, x_5 \oplus x_7 \,||\, x_4 \oplus x_6 \,||\, x_3 \oplus x_5, \quad (12)$$

where $X = x_8|| \dots ||x_1 \in V_{64}, x_8, \dots, x_1 \in V_8$.

Now we need to define a one-step compression function. It makes no sense to make major changes here, other than to reduce the use of mixing transforms, since the number of mixing blocks has changed.

Finally, given the above, let's redefine the one-step hashing function as:

$$\chi(M, H) = \Psi^{61}(H \oplus \Psi(M \oplus \Psi^{12}(E_K(H)))), \quad (13)$$

where $M, H \in V_{64}$.

The algorithm for calculating the value of the hash function h will change only by the dimensions of the blocks and the use of a new one-step hashing function. Specifically, the algorithm is defined as follows.

Input: $M \in V^*$, $H \in V_{64}$
1. $\Sigma = 0^{64}$, $L = 0^{64}$.
2. If the condition $|M| > 64$ is satisfied, then go to step 3, otherwise go to step 8.
3. Calculate M_p, M_s: $M = M_p||M_s$, where $M_s \in V_{64}$.
4. $H = \chi(M_s, H)$.
5. $\Sigma = \Sigma \boxplus M_s$, $L = L \boxplus 64$.
6. $M = M_p$.
7. Go to step 2.
8. $M' = 0^{64-|M|}||M$.
9. $\Sigma = \Sigma \boxplus M'$, $L = L \boxplus |M|$.
10. $H = \chi(M', H)$.
11. $H = \chi(L, H)$.
12. $H = \chi(\Sigma, H)$.
Output: H – hash value.

3. Optimization score

Let's consider the optimization of the GOST R 34.11-2012 algorithm.

First, let's look at the total amount of memory saved. Saved 1728 bits of dynamic memory and 10240 bits of static memory. This is a very large volume and if we translate this into the size of the algorithm, then about 65024 gate equivalent (GE) will be released.

In terms of speed, not such results have been achieved. However, a plus can be noted in that the processing speed of short data is several times higher than that of the original. Thus, if the input messages are less than 64 bytes, then the average data processing speed is twice that of the original. Note that if the input messages are less than 16 bytes in length, then the processing speed is 600000000 bytes per second, when the original has 100000000 bytes per second. This is of course short data, but such data includes passwords, which are usually 8 to 16 bytes long. Also, the flogs are constantly compared, so this algorithm is very suitable for them. However, if we take a conditionally infinite string as input, then for the original the data processing speed is 2500000000 bytes per second, while the new algorithm has 1250000000 bytes per second. That is, twice as slow.

The new algorithm meets the requirements of lightness in that it has a small size and a high speed of processing short input data. But hash function is not suitable for other purposes, since it has a low processing speed for big data.

Let us evaluate the optimization of the GOST R 34.11-94 algorithm.

First, let's estimate the data processing speed. It increased greatly, but, as for the previous algorithm, it turned out to be different for different lengths. If we consider conventionally infinite lines, then the speed of the original algorithm is 3000000000 bytes per second, while the speed of the new algorithm is 6000000000 bytes per second. That is, even for a conditionally infinite string, the new algorithm is twice as efficient in time as the original.

If we consider the input data of the order of 16 bytes, then the speed of the original algorithm is 850000000 bytes per second, and the new algorithm is 3830000000 bytes per second. That is, for small input data, the speed of the algorithm has increased by about 4.5 times.

Now let's look at optimization for the amount of stored information. 768 bits of dynamic memory was saved and 128 bits of static memory were added, thus reducing the algorithm size by about 6272 GE.

As a result, we can say that the optimization of this algorithm is successful and is suitable both for processing long messages and for processing a large number of short messages. It can also be noted that, although not by a little, the amount of stored information has decreased, which still slightly affects the size of the hardware implementation of the algorithm and, accordingly, the cost.

4. Safety assessment

It remains only to consider the safety of the developed algorithms. We have considered algorithms for generating a hash, so we will consider the possibility of selecting collisions and modifying data, as well as finding a second preimage.

It is worth adding a number of statements about hash functions that were not covered in the first chapter. First, if the hash function is built on a one-step compression function f, then the collision resistance of the function f implies the collision resistance of the entire hash function. The second statement is that if the function is resistant to collisions, then it is resistant to finding the second preimage. The third statement - the condition of resistance to finding the inverse image of a function is stronger than the condition of collision resistance.

Let's start by considering the cryptographic strength of the GOST R 34.11-2012 algorithm [4].

Before starting the research, we note that the algorithm has an avalanche effect, as a result of which, even with small changes in the initial data, serious changes occur in the final result. It is also worth noting that the end result is equally likely. Consider the estimate of the complexity of finding the second preimage. Since the algorithm is equally probable, the probability that a random input value will have a specific desired hash is $\frac{1}{|Y|} = \frac{1}{2^{64}}$. The probability that out of n randomly taken messages at least one of the messages will coincide in the hash value with the original is equal to:

$$P = 1 - \left(\frac{2^{64}-1}{2^{64}}\right)^n = 1 - \left(1 - \frac{1}{2^{64}}\right)^n \approx 1 - e^{-\frac{n}{2^{64}}} \quad (14)$$

Thus, in order to find the second preimage with a given probability P, it is necessary to take the following amount of input data:

$$n = -2^{64}\ln(1-P) \quad (15)$$

That is, in order to find the second preimage with a probability of 0.5, it is necessary to take approximately $n = 2^{63}$ input data. And in order to almost guaranteed to find the second preimage, you need to take $n = 2^{70}$ input data.

Consequently, the algorithm lost its resistance to the search for the second preimage, due to the decrease in the hash block, but it still remained acceptable.

Next, we will consider the stability of this algorithm to the "Birthdays" paradox. Following the paradox of "Birthdays", we can say that to find a collision with a probability of 0.5, you need to take 2^{32} inputs. Hence the algorithm is not very collision resistant. This is a small hash length problem.

All of the above applies only to hash length, but not to the algorithm itself. The security of the algorithm itself has great cryptographic strength, since in the lightweight version, almost everything remains from the original.

In general, the functions used in the algorithm make it quite cryptographically strong. But the algorithm is weak for collision detection and a list attack due to a small hash value. Again, this was expected and justified, because lightweight algorithms are used on low-power devices and, as a rule, do not use expensive information and the cost of cracking them is more expensive than the information itself, even with such a low cryptographic strength. In general, this algorithm demonstrates the main advantage of constructing a lightweight algorithm based on the existing one - it is high cryptographic strength due to proven functions.

Now let's move on to studying the safety of the optimized algorithm GOST R 34.11-94 [5,6].

To begin with, we note that the algorithm has a hash dimension of 64 bits. The Stribog-64 algorithm, considered in the previous paragraph, has the same hash dimension. This means that the entire estimate of the optimized GOST R 34.11-94 algorithm with respect to the length of its hash will be the same as for the Stribog-64 algorithm.

Similar to the previous considered algorithm, this algorithm is not sufficiently cryptographically strong, since it has a small hash length, but is resistant to attacks of search for a preimage. In principle, this robustness is sufficient for lightweight algorithms. Perhaps, for further work on this topic, we can consider blocks of greater length than 64 bits to increase the cryptographic strength of the hash function.

These algorithms turned out to be not resistant to collision search, but rather resistant to searching for images.

5. Conclusion

In this paper, issues related to lightweight cryptography, methods for the implementation of lightweight hashing algorithms, and methods for their analysis were considered.

These algorithms have shown good results in terms of lightness. That is, these algorithms require less stored information and perform operations faster than the originals. At the same time, the cryptographic strength of these algorithms turned out to be low, but sufficient for lightweight algorithms.

References

[1] Weis S A, Sarma S E, Rivest R L and Engels D W 2004 Security and privacy aspects of low-cost radio frequency identification systems (*In Security in Pervasive Computing 2003, Lecture Notes in Computer Science*) v 2802 pp 201-212

[2] Leander G, Paar C, Poschmann A and Schram K 2007 New Lightweight DES Variants (*FSE 2007, LNCS*) A. Biryukov (Ed.) **4593** pp 196–210

[3] Bard G V, Courtois N T, Nakahara J, Sepehrdad P and Zhang B 2010 Algebraic, AIDA/Cube and Side Channel Analysis of KATAN Family of Block Ciphers (*INDOCRYPT 2010, LNCS*) G. Gong and K.C. Gupta (Eds.) **6498** pp 176–196
[4] Sedov G K 2015 The security of GOST R 34.11-2012 against preimage and collision attacks (*Mathematical Aspects of Cryptography*) vol 6 pp 79–98
[5] Levin V Y 2011 Increasing the security of hash functions (*J Math Sci*) **172** pp 734–739
[6] Poschmann A, Ling S and Wang H 2010 256 Bit Standardized Crypto for 650 GE GOST Revisited (*In CHES 2010, Lecture Notes in Computer Science*) v 6225 pp 219-233

General approach to automating the process of responding to computer security incidents

A V Pilitsky, O E Prokopenko and V N Halizev

Kuban State Technological University, Krasnodar, Moscovskaya str., 2, 350072, Russia

E-mail pilitsky.alex@yandex.ru

Abstract. The article considers the process of responding to information security incidents as a cyclical process divided into several stages. The main approaches to automating response processes are presented, as well as cases in which automation is not effective.

Keywords: information security; investigation; response; automation; information security incident.

1. Introduction

Now it is impossible to imagine any organization without its own information system, whether it is a large Corporation or a small private business, the only difference is in the scale of this system. And every year the economic losses of enterprises due to computer security incidents only grow. We can't say that there were no such incidents before. No, they were, but not in such quantity.

At the moment, there are no organizations or people who are not victims of hacking or, for example, a computer virus. Yes, now, probably, everyone has an idea about an antivirus program, and it is possible that almost every enterprise has it installed, but there is no such thing as "absolutely protected system". Any system can be hacked, regardless of the degree of its security, it only needs funds, sometimes not only of an intellectual nature, but also financial. With these two components, you can get around everything. This is when computer security incidents occur. If your system does not provide protection against this type of incident, then you can only protect yourself from it in the future.to do this, you need to investigate the computer security incident that occurred, and take appropriate measures so that the security system will prevent such a case in the future.

Now we live in the age of information technology, and where the means and methods of protecting information are being strengthened, the capabilities of hackers (intruders) are also growing. It is worth noting that the development of the second is faster. Every year there are more and more targeted attacks that require a large amount of resources, both material and human, to investigate them. This is due to the fact that, as a rule, such attacks affect a large number of enterprise assets using advanced penetration techniques and anchoring in the information infrastructure. Investigating these types of incidents requires experienced specialists who take years to prepare. At the same time, in addition to complex attacks, there are also simple attacks, but also require the response of specialists. It becomes impossible to provide sufficient incident response specialists. The solution to this problem was the use of automation tools for investigating computer security incidents.

IOP Conf. Series: Materials Science and Engineering 1069 (2021) 012023 doi:10.1088/1757-899X/1069/1/012023

This approach allows you to reduce the number of specialists involved in the investigation of an incident, increase the speed of response, and allow automation tools to resolve simple incidents that do not require human intervention on their own.

As mentioned above, first of all, automation processes allow specialists to focus their attention on cases where human intervention is required, but this approach is not always correct. Automation of some processes can lead to an increase of cases in false positives, and sometimes cause important indicators that show a more serious and complex attack to be missed. Just like any other software, automation tools may contain errors in the code that may lead to the leakage of critical information [1].

General approaches to investigating computer security incidents.

There is no one-size-fits-all method for investigation, each organization must adapt standard recommendations based on its own characteristics and priorities. The approach to investigating computer security incidents in a particular company will depend on many factors, whether it is the role of the Information security Department, Information security policy, acceptable risk level, and others. But if you study international standards and best global practices, the General approach to responding to computer security incidents can look like the one shown in the Figure 1 [2].

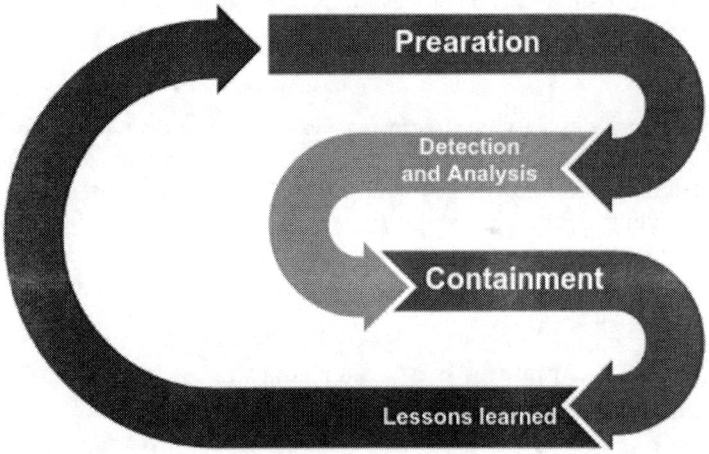

Figure 1. General approach to incident response

2. The need to automate the investigation of computer security incidents

SOC (Security Operations Center) analysts perform two types of work: investigating template incidents that often do not deviate from the standard scenario, or reviewing tasks that require special analysis and often a lot of experience to investigate such an incident. Based on this, a problem arises: the necessary specialists with sufficient skills are not enough to solve the second type of work, while the first requires a large number of employees, which is very expensive.

Routine analysis can be automated, which is advisable on the bases of financial and human costs. This process of automating the investigation of computer security incidents has a number of advantages and disadvantages.

Based on the fact that many organizations have huge information systems, it will be impossible to hire the necessary number of employees to effectively investigate incidents. Firstly, it is economically inefficient, and secondly, a person processes simple requests slower than a machine.

Thus, the advantages of automating the investigation of incidents will be:

1. more appropriate (logical) response to notifications and requests;
2. Higher volume of application closing and incident response;
3. more careful attention of analysts to higher priority issues;
4. Improved visibility of what is happening;

5. Coverage of a larger area and a larger number of applications [3].

But also in any situation, in addition to the pros, there are a number of disadvantages. If you take, for example, critical infrastructure systems, then false positives can lead to disastrous results, especially if you take healthcare or communications. In such systems, disabling critical components will lead not only to economic damage, but also to the loss of human lives.

Disadvantages of automation:

1. Incorrect classification of the attack, so the wrong actions are taken;
2. automation of requests that should have been processed manually;
3. Omission of key information or data;
4. The adoption of incorrect or inappropriate decisions;
5. The creation of excessive and expensive dependencies [3].

These days, organizations use the automation process because, in their opinion, without it, the probability of missing an attack that will cause huge damage is worse than the possibility of negative consequences of the automation process.

Basically, the process of automating the investigation of computer security incidents is considered as automation of everything. But companies can only upgrade some of the most relevant and secure stages of the incident investigation process. The idea is to find a balance between the elements that are best automated and the parts that will be processed by analysts.

3. Automating investigation of incidents of computer security

3.1. Preparation

The preparation phase includes activities to create a system for protecting the information infrastructure from attacks. At this stage, security policies are formed, information security tools and monitoring systems are adjusted based on knowledge about the types of possible attacks and potential violators. Collecting and correlating information about threats and sources, as well as information about past incidents, can greatly simplify the task of specialists in the field of building a security system, as well as increase efficiency by using the most relevant information from a large number of sources. The main types of such information are:

1. SIEM data and log Files
2. checking the reputation of IP, URLS, files, and domains
3. Search in the white and black list
4. IP Geolocation
5. Similar events, same goals, and critical resource checks
6. Threat analysis

Threat analysis platforms collect data about cyber threats and make it available to your organization. There are many useful channels, including HPE Threat Central, FireEye iSIGHT, and IBM X-Force. Integrating with one or more of these channels for automatic enrichment is a great way to immediately put each incident in context.

3.2. Detection and analysis

The promise of security automation lies in the ability to automate not only the collection of information, but also entire processes. Some automation providors have a vision of a SOC that addresses threats and responds to warnings without human intervention. This type of automation can be useful, but there are also negative aspects to this approach. They will be discussed further.

As a rule, at this stage, detection is performed by automated tools that use certain rules formed by the analyst at the preparation stage. The success of the response depends on how well these rules were drawn up, and whether they correspond to current threats.

Among the automated analysis tools, it is worth highlighting sandboxes. These tools allow you to dynamically analyze files that enter the system and identify malicious software based on signatures and behavioral analysis. Using such solutions allows you to switch the attention of analysts to cases where atypical exploitation and infection techniques are used.

When a particular trigger has worked, the system must perform a certain number of actions to prevent an attack. These actions are also prescribed by a specialist. However, it is important that not all actions are performed automatically. If the system sees deviations from the specified pattern, this incident should be passed to analysts, so that they can analyze the data manually to understand the reason for its occurrence, and form a new rule or exception based on the received data.

3.3. Prompt response to an incident

At the incident response stage, it is important that tasks are quickly distributed so that the workload of specialists is uniform. It is also important to establish communication between departments so that employees do not spend extra time providing and receiving information.

Automatic notifications make status updates, collaboration, reporting, and task management more streamlined and error-free. This provides a big operational advantage over manual processes that force analysts to search for contact information, prepare reports, and delegate responsibilities manually-all during the chaos of a security breach. Examples of valuable automated communications include:

- Reports
- Assign tasks on the basis of the playbooks
- Notification of the action

Warning fatigue is a serious problem for most corporate SOCS. Analysts are clogged with thousands of warnings, and most of them are false. Automation allows analysts to focus on warnings that pose a real threat to the company. Using prediction criteria, the system can automatically set priority levels in analyst queues to reduce the number of most likely false positives. This can take the form of a percentage of the probability of a false positive assigned by the system, or a more General risk assessment. If designed correctly, this type of incident response automation can significantly increase the productivity of your SOC.

3.4. Securing the results of the investigation

At this stage, new or old security policies are formed based on the received information about the incident. Here, automating the collection of statistics and correlating the results obtained with existing ones will allow you to look at the situation much more broadly and may allow you to identify larger chains of attacks on specific resources, which in the future will allow you to form a strategy for protecting specific assets based on the context of possible previous attacks.

4. Processes that do not require automation

Despite all the advantages of automation, there is a number of cases in which this approach may not only be ineffective, but also lead to a threat to the security of the infrastructure.

When automating, it is important that all necessary actions that can affect the interaction of infrastructure elements, or their integrity, are confirmed by a specialist. An error in the processing of input parameters or the logic of the system can paralyze the operation of the enterprise by, for example, blocking critical resources. Such an error can cause damage equal to the actual attack.

It is also important that the investigation is not fully automated. The participation of specialists is due not only to the existence of complex cases, but also to the fact that the incident response system must be flexible and meet both information security trends and regulatory requirements of global and local regulators. And it is the specialists who ensure the updating of the system.

It is important that automation tools do not restrict the incident response system in any way in the field of its modernization. The system should not only solve strictly typed tasks, but also maintain a database of past incidents and decisions. All this is necessary so that in the future I can dynamically make decisions based on similar cases.

Integration with third-party sources is a significant aspect in building an automated response system. Some automation platforms get all their functionality by connecting to external data sources, which often overlap with each other or with data from existing internal sources. Not to mention that these external data sources often require expensive subscriptions, and in some cases even require creating or maintaining integration. The cost of this approach can scale very quickly, while your ability to respond to incidents depends on the uptime of a dozen different integrations and companies. Therefore, you should avoid this type of dependency if it doesn't pay off.

5. Conclusion

Thus, we can say that large and medium-sized companies cannot do without automating the investigation of computer security incidents. This will allow companies not only to compensate a huge number of employees, but also to quickly respond to incidents that cause material and sometimes reputational damage. But automation should not be 100%. every step of this process should be considered - whether there is a need to improve this stage of the investigation. It is also necessary to take into account the individual factors of your company when automating a particular step.

References

[1] Securit y Orchestration and Automation Benefits: How SOAR Helps Improve Incident Response: https://securityboulevard.com/2020/04/9-security-orchestration-and-automation-benefits-how-soar-helps-improve-incident-response/
[2] Incident Response Steps and Frameworks for SANS and NIST: https://cybersecurity.att.com/blogs/security-essentials/incident-response-steps-comparison-guide
[3] SOC Automation-Deliverance or Disaster (A SANS Spotlight Paper Written by Eric Cole, PhD December 2017)
[4] Computer Security Incident Handing Guide (NIST Special Publication 800-61 Revision 2)
[5] Automating security events management (FabianLibeau(EMEA marketing director, ArcSight)) https://doi.org/10.1016/S1353-4858(08)70139-9
[6] Automated systems only: why CISOs should switch off their dumb machines (Spencer Young) https://doi.org/10.1016/S1353-4858(19)30106-0
[7] The future of information security incident management training: A case study of electrical power companies (Maria Bartnes, Nils Brede Moe, Poul E.Heegaarda) https://doi.org/10.1016/j.cose.2016.05.004
[8] Linux Forensics (Published by Pentester Academy, a division of Binary Security Innovative Solutions Pvt. Ltd)
[9] Actionable Information for Security Incident Response (European Union Agency for Network and Information Security, November 2014)
[10] A methodology for improving informationsecurity incident identification and response (Master Thesis Informatics & Economics By Nanno Zegers)

A version of the voting protocol based on blockchain technology

V N Killer [1], A E Borshevnikov[1], K S Soldatov [1]

[1]Far Eastern Federal University, Vladivostok, Russky Island, 10 Ajax Bay, 10, 690922, Russia

E-mail killervlad997@gmail.com

Abstract. The paper describes a generalized version of the voting protocol based on blockchain technology. The analysis of the proposed protocol is also carried out. The proposed protocol is based on such cryptographic primitives as digital signatures, asymmetric ciphers, and zero knowledge proof protocol. Safety analysis was carried out using the AVISPA software package. Anonymity was analyzed separately as part of this protocol by analyzing differential privacy.

Keywords: Voting; Blockchain; AVISPA; Differential privacy

1. Introduction

Information technology is developing more and more, year after year there are more large-scale breakthroughs. So in the 21st century, cryptography, neural networks, biometrics and other areas related to information technology are developing rapidly. With the development of these industries, people's lives begin to change.

Everywhere people have been trying for centuries to achieve the right to vote, the right to change their lives for the better. Most countries hold presidential elections, elections to the State Duma, and even schools in some countries elect the president by voting. However, with an incorrectly structured voting procedure, there remains the possibility of manipulating as a result of voting, which can lead to an illegitimate victory for one of the candidates.

One of the results of the development of information technology is blockchain technology. If we consider this technology in relation to the voting procedure, then we can note its promising application in the construction of voting protocols. Within the framework of this work, the applicability of blockchain technology in voting protocols has been studied using a separate example.

The voting protocol will never be applied in an election procedure unless the protocol has been shown to be capable of resist to fraud and protecting anonymity. An ideal protocol should have at least the following six properties [1, 2].

1. Only those who have the right to vote can vote;
2. Everyone can vote no more than once;
3. No one can find out for whom a particular voter voted;
4. No one can vote instead of another;
5. Each voter can check that his vote was taken into account when summarizing the voting results;
6. Everyone knows who voted and who did not.

It is usually for voting protocols to use a number of cryptographic primitives and specially proposed methods to meet the requirements described above.

For example, to implement voting protocols, it was proposed to use blind signatures [1] and mix networks [2].

Another method that has been proposed to reduce the possibility of manipulating the voting results is based on the division of the functions of the central election commission between two separate bodies [3]. This approach was developed in works [4-6].

At the moment, there are a number of works in which voting protocols are based on the use of blockchain technology [7,8]. However, the mentioned works do not study the security of the proposed protocols.

2. Formalization of the voting protocol based on blockchain technology

Let here and below: A is the sender; B is the recipient; T is the trusted source (blockchain); Tr is the transaction (operation); $S_k(Tr)$ is the digital signature of transaction Tr with key k; $Anon()$ is the anonymization procedure; $Block_i$ is the *i*-th data block; $h()$ is the hash function; Users are voters; k k_{pub_A} is the public key of the asymmetric encryption / digital signature verification algorithm for user A; k_{sec_A} is the private key of the asymmetric encryption / digital signature generation algorithm for user A; $AnID$ is an anonymous identifier; datum -is the set of transactions; $HEAD$ is the header in the start block of blockchain; IV is an initialization vector.

Let's set the following formalization for the blockchain:

1. $Tr(A \rightarrow B: data)$

$A: \{k_{pub_A}, k_{sec_A}\}$

$A \rightarrow T: Anon(S_{k_{sec_A}}(Tr(A \rightarrow B: data), h(Block_{i-1}), \{AnID, k_{pub_A}\}))$

2. $T(Users): Block_i = \{HEAD, datum, h(datum)\}$

$datum = \{Tr_1, Tr_2, \ldots, Tr_m\}$

$Block_0 = \{HEAD, IV, h(IV)\}$

3. $T \rightarrow B: data$

In this formalization, the following steps were described:

1. The sender wants to transfer funds to the recipient. The sender generates a transaction, public and private keys. The sender writes an anonymous signed transaction to the blockchain, which includes data, a hash of the previous block, an anonymous user ID along with a public key.

2. In the blockchain, users form blocks. Also, a lot of transactions and a zero block are recorded in the blockchain.

3. The recipient receives the funds.

The above formalization of the blockchain technology did not exist before, which makes it unique.

Let's describe a voting protocol based on blockchain technology. The connection between the Central Election Commission (hereinafter - the CEC) and voters is carried out through blockchain technology (T).

Based on the blockchain technology formalization described above, we present the following voting protocol:

1. $CEC: \{k_{pub_{CEC}}, k_{sec_{CEC}}\}$

$CEC \rightarrow T: Param, k_{pub_{CEC}}$

1.1. $CEC \rightarrow T: S_{k_{sec_{CEC}}}(Tr(CEC \rightarrow T: Param, k_{pub_{CEC}}), ID(CEC), k_{pub_{CEC}}, h(Block_{l-1}))$

1.2. $T(Miners): BLOCK_l\{HEAD_l, datum1, h(datum1)\}$,

$datum1 = \{Tr(CEC \rightarrow T: Param, k_{pub_{CEC}})\}, Block_0 = \{HEAD_0, IV, h(IV)\}$

2. $T: \{k_{pub_T}, k_{sec_T}\}$

$T \rightarrow A: k_{pub_T}$

$T \rightarrow CEC: k_{pub_T}$

2.1. $T \rightarrow A: Anon(S_{k_{sec_T}}(Tr(T \rightarrow A: k_{pub_T}), Tr(T \rightarrow CEC: k_{pub_T}), D(T), k_{pub_T}, h(Block_{y-1})))$

2.2. $(Miners): BLOCK_y\{HEAD_y, datum2, h(datum2)\}$
$datum2 = \{\{Tr(i \to T: k_{pub_i})\}_{i=1}^{n}, Tr(T \to CEC: k_{pub_T})\}$
3. $A: \{k_{pub_A}, k_{sec_A}\}$
$A \to T: Anon(E_{k_{pub_T}}(rand_A, k_{pub_A}))$
3.1. $A \to T: Anon(S_{k_{sec_A}}(Tr(A \to T: Anon(rand_A, k_{pub_A})), k_{pub_A}, h(Block_{z-1})))$
3.2. $T(Miners): BLOCK_z\{HEAD_z, datum3, h(datum3)\}$
$datum3 = \{\{Tr\left(i \to T: Anon(rand_i, k_{pub_i})\right)\}_{i=1}^{n}\}$
4. $CEC \to T: E_{k_{pub_T}}(List(ID)),\ ID = \{ID(i)\}_{i=1}^{n}$
4.1. $CEC \to T: S_{k_{sec_{CEC}}}(Tr(CEC \to T: E_{k_{pub_T}}(List(ID))), ID(CEC), k_{pub_{CEC}}, h(Block_{x-1}))$
4.2. $T(Miners): BLOCK_x\{HEAD_x, datum4, h(datum4)\}$
$datum4 = \{Tr\left(CEC \to T: E_{k_{pub_T}}(List(ID))\right)\}$
5. $A \to T: Anon\left(ZKP\left(k_{sec_A}, rand_A, ID(A)\right)\right)$
5.1. $A \to T: Anon(S_{k_{sec_A}}(Tr(A \to T: Anon(ZKP(k_{sec_A}, rand_A, ID(A))), k_{pub_A}, h(Block_{v-1})))$
5.2. $T(Miners): BLOCK_v\{HEAD_v, datum5, h(datum5)\}$
$datum5 = \{Tr(i \to T: Anon\left(ZKP\left(k_{sec_i}, rand_i, ID(i)\right)\right)\}$
6. $T \to CEC: E_{k_{pub_{CEC}}}(List(Rand)),\ Rand = \{rand_i\}_{i=1}^{n}$
7. $A \to T: Anon\left(E_{k_{pub_T}}(B, rand_A)\right)$
7.1. $A \to T: Anon(S_{k_{sec_A}}(Tr(A \to T: Anon(E_{k_{pub_T}}(B, rand_A)),\ k_{pub_A}, h(Block_{w-1})))$
7.2. $T(Miners): BLOCK_w\{HEAD_w, datum6, h(datum6)\}$
$datum6 = \{Tr(i \to T: Anon(E_{k_{pub_T}}(B, rand_A)))\}$
8. $T: List\left(\{B_i, k_{pub_i}, E_{k_{pub_i}}(rand_i)\}_{i=1}^{voter}\right)$
8.1. $T: S_{k_{sec_T}}(Tr(T: List(\{B_i, k_{pub_i}, E_{k_{pub_i}}(rand_i)\}_{i=1}^{voter}), k_{pub_T}, h(Block_{a-1}))$
8.2. $T(Miners): BLOCK_a\{HEAD_a, datum7, h(datum7)\}$
$datum7 = \{Tr(T: List(\{B_i, k_{pub_i}, E_{k_{pub_i}}(rand_i)\}_{i=1}^{voter}))\}$
9. $T \to CEC: E_{k_{pub_{CEC}}}\left(\sum(B), List(\{B_i, rand_i\}_{i=1}^{voter})\right)$
10. $CEC \to T: \sum(B)_{CEC}$
10.1. $CEC \to T: S_{k_{sec_{CEC}}}(Tr(CEC \to T: \sum(B)_{CEC}),$
$ID(CEC), k_{pub_{CEC}}, h(Block_{b-1}))$
10.2. $T(Miners): BLOCK_b\{HEAD_b, datum8, h(datum8)\},\ datum8 = \{Tr(CEC \to T: \sum(B)_{CEC})\}$

The formalization described above can be presented as follows.

1. The CEC sends parameters and the public key that was generated by the CEC to the blockchain.

1.1. The CEC signs the transaction that was described in step 1, the CEC ID, the CEC public key and the hash from the previous block of the blockchain.

1.2 . Miners form a block, which includes a block header, a set of transactions and a hash value from this set of data. The zero block consists of the header of this block, the initializing vector and the value of the hash function of the initializing vector.

2. The blockchain generates public and private keys and sends the public key to voters and the CEC.

2.1. The blockchain sends anonymized signed transactions to the voter along with its identifier, public key, and hash value from the previous block.

2.2. Miners form a new block, which includes a block header, a set of transactions and a hash value from this set of data.

A set of transactions consists of transactions,

- voters send their public keys to the blockchain;
- the blockchain sends its public key to the CEC.

3. The voter sends an anonymized and encrypted random number and public key to the blockchain.

3.1. The voter sends to the blockchain the anonymized signed transaction that was described in step 3, along with the public key and hash value from the previous block.

3.2. Miners form a new block, which includes a block header, a set of transactions and a hash value from this set of data. Many transactions include voter transactions that send an anonymized random number and public key to the blockchain.

4. The CEC sends an encrypted voter ID list to the blockchain.

4.1. The CEC sends to the blockchain the signed transactions described in step 4 along with the CEC ID, the CEC public key, and the hash value from the previous block.

4.2. Miners form a new block, which includes a block header, a set of transactions, and a hash value from this set of transactions. The set of transactions includes the transaction described in step 4.

5. The voter sends an anonymized secret key, random number and identifier to the blockchain using a zero knowledge proof protocol.

5.1. The voter sends the anonymized signed transaction described in step 5 to the blockchain along with the public key and hash value from the previous block.

5.2. Miners form a new block, which includes a block header, a set of transactions, and a hash value from this set of transactions. Many transactions include transactions in which voters send an anonymized secret key, random number, and identifier to the blockchain using a zero knowledge proof protocol.

6. The blockchain sends to the CEC an encrypted list of random numbers chosen by the voters.

7. The voter sends an anonymized encrypted ballot with his vote and a random number to the blockchain.

7.1. The voter sends the anonymized signed transaction described in step 7 to the blockchain along with the public key and hash value from the previous block.

7.2. Miners form a new block, which includes a block header, a set of transactions and a hash value from this set of data. Many transactions include transactions in which voters send an anonymized encrypted ballot and a random number to the blockchain.

8. The blockchain generates a list that includes ballots, public keys, encrypted random numbers of all voters.

8.1. The blockchain generates the signed transaction described in step 8, the public key and the hash value from the previous block.

8.2. Miners form a new block, which includes a block header, a set of transactions and the value of hash functions from this set of transactions. The set of transactions includes the transactions described in step 8.

9. Blockchain sends encrypted ballot counts to the CEC, as well as a list of voter ballots and their randomly selected numbers.

10. The CEC sends the counting result received by the CEC to the blockchain.

10.1. The CEC sends the signed transactions described in step 10 to the blockchain along with the CEC ID, the CEC public key, and the hash value from the previous block.

10.2. Miners form a new block, which includes a block header, a set of transactions, and a hash value from this set of transactions. The set of transactions includes the transaction described in step 10.

3. Protocol analysis

Let's analyze the proposed protocol.

Step 10 compares the values of the votes counted by the CEC and the votes counted in the blockchain. If they match, then the voting results can be considered correct.

The protocol described in this work was studied using the AVISPA software package. To analyze the protocol using AVISPA, the proposed protocol was formalized in the high-level protocol specification language.

A study using the AVISPA software package showed that the proposed protocol can be considered safe from the point of view of users' trust in the data sent between them.

We will investigate the anonymity of voters within the framework of the developed protocol. For this, we will use such a concept as differential confidentiality [9].

Let ε be a non-negative real number and A be some probabilistic algorithm that takes a data set as input. We denote the image of A by imA. Algorithm A is ε-differentially private if for all datasets D_1 and D_2 that differ by one element (that is, the data of one person), as well as all subsets S of the set imA [9]:

$$P\{A(D_1) \in S\} \leq e^{\varepsilon} \cdot P\{A(D_2) \in S\}, \quad (1)$$

where P is the probability

Let's define the next ordered element - *{IP, ID, rand, k, vote}*, where *IP* is the user's IP address; *ID* is the user identifier; *rand* is a random number used within the protocol; *k* is the voter's public key used within the protocol; *vote* is the voter's vote. Let us denote by Ø is the absence of data transmitted to the voter. Let's describe the voting protocol in the following notation:

1. $CEC \rightarrow T: \emptyset$
2. $T \rightarrow A: \{IP, \emptyset, \emptyset, \emptyset, \emptyset\}$
   $T \rightarrow CEC: \emptyset$
3. $A \rightarrow T: \{AnIP, \emptyset, rand, k, \emptyset\}$
4. $CEC \rightarrow T: \{\emptyset, ID, \emptyset, \emptyset, \emptyset\}$
5. $A \rightarrow T: \{AnIP, ZKP(ID), ZKP(rand), ZKP(k), \emptyset\}$
6. $T \rightarrow CEC: \{\emptyset, \emptyset, rand, \emptyset, \emptyset\}$
7. $A \rightarrow T: \{AnIP, \emptyset, rand, \emptyset, vote\}$
8. $T: \{\emptyset, \emptyset, rand, k, vote\}$
9. $T \rightarrow CEC: \{\emptyset, \emptyset, rand, \emptyset, vote\}$
10. $T \rightarrow CEC: \emptyset$

Consider two random voters in a voting protocol. You can see that during the protocol, the data of the two users is indistinguishable.

However, the following should be noted. At step 5, it is possible to establish a connection between the data provided by the user and the user's identity. The binding is eliminated by using a zero knowledge knowledge proof protocol that is additionally anonymized (for example, using a set of random proxies). The anonymization of data, in particular the IP address, also leads to the provision of anonymity in step 3.

The only thing that can be established within the framework of step 3 is the willingness of the voter to vote within this voting. To eliminate this knowledge, this step can be anonymized, for example, using a zero knowledge proof protocol.

Hence, we can assume that the proposed protocol is 0-differentially confidential, i.e. the protocol is anonymous.

4. Conclusion

The paper proposes a version of the voting protocol based on blockchain technology. As a result of the analysis of this protocol, the following results were obtained:

1. the proposed protocol is anonymous,
2. the trust in the transmitted data is observed among the participants of this protocol,
3. Confidentiality of data transmitted by voters is ensured using asymmetric encryption algorithms, i.e. data confidentiality depends on the strength of the protocols used.

The results obtained show that blockchain technology can be used as a reliable basis for constructing voting protocols.

References

[1] Chaum D L 1982 Blind Signatures for Untraceable Payments (*Advances in Cryptology Crypto 82 : Proceedings*) (Springer US) pp 199–203
[2] Chaum D L 1981 Untraceable electronic mail, return addresses, and digital pseudonyms (*Communications of the ACM*) vol 24 pp 84–90
[3] Nurmi H, Salomaa A and Santean L 1991 Secret ballot elections in computer networks (*Computers and Security*) vol 10 pp 553-560
[4] Fujioka A, Okamoto T and Ohta K 1993 A practical secret voting scheme for large scale elections (*Lecture Notes in Computer Science*) vol 718 pp 244-251
[5] Cranor L F and Cytron R K 1997 Sensus: a security-conscious electronic polling system for the Internet (*Proceedings of the Thirtieth Hawaii International Conference on System*) pp 561-570
[6] Sako K and Kilian J 2001 Secure Voting Using Partially Compatible Homomorphisms (*Lecture Notes in Computer Science*) vol 839 pp 411–424
[7] Li P, Peng J, Yang L, Zheng Q and Pan G 2018 Crux – A New Fast, Flexible and Decentralized Consensus Algorithm with High Fault Tolerance Rate (*Lecture Notes in Computer Science*) vol 11373 pp 66–76
[8] Wang L, Hu M, Jia Z, Gong B, Yang Y, Liu X and Cui W 2019 A Voting Scheme in Blockchain Based on Threshold Group Signature (*Trusted Computing and Information Security*) pp 184–202
[9] Dwork C and Roth A 2014 The Algorithmic Foundations of Differential Privacy (*Foundations and Trends in Theoretical Computer Science*) vol 9 pp 211–407

Analysis of the Use of Artificial Immune Systems

R M Mikherskii[1], M R Mikherskii[1]

[1]V.I. Vernadsky Crimean Federal University, Simferopol, Prospekt Vernadskogo, 4, 295007, Russia

E-mail mrm03@mail.ru

Abstract. Artificial immune systems are systems of artificial intelligence at the basis of work, which are based on the principles of the functioning of the biological immune system. These systems are of great interest from researchers developing models and algorithms in the field of machine learning for solving complex computational and engineering problems. In the presented work, a review of modern methods for the implementation of artificial immune systems is carried out, taking into account the results obtained recently.

Keywords: Artificial intelligence; Machine learning; Pattern recognition

1. Introduction

Recently, machine learning systems are increasingly used in the study of a wide range of problems, the solution of which until recently was available only to humans. The most used machine algorithms are now algorithms of convolutional neural networks [1-10].

This popularity is due to two factors. First, convolutional neural networks currently demonstrate the best quality of pattern recognition and data classification among all existing methods. Secondly, a large number of software libraries have been implemented that allow a wide range of users to use convolutional neural networks.

However, it should be noted that convolutional neural networks also have a number of disadvantages, the main one of which, perhaps, is the long training time of such networks. In some tasks, this time can be tens of hours [11]. In addition, if such a network is trained to classify data according to a certain set of classes, then adding a new class requires a complete retraining of the network.

The artificial immune systems studied in this work do not have the disadvantages described above. Artificial immune systems are computing systems based on the principles of the biological immune system [12-17].

2. Review

The field of application of artificial immune systems is quite extensive. Currently, these networks are used for processing data without a structure [18], searching for anomalous data [19], control [20], solving problems of classification and optimization [21, 22, 23], information compression [24], security computer systems [25], machine learning [26,27].

In the article [28], an implementation of an artificial immune system is presented, which makes it possible to recognize from 77.55% to 99.78% of the images of human faces presented to this system for identification.

The use of artificial immune systems for recognition of visual images is investigated in works [29,30]. In these works, a new algorithm for an artificial immune system is presented, which makes it possible to recognize such images in real time using a webcam. As a result of the experiment, it was determined that this artificial immune system can be successfully used not only to recognize the faces of people, but also any other objects. These works also investigate the possibility of using artificial immune systems in parallel computing systems. The developed artificial immune system can be trained in a new image at any time of its operation. It should also be noted that this system learns new images rather quickly. The advantages described above allow you to create artificial intelligence systems that learn in real time.

The article [31] presents a new set of multiple artificial immune system detectors for detecting malicious programs on mobile devices running the Android operating system. An artificial immune system with multiple detectors differs from the commonly used artificial immune system in that several sets of detectors are developed simultaneously through negative selection. The first set of detectors includes detectors associated with protected applications, the second set - related to malicious applications. An artificial immune system with multiple detectors also includes, along with the negative selection method, feature selection, known as the detector separation technique. The approach presented in this work allowed achieving recognition accuracy of 93.33%.

It should be noted that artificial immune systems are also used in desktop computers to protect against malicious attacks over the Internet. The article [32] investigated security systems based on an artificial immune system used in such computers.

In works [33, 34], a fairly successful attempt was made to use an artificial immune system to solve the problem of identification and authentication of users using biometric data. For this purpose, iris recognition was used.

The works [35,36] consider the use of an artificial immune system in e-mail to solve the problem of detecting messages containing spam. To solve this problem, the authors proposed to use two classes of antibodies. The first class included antibodies that multiply when interacting with email messages in which spam is absent. The second class is those that multiply when interacting with mail messages that contain spam.

In [37], a detailed analysis of algorithms used in pattern recognition and intelligent analysis is made. An overview of the use of artificial immune systems in problems of authentication and identification of people using their biometric data is carried out. It is shown that these systems have advantages in use in comparison with "fuzzy extractors" and artificial neural networks. Possible solutions to the problem of secure storage of biometric data using an artificial immune system were also considered.

In [38], it was proposed to use an artificial immune system to detect information security incidents in intrusion detection systems. In this system, a modification of the clonal selection algorithm is considered, in which an external optimization structure is used, the principle of which is based on the use of the strategy of evolutionary algorithms.

In article [39], it is proposed to use an artificial immune system to identify the author of a Russian handwritten text based on his handwriting sample. When implementing this system, methods of negative selection, Erne idiotypic networks and risk theory were used.

The paper [40] presents a method for analyzing handwritten text, which is based on the principle of functioning of an artificial immune system. To solve this problem, a negative selection algorithm was used. The initial data for this system are the coordinates of the pen taken from the output of the digitizer, as well as the value of the pen pressure. This method allows for a text-independent analysis of the manuscript. According to the authors, it potentially has a higher analysis accuracy and, accordingly, expands the scope of online manuscript analysis systems.

The paper [41] considers the possible use of an artificial immune system to create a search system for design solutions in the electronic archive of design documentation. An immune multimodal algorithm was used as a search method. The descriptions of software implementation tools are given.

In article [42], a model of an intrusion detection system based on an artificial immune system is proposed. This model examines the abnormal behavior of packets in network traffic during transmission.

It combines machine learning rules and negative selection theory to improve the performance of an intrusion detection system. The article tested the proposed model on the DARPA1999 dataset.

In [43], a hybrid approach to predicting the temporal change in the parameters of an on-board automated control system is considered. To solve this problem, a modified immune clonal selection algorithm was proposed. The theoretical results obtained in this work can be used in the development of intelligent decision support systems for monitoring the state of complex technical objects of the rocket and space industry. Application of the proposed approach makes it possible to increase the operational efficiency and reliability of the operation of these objects.

In [44], an algorithm for an artificial immune system is proposed that allows solving problems of controlling autonomous systems, symbolic regression, and recognition of single characters. Based on the presented algorithm in the C # programming language, a software system has been developed that has shown its efficiency in solving the assigned tasks. Computational experiments were carried out for each considered problem. The accuracy of the results obtained was compared with that for such methods as: linear classifier (1st level neural network); k-nearest neighbors method; principal component analysis and quadratic classifier; support vector machine (Gaussian kernel); two-layer, three-layer and six-layer neural networks. In general, compared with these methods, the artificial immune system showed good results.

In [45], the theoretical foundations of the three main types of algorithms for the artificial immune system, such as the clonal selection algorithm, the negative selection algorithm, and the immune network algorithm, are studied in detail. The paper also discusses the possible benefits of artificial immune systems for understanding how real biological immune systems work.

To solve the problem of resource allocation when using an artificial immune system, an optimal approach based on control was proposed in [46].

In [47], studies were carried out on the use of dendritic cells to detect abnormal or dangerous events in the data stream.

In [48], a type of data space representation was proposed that can evolve and adapt over time, providing a link between the data and the immune algorithm itself.

In [49], it was proposed to use a model of an artificial immune system for early diagnosis of breast cancer. This model is called the artificial immune system for associative classification.

The goal of [50] was to develop a decentralized adaptive system capable of studying the model of incoming problems and being able to choose the best from a given set of solutions. The system presented in this work is based on the integration of three main theories in immunology: the theory of the immune network, the theory of danger and the theory of clonal selection. The efficiency of the proposed mechanism in real scenarios was tested experimentally on a physical network of real robots.

The application of the principles of artificial immune systems to control mobile robots is discussed in [51]. In this article, an algorithm for an artificial immune system was proposed by combining two models: clonal selection and the immune network. This hybrid mechanism has been tested on a real robot located in an unknown environment including obstacles. The main goal of the robot was to detect safe areas in the environment. The results obtained in various test cases indicate that the proposed model is viable and able to cope with different environmental conditions.

The article [52] presents a model of an artificial immune system using the CNN algorithm. The model is based on the bone marrow and thymus model. The results obtained and the time measurement carried out at the same time show that the model presented in this work can be implemented in applications in real time.

3. Conclusion

The literature review presented in this article is, of course, incomplete. This is due to the fact that artificial immune systems are a rapidly developing and promising direction in the development of machine learning and the scope of their application is extremely wide. Nevertheless, we very much hope that this work will be useful both for specialists in the field of artificial intelligence and machine learning, and for scientists working in related fields of knowledge.

4. Acknowledgment
The authors would like to express their gratitude to Marina Vsevolodovna Glumova, Director of the Physico-technical Institute of the V. I. Vernadsky Crimean Federal University, and Viktor Milyukov, head of the Department of computer engineering and modeling of the Physico-technical Institute of the V. I. Vernadsky Crimean Federal University, for their comprehensive assistance in organizing research work.

References
[1] Bulat A and Tzimiropoulos G 2020 *IEEE Transactions on Pattern Analysis & Machine Intelligence* vol 42 no 02 pp 343-356
[2] Farrugia and Guillemot C 2020 *IEEE Transactions on Pattern Analysis & Machine Intelligence* vol 42 no 05 pp 1162-1175
[3] Rocco I, Arandjelovic R and Sivic J 2019 *IEEE Transactions on Pattern Analysis & Machine Intelligence* vol 41 no 11 pp 2553-2567
[4] Lian C, Liu M, Zhang J and Shen D 2020 *IEEE Transactions on Pattern Analysis & Machine Intelligence* vol 42 no 04 pp 880-893
[5] Kalayeh M and Shah M 2020 *IEEE Transactions on Pattern Analysis & Machine Intelligence* vol 42 no 06 pp 1483-1500
[6] Lin S, Ji R, Chen C, Tao D and Luo J 2019 *IEEE Transactions on Pattern Analysis & Machine Intelligence* vol 41 no 12 pp 2889-2905
[7] Sindagi V A and Patel V M 2017 *Pattern Recognit. Lett.* vol 107 pp 3–16
[8] Zhou Z, Shin J, Zhang L, Gurudu S, Gotway M and Liang J 2017 *Proc. IEEE Conf. Comput. Vis. Pattern Recognit.* pp 7340–7349
[9] Li Y, Zhang X and Chen D 2018 *Proc. IEEE Conf. Comput. Vis. Pattern Recognit.* pp 1091–1100
[10] He K, Gkioxari G, Dollar P and Girshick R 2020 *IEEE Transactions on Pattern Analysis & Machine Intelligence* vol 42 no 02 pp 386-397
[11] Nemkov R M 2015 *Development of neural network algorithms for invariant pattern recognition* (The thesis for the Candidate's degree in Technical Sciences, Stavropol)
[12] Dasgupta D 1999 *Artificial Immune Systems and Their Applications* (Springer, Berlin Heidelberg) pp 3–21
[13] Dasgupta D and Gonzalez F 2002 *IEEE Trans Evol Comput* vol 6 no 3 pp 281–291
[14] Timmis J, Knight T, LN de Castro and Hart E 2004 *An overview of artificial immune systems, in Computation in Cells and Tissues: Perspectives and Tools for Thought ed. by R Paton, H Bolouri, M Holcombe, JH Parish, R Tateson. Natural Computation Series*
[15] Hofmeyr S 1999 *An Immunological Model of Distributed Detection and Its Application to Computer Security* (Ph.D. Dissertation, Department of Computer Science, The University of New Mexico, Albuquerque)
[16] Hofmeyr S and Forrest S 1999 *The Proceedings of the 1999 Genetic and Evolutionary Computation Conference (GECCO-1999)* (Morgan-Kaufmann, San Francisco)
[17] Al-Enezi J R, Abbod M F and Alsharhan S 2010 *International Journal of Research and Reviews in Applied Sciences* vol 3 no 2 pp 118 – 131
[18] Knight T and Timmis J 2001 *IEEE Intern. Conf. on Data Mining* pp 297– 304
[19] Garrett S M 2005 *Evolutionary Computation* vol 13 pp 145–178
[20] Krishna K K and Neidhoefer J 1999 *Immunized adaptive critic for an autonomous aircraft control application* ed. by Dasgupta D (Springer-Verlag Inc.) vol 20 pp 221–240
[21] Bryukhovetskiy A A and Skatkov A V 2010 *Optimization of viral processes* vol 7 pp 119–122
[22] Gao X Z, Ovaska S J, Wang X and Chow M.Y 2009 *Neural Computing and Applications* vol

18 no. 7 pp 719–729
[23] Bardachev Yu N and Didyk A A 2007 *Automation, automation, electrical complexes and system* vol 2 pp 107-111
[24] Stankevich L A and Kazansky A B 2006 *Actual problems of protection and security: tr. 9th All-Russia. scientific-practical conf.* vol 5 pp 145-152
[25] Kim J and Bentley P 2002 *In Proc. Congress on Evolutionary Computation* (Honolulu, HI, USA) pp 1244–1252
[26] Hunt J E and Cooke D E 1996 *Journ. of Network Computing Applications* vol 19 pp 189–212
[27] Dasgupta D, Yu S and Nino F 2011 *Applied Soft Computing* vol 11 no 2 pp 1574-1587
[28] Luh G 2014 *Immune Computation* vol 2 no 1 pp 21-34
[29] Mikherskii R 2018 *Computer Optics* vol 42 no 1 pp 113 – 117
[30] Mikherskii R 2017 *Information technologies and nanotechnologies (ITNT-2017): III international Conf. and youth school* (Samara) pp 518-521
[31] Brown J, Anwar M and Dozier G 2017 *EURASIP J. on Info. Security 2017* 7
[32] Ramakrishnan S and Srinivasan S 2009 *Artif Intell Rev* vol 32(1-4) pp 13–43
[33] Hou H and Dozier G 2005 *Proceedings of the ACM Symposium on Applied Computing* pp 952–956
[34] Hou H 2006 GENERTIA: *A System for Vulnerability Analysis, Design and Redesign of Immunity-Based Anomaly Detection Systems* (Ph.D. Dissertation, Computer Science & Software Engineering Department, Auburn University, Auburn)
[35] Idris I and Abdulhamid SM 2014 *An Improved AIS Based E-mail Classification Technique for Spam Detection arXiv preprint arXiv:1402.1242*
[36] Idris I 2012 *International Journal of Artificial Intelligence & Applications (IJAIA)* vol 3 no 1 pp 83–94
[37] Sulavko A E, , Shalina E V, Stadnilov D G and Choban A G 2019 *Information security issues* vol 1 no 124 pp 38 - 46
[38] Zhukov V G and Salamatova T A 2014 *Vestnik SibGAU* vol 4(56) pp 41–47
[39] Eremenko Yu I and Shatalov 2013 A A *Scientific Bulletin of the Belgorod state University. Series: Economics. Informatics* vol 28 no 22-1 (165) pp 218-224
[40] Bryukhomitsky Yu 2015 A *Proceedings of the southern Federal University. Technical Sciences* no 5 (166) pp 174-183
[41] Eremenko Yu I, Melnikova I V and Shatalov A A 2011 *Bulletin of the Voronezh state technical University* vol 7 no 7 pp 115-121
[42] Burlakov M E and Ivkin A N 2019 *Bulletin of Perm national research Polytechnic University. Electrical engineering, information technology, control systems* no 29 pp 209-224
[43] Smirnov V A and Smirnov D V 2018 *Science-intensive technologies in earth space research* vol 10, no 1 pp 66-78
[44] Astachova I F, Zolotukhin A E, Kurklinskaya E Yu and Belyaeva N V 2019 *J. Phys.: Conf. Ser.* 1203 012036
[45] Timmisa J, Honec A, Stibord T and Clarka E 2008 *Theoretical Computer Science* 403, pp 11–32
[46] Hone A and Van den Berg H 2007 *Silico Immunology* (Springer) pp 351–374
[47] Greensmith J, Aickelin U and Twycross J 2006 *Proceedings of the 5th International Conference on Artificial Immune Systems*, vol 4163 pp 404–417
[48] Bentley P J, Greensmith J and Ujjin S 2005 *Proc. of the 4th International Conference on Artificial Immune Systems, ICARIS, in: Lecture Notesin Computer Science* (Springer) vol 3627 pp 139–152
[49] González-Patiño D, Villuendas-Rey Y, Argüelles-Cruz A J, Camacho-Nieto O and Yáñez-Márquez C 2020 *Appl. Sci.* vol 10 515
[50] Semwal T and Nair S B 2020 *Applied Soft Computing* vol 86 105920
[51] Shrivastava K, Jha S S and Nair S B 2013 *Proceedings of Conference on Advances in Robotics,*

ACM pp 1-7
[52] Cserey G, Falus A, Porod W and Roska T 2004 *IEEE International Symposium on Circuits and Systems (IEEE Cat. No.04CH37512)* (Vancouver, BC) pp III-17

Detection of compromising radiation from modern data transfer interfaces using the example of high definition multimedia interface

A P Durakovskiy[1], L N Kessarinskiy[1], E A Simakhin[1]

[1] National Research Nuclear University MEPhI, Moscow, Kashirskoe hwy, 31, 115409, Russia

E-mail EASimakhin@mephi.ru

Abstract. Recently the technical channel of information leakage from electromagnetic emanation more and more interesting for specialists in the field of information security. At the same time, new versions of data transfer interfaces are being developed, that quickly distribute around the world. The use methods to detecting informative frequencies are based on the principles of operation of old data transfer interfaces, consider only the harmonics of the signal, excluding information about its type and interfaces that transmit information to. Accordingly, these methods to detecting informative frequencies of modern data transfer interfaces, such as high definition multimedia interface, do not reflect the completeness and may contain errors, which negatively affects the measurement stage. The article suggests a method to detecting informative signals of compromising emanation from modern data transmission interfaces, such as high definition multimedia interface, using an anechoic chamber, special test program and control program for measuring equipment. The proposed method will be useful both for educational purposes and for beginners in the field of technical information security.

Keywords: information security, compromising emanation, anechoic chamber, HDMI interface, LVDS interface

1. Introduction

In the processing of information data transfer interfaces emit informative electromagnetic waves into the surrounding space. Using special equipment and analysing the parameters of this radiation, it is possible to non-invasively intercept information and restore it [1], [2]. The emergence and widespread use of new technologies in information processing increases the risk of loss of privacy. In this regard, the information content of interfaces [3] and the complexity of information recovery [4] are analyzed, and situations of interception of informative signals of modern data transfer interfaces are modeled [5]. The characteristics of modern data transfer interfaces are analyzed in detail [6], new methods of information recovery are developed [7-8]. Extensive research in this field of information security shows the relevance of protecting information from its leakage through this technical channel. Specialists in technical data protection are particularly interested in works related to the practical detection of frequencies of modern data transfer interfaces.

2. Test signal

One of the main steps in analyzing the security of a data interface is to determine the frequencies at which it operates. To do this, specialists in the field of technical security information use specialized programs that transmit a predetermined sequence of data to the interface under study. To form such a sequence, it is necessary to accurately determine its characteristics.

2.1 Characteristics of the test signal

Due to the fact that the electromagnetic field in the conductor appears when the current or voltage changes, a periodic and time-stable test signal must be transmitted to clearly detect compromising radiation from the interface and configure the equipment. The result of the analysis is a set of security indicators that are numerically equal to the signal-to-noise ratio at each harmonic of the interface under study in a given frequency range. For the technical channel of information leakage from side electromagnetic radiation and interference, the signal-to-noise ratio for a pulse sequence in the frequency range is determined by the formula, the General form of which is given in [9]:

$$\Delta = \frac{E_S}{E_N} = \frac{\sqrt{\frac{1}{2\tau * F_C} \sum_{j=0}^{K} \left(E_{S+N}^2 - E_N^2\right)}}{\sqrt{\int_{\Delta F} E_N^2(f)df}}, \quad (1)$$

where F_C is a clock frequency, τ is a pulse duration, K is a number of harmonics in range, E_{S+N} is a "pure" signal and noise intensities, E_N is a noise intensities, f_h and f_l are high and low frequency range, $E_N^2(f)$ is a "dependence" of the noise field strength on the frequency in the range.

When conducting tests related to determining the maximum distance at which it is possible to restore information, the critical operating conditions of the interface under study are evaluated. Critical conditions are determined by the minimum level of noise or interference and the maximum signal level in the surrounding space of the interface under study. The minimum noise level is achieved by choosing the time of testing or using a specialized measuring area – an anechoic chamber.

Based on the formula (1), the maximum signal level is achieved if the duty cycle of the test periodic pulse signal $D = \tau * F_c$ is less than or equal to 1/2. Otherwise, when calculating security indicators, we will not get the maximum values. If the duty cycle of such a signal is less than 1/2, the pulse emission time will decrease, and therefore the amount of radiated energy will decrease, which contradicts the achievement of critical conditions. Accordingly, the duty cycle of the test signal should be equal to 1/2. In addition, this characteristic of the test signal shows an equally probable distribution of zeros and ones in the sequence transmitted to the interface input, and accordingly bring it closer to real operating conditions.

2.2 Generating a test signal

In accordance with the specification [10], signal transmission in the HDMI interface is based on transition-minimized differential signaling (TMDS) technology, where three channels (Ch0, Ch1, Ch2) transmit sub – pixel information, and the fourth channel is the clock channel (figure 1).

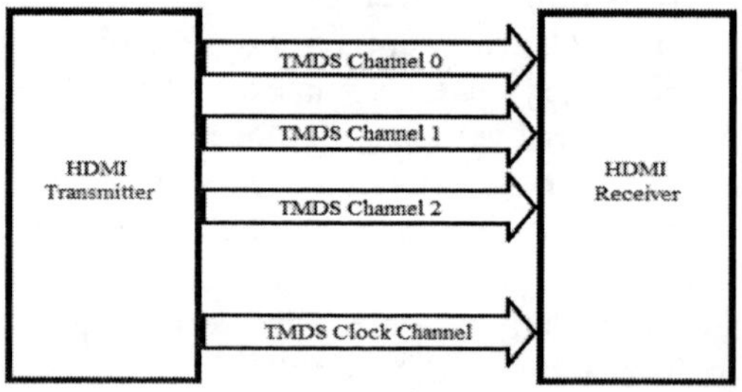

Figure 1. Block diagram for transmitting an informative signal via HDMI

To confirm the assumption in point 2.1, we used the Rohde Schwarz rtb2004 digital oscilloscope. Using an oscilloscope, the differences between the maximum and minimum voltage values of the transmitted signal were obtained in the channels of the HDMI interface of type "A", corresponding to the transmission of an informative signal with numbers: 1 and 3, 4 and 6, 7 and 9.

The signal transmitted via the HDMI interface is differential, so we can only evaluate 3 of the 6 channels corresponding to different sub-pixels, for example, channels 1, 4, and 7. If we refer to the RGB palette, the boundary colors for it are white - (255, 255, 255) and black - (0, 0, 0). When these colors alternate, we get the longest radiation time. The values of all three pixel components for white and black are the same, so we can only evaluate 1 channel, for example, the 4th (figure 2).

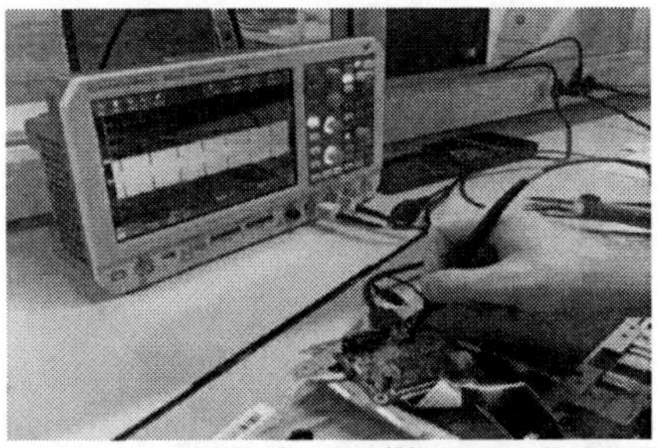

Figure 2. Working with the oscilloscope R&S RTB2004

The results of working with the oscilloscope are shown in table 1.

Table 1. The voltage difference in the test signal

Title	Image on the monitor display	Oscillogram	ΔV, mV
White screen			791,02
Black screen			561,52
Point-to-point	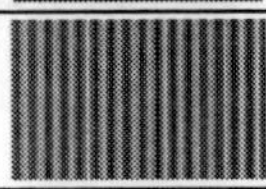		859,37
2 points through 2 points			830,47
4 points through 4 points			805,21
Half of the screen			66,48
3 lines		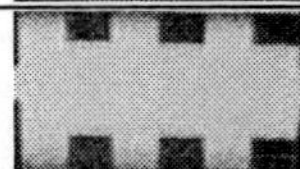	83,64
5 lines			80,78

A point-to-point test signal is a repeating sequence of one white pixel and one black pixel. A point is defined as a single pixel. Similarly for the "2 points through 2 points" and "4 points through 4 points" tests. Table 1 shows that the point-to-point test sequence has the largest voltage difference, and as the time between the back and front edges of the two pulses increases, the signal energy decreases.

Thus, it can be established that the point-to-point test has more relevant characteristics for determining the interface frequencies. To show the difference in the energy of test sequences that differ in characteristics from sequences with a duty cycle equal to 1/2, sequences that do not meet the requirements of paragraph 2.1 were additionally considered. As can be seen from table 1, their energy is much less. To create a test sequence we was written a C++ program using the Windows API.

3. Description of the detection method

Important criteria for detecting an informative signal for the researcher are the conditionally stable electromagnetic environment and the type of measuring device that can most accurately determine the presence of the signal. In real conditions of information leakage through side electromagnetic radiation, the attacker has more sensitive and expensive equipment and searches for information with minimal noise, for example, at night. To approach such conditions, it is necessary to use a specialized measuring area - an anechoic chamber (AC) for the electromagnetic frequency range.

3.1 Measuring area

In addition to the characteristics of the measuring area, the location of the object under study is also important. The location of the interface or computer is also important for actual operating conditions. To do this, place the monitor in the center of the turntable at a height of approximately 80-90 cm from the floor. The distance from the floor is selected based on the average height of the office desk (figure 3). Also need to exclude other monitors or computers from working, which will increase the accuracy of detecting the interface's own signal.

Figure 3. The location of the monitor

In addition, there should be no devices with the analyzed interface inside the AC, even if they are turned off. This is necessary because if the wavelengths of the conductor and the data interface match, the digital signal can be modulated by various generators in the computer. Thus, the measured energy of the signal from the analyzed interface may change, which will negatively affect the accuracy of the research results. The described location measures are also used to reduce noise.

3.2 Detection

In process of analyzing the specification of HDMI and the monitor architecture, we paid attention to the adjacent informative data transfer interfaces. Next, we will call an adjacent informative interface such that it receives data from the HDMI and carries the information necessary to restore the original signal. In the case of HDMI, the following is an informative interface with low-voltage differential signaling (LVDS). Due to spread spectrum modulation (SSM), the LVDS spectrum is broadband. The spectrum of the LVDS interface for different SPAN (figures 4 and 5).

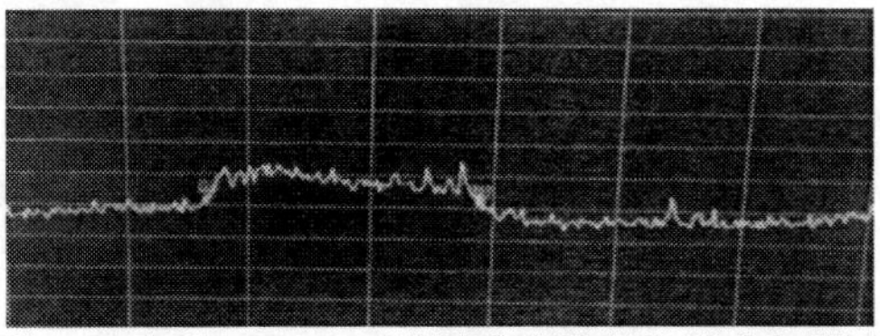

Figure 4. Spectrum LVDS (SPAN 40.0 MHz)

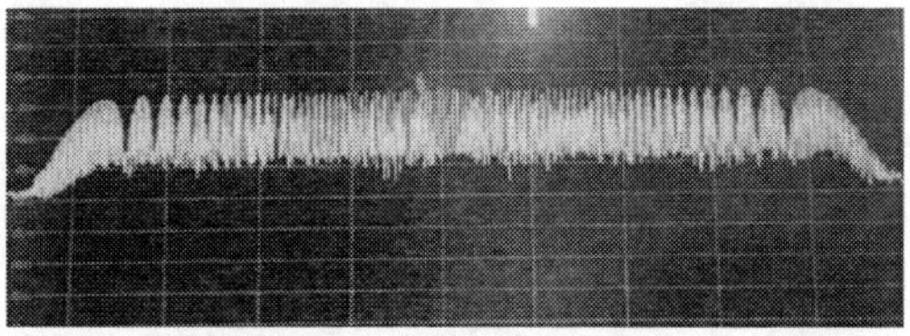

Figure 5. Spectrum LVDS (SPAN 15.5 MHz)

As the research has shown, the width of the LVDS spectrum increases depending on the harmonic number (figures 6, 7 and 8). Further, the LVDS harmonic will be called the central frequency of the spectrum. This does not contradict generally accepted terms, since before applying modulation transformations, the LVDS spectrum looks like the spectrum of a harmonic signal.

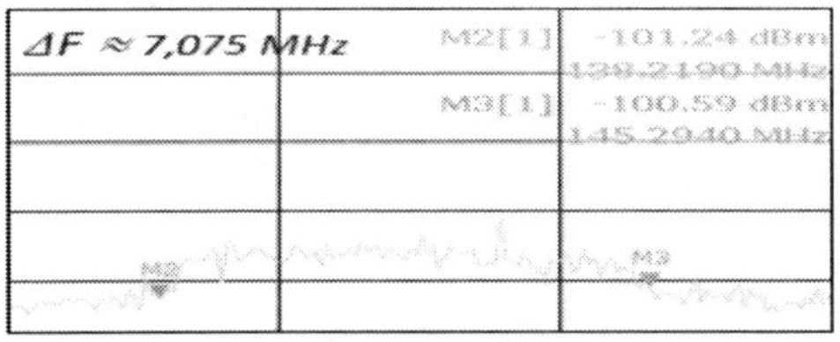

Figure 6. The 2nd "harmonic" of LVDS

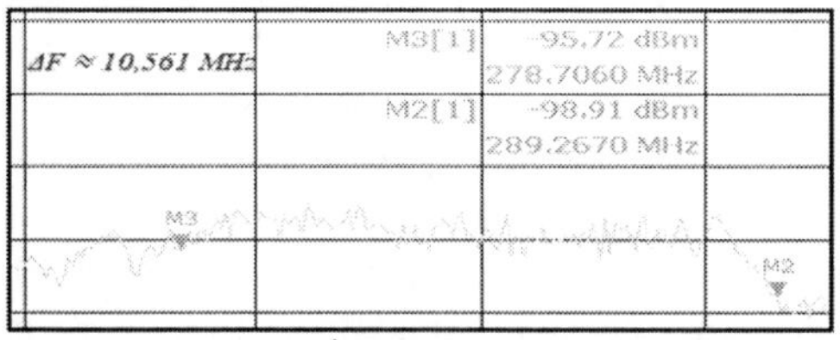

Figure 7. The 3rd "harmonic" of LVDS

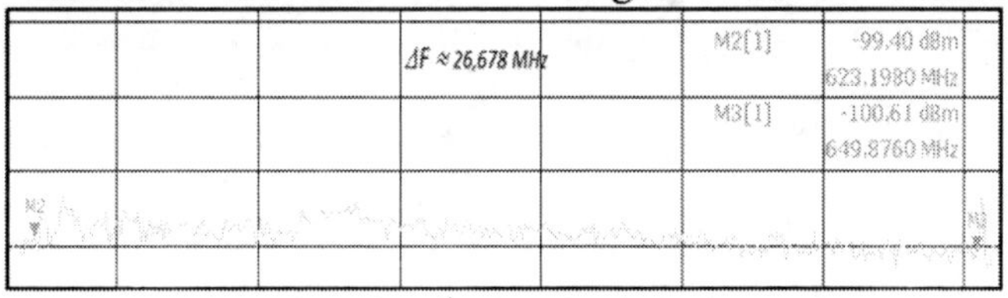

Figure 8. The 9th "harmonic" of LVDS

Also we revealed that the LVDS interface operates at the same frequencies as the HDMI interface. Thus, we observe an overlap of the spectrum of the two signals. And since the signal power from the LVDS interface is greater than the signal power from HDMI, during normal operation of the LCD monitor, we will always observe the spectrum of the LVDS interface on the measuring receiver. Thus, it will not be sufficient to simply use the receiving antenna to explicitly detect an informative signal from HDMI. To solve this problem, use a voltage probe for destructive testing or a current collector for non-invasive testing. In this case, destructive research that leads to a violation of the shell of the HDMI cable. After such an impact on the cable, it cannot be used for measurements as this is a deviation from the normal (factory) configuration of the interface and reduces the level of noise immunity. To avoid this, we can use a current collector that has a verification certificate for the search range. In this case, to measure the maximum value of the signal intensity, it is necessary to determine the wavelength. After the calculation, move the device along the cable at a distance of $\lambda/4$ from its beginning, where λ is a wavelength in meters. Based on this, the minimum will be at a distance of $\lambda/2$ (figure 9 and 10).

Also, to solve this problem, we can turn off the monitor by pressing the button on its case. Because of this, the transmitted frames will not be displayed on the monitor screen. However, the search for the informative interface of the frequency or measuring the intensity of the receiving antenna is also incorrect. We know that an electromagnetic field propagates through space with the property of attenuation or scattering. At short distances or in the near zone from the source, the electromagnetic field is weakened in proportion to the cubic function of the distance. The distance from the cable axis to the receiving antenna is greater than to the current collector housing. The difference in distance depends on the size of the receivers and the location of the receiving antenna. From this point of view, searching for dangerous frequencies using a current collector is the right solution.

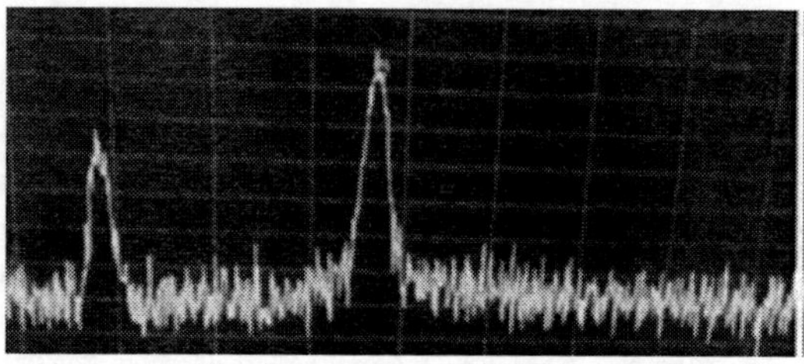

Figure 9. The intensity of HDMI signal ($\lambda/4$)

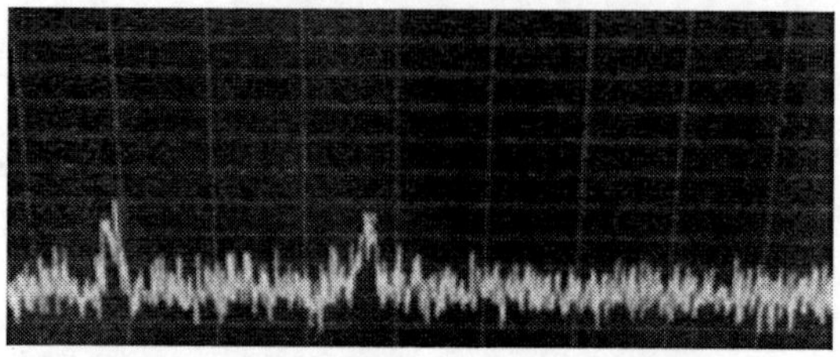

Figure 10. The intensity of HDMI signal ($\lambda/2$)

3.3 Experiment

The operating detection range from 9 kHz to 300 MHz was selected for the experiment. The Rohde Schwarz FSW 13 spectrum and signal analyzer with a frequency range from 2 Hz to 13.6 GHz was used as a measuring receiver. To search for informative frequencies of the HDMI interface, a TI2-3 current collector was used, operating in the range from 9 kHz to 300 MHz. The AI5-0 dipole antenna was used, operating in the range from 9 kHz to 2 GHz. An HDMI cable version 1.3 was used as the object of research. To control the measurement system, a program was developed in the graphical programming language "G" in the LabVIEW development environment. Using this program, we can control the measuring receiver in real time, automatically search for signals relative to noise and threshold values of intensity in a given frequency range, adjust the central frequency of the signal depending on the maximum intensity, automatically identify informative frequencies, measure the electromagnetic field strength at detected frequencies, and output a measurement report in the "*.docx" format.

The signal search is performed as follows. The first step in the search is to determine the clock frequency of the HDMI interface. Depending on the cable category, it may be different. We chose the category "A" cable, so the clock frequency is 74.25 MHz. Set the center frequency equal to the clock frequency and SPAN at least 5-10 MHz on the measuring receiver. Turn on the test signal, set the current collector to the initial position (next to the monitor connector). Then slowly move the current collector along the cable from the monitor, observing the increase in tension. To speed up the process of detecting informative frequencies, we used the reception of cyclic and sequential switching on and off of the test signal, changing the values of RBW and SPAN. After determining the expected frequency, go to the time domain (SPAN is 0 Hz) and send a test signal that is not typical for the interface, for example, 5 lines (figure 11) When the signal intensity is high, a special test for visual verification is not required, since the transmitted signal has the characteristic appearance (figure 12).

Figure 11. Example of not typical test signal

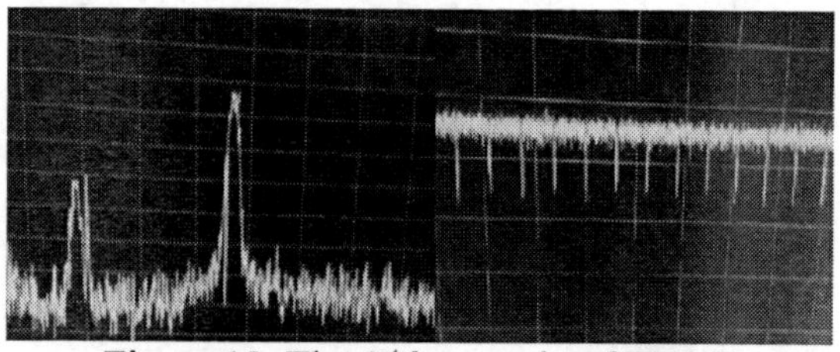

Figure 12. The 1st harmonic of HDMI

However, this method is relevant for detecting only the particular harmonic. To detect the others harmonic, we need to turn off the monitor, put the dipole antenna to the monitor connector, and select vertical polarization. Turn on the test signal and move away from the antenna at a distance of more than 1 meter to avoid interference. The accuracy and execution time of the control program search module depends on the number of samples in SPAN. A large number of samples increases the accuracy of detection, but also increases the detection process. To minimize the number of such samples, the center frequency correction mode was used. To do this, we check the small frequency region around the detected frequency equal to $2 * RBW$ and refine the central frequency. Then we run a selection of informative frequency, not using a typical test as a reference. The program compares the signal at each frequency and calculates the correlation coefficient, which determines the “informative” of the frequency.

4. Conclusion

This article describes a General approach to detecting compromising radiation from modern data transmission interfaces using the example of HDMI using a dipole antenna and a current collector, and provides experimental confirmation. It is also shown that the use of an anechoic camera, a test program written in C++ using the Windows API, and a special measurement receiver control program written in g in the LabVIEW development environment significantly increases the detection accuracy and speeds up this process. The proposed method will be useful both for educational purposes and for beginners in the field of technical information security. Further work will be aimed at automating the process of analyzing broadband signals and working with LVDS and RSDS interfaces.

5. Acknowledgment

The authors express their gratitude to the management of the NRNU MEPhI for organizing work in the anechoic chamber, the center for the development of automated systems and rental of measuring devices "Priboroteka" for providing specialized equipment and consulting high-level specialists.

References

[1] Eck W V 1985 Electromagnetic Radiation from video display units: An eavesdropping risk? *Comput. Security* vol 4 no 4 (Amsterdam: North-Holland/American Elsevier) p 269
[2] Wang L and Yu B 2012 Research on the compromising electromagnetic emanation from digital signals *Proc. Int. Conf. on Automatic Control and Artificial Intelligence(Xiamen)* p 1761
[3] Kuhn M G 2003 Compromising emanations: Eavesdropping risks of computer displays, *Technical report UCAM-CL-TR-577* (Cambridge: University of Cambridge Computer Laboratory) chapter 4 pp 77–84
[4] Kuhn M G 2013 Compromising emanations of LCD TV sets *Proc. IEEE Transaction on electromagnetic compatibility* vol 55 no 3 (London: IET) p 564
[5] Song T, Jeong Y and Yook J 2015 Modeling of Leaked Digital Video Signal and Information Recovery Rate as a Function of SNR *Proc. IEEE Transactions on Electromagnetic Compatibility* vol 57 no 2 (New York: IEEE) p 164
[6] Kubiak I and Przybysz A 2019 DVI (HDMI) and DisplayPort digital video interfaces in electromagnetic eavesdropping process *Proc. Int. Symposium on Electromagnetic Compatibility – EMC EUROPE 2(Barcelona)* (New York: IEEE) p 388
[7] Meulemeester P Scheers B and Vandenbosch G A E 2019 A quantitative approach to eavesdrop video display systems exploiting multiple electromagnetic leakage channels *Proc. IEEE Transactions on Electromagnetic Compatibility* vol 62 no 3 (New York: IEEE) p 663
[8] Meulemeester P Scheers B and Vandenbosch G A E 2020 Differential signaling compromises video information security through AM and FM leakage emissions *Proc. IEEE Transactions on Electromagnetic Compatibility* (New York: IEEE) p 1
[9] Rembovskiy A A Ashihmin A V and Kozmin V A 2006 Radio monitoring: tasks, methods, tools (Moscow: Hot line - Telecom) chapter 11 pp 444–451
[10] Hitachi, Matsushita Electric Industrial Co, Philips Consumer Electronics, Silicon Image, Sony Corporation, Thomson Inc and Toshiba Corporation 2006 High-Definition Multimedia Interface Specification Version 1.3a chapter 4 pp 10–48

Parallel implementation of algorithm for creating and verifying digital signature based on elliptic curves

P T Burtseva, L I Zelenina, L E Khaymina, E S Khaymin, I M Zashikhina

Northern (Arctic) Federal University named after M.V. Lomonosova, Arkhangelsk, Naberezhnaya Sev. Dviny str., 17, 163002, Russia

E-mail l.zelenina@narfu.ru

Abstract. The article discusses the use of parallel data processing in the implementation in the C ++ programming language using the Visual Studio integrated development environment of an algorithm for information protection and authentication. Data processing is based on the method of discrete logarithm in a group of points of an elliptic curve. The studied case has been developed as *Cost Optimization Program for an Oil and Gas Company* and includes creating an *Optimization Map* information system. A part of the system is a module generating and verifying a digital signature. Due to the use of parallel data processing and the ECDSA algorithm for digital signature creation, oil and gas company will be able to increase its working efficiency and information security.

Keywords: Information security, Electronic signature, Discrete logarithm, ECDSA algorithm, Parallel data processing, Company's work optimization

1. Introduction

Today, companies all around the globe strive to enhance their business potential and staff efficiency. Company managers, data security and R&D specialists, employ information technologies of the latest generation to ensure working processes' congruence, enterprise's safety and swift personnel reactions. One of the solutions is the use of a digital signature as a means of verifying and protecting information [1, 2, 3]. In Russia, to legalize the procedure of a digital signature use, a new standard for the electronic signature has been adopted, and the Law on Digital Signature has been issued. Western companies and regulatory agencies like Institute of Electrical and Electronics Engineers and National Institute of Standards and Technology (USA) adopted a new standard for digital signature ECDSA (Elliptic Curve Digital Signature Algorithm), based on the use of elliptic curves.

The digital signature of files or e-mail messages is performed following cryptographic algorithms that employ asymmetric keys, such as the "secret key", used for the signature itself, and the "public" key, used to verify another person's signature. The most famous schemes for creating an electronic digital signature are schemes like DSA, RSA, ElGamal, Rabin, Schnorr, Diffie-Lamporta [4, 5]. A special advantage of DSA scheme, which represents elliptic curve cryptosystems is that they provide significantly higher security with equal labor intensity, or, conversely, significantly lower labor intensity with equal resistance as compared to the other algorithms [6, 7]. That is why DSA or ECDSA has attracted special attention of the researches worldwide [8, 9, 10, 11].

For a signature to be solid, it needs to be based on a hard-to-calculate mathematical problem. One of such problems is the problem of discrete logarithm in a group of points of an elliptic curve. Application of methods of parallel data processing to an algorithm that implements labor-intensive stages of discrete logarithm methods allow for more efficient use of multiprocessor computing systems in the process of creating and verifying a digital signature [12]. The discrete logarithm problem, which determines the stability of cryptography algorithms, is complex; its solution requires rather large computing power. The larger the size of the initial data, the higher the complexity of the problem. Parallel algorithms that ensure effective use of multiprocessor computing systems in the process of creating an electronic signature can successfully implement time-consuming steps of mathematical methods of discrete logarithm [13, 14].

The study examines the case of parallel implementation of an algorithm for creating a digital signature using discrete logarithm in a group of points of an elliptic curve and its application to protect the data of optimization maps in an information system to balance the expenses of an oil and gas company. In this paper, we are studying the functionality of signing and verifying an electronic signature. We intend to present the results of an optimization map development in order to investigate the potential of cost optimization program in multiprocessor computing systems.

2. Information protection by means of digital signature, discrete logarithm problem and elliptic curve

One of the means of protecting information is a digital signature. This is the requisite of an electronic document obtained as a result of cryptographic information transformation, which allows checking the integrity of information in an electronic document. According to the GOST R 34.10-2018, it represents a string of bits resulting from the signature generation process. The discrete logarithm is a mathematical problem that uses an asymmetric encryption scheme. Asymmetric encryption scheme provides a basis for a digital signature. Its idea is based on the high computational complexity of the function inversion F_ (n, a) (x) = a ^ x mod (n).

An elliptic curve E over a field F is the set of points (x; y) whose coordinates belong to the field and satisfy the following cubic equation:

y ^ 2 + a_1 xy + 1a_3 y = x ^ 3 + a_2 x ^ 2 + a_4 x + a_6, a_i∈F.

Elliptic curves are used as a basis for constructing cryptosystems.

Since the initial standardization of Elliptic Curve Cryptography, significant progress has been made in terms of the efficiency and security of curves. Some elliptic curves allow faster calculations in modular arithmetic. As part of the implementation of the parallel algorithm, the Weierstrass elliptic curve P-256 can be used to create an electronic signature and its verification. This curve provides an adequate level of security and software efficiency.

3. Cost optimization program and optimization activity map

For the practical application of the algorithm, the information system *Cost Optimization Program for an Oil and Gas Company* was chosen. The software aims at the improvement of costs' regulation and increasing the efficiency of oil production processes. Within the *Cost Optimization Program for an Oil and Gas Company* information system, the section *Optimization Map* has been developed.

Optimization Map (OM) is a plan for finding and putting into operation the optimal solution in order to increase the efficiency of oil production processes.

The life cycle of *Optimization Map* is divided into 4 main blocks:

- Creation of *Optimization Map*, its correction;
- Negotiation of *Optimization Map*. At this stage, the map is signed using the digital signature;
- Approval and implementation of *Optimization Map*;
- Completion of the work cycle.

The matching block in the system is divided into several stages. At each stage, the official will work only on the part of the map which belongs to their field of responsibility.

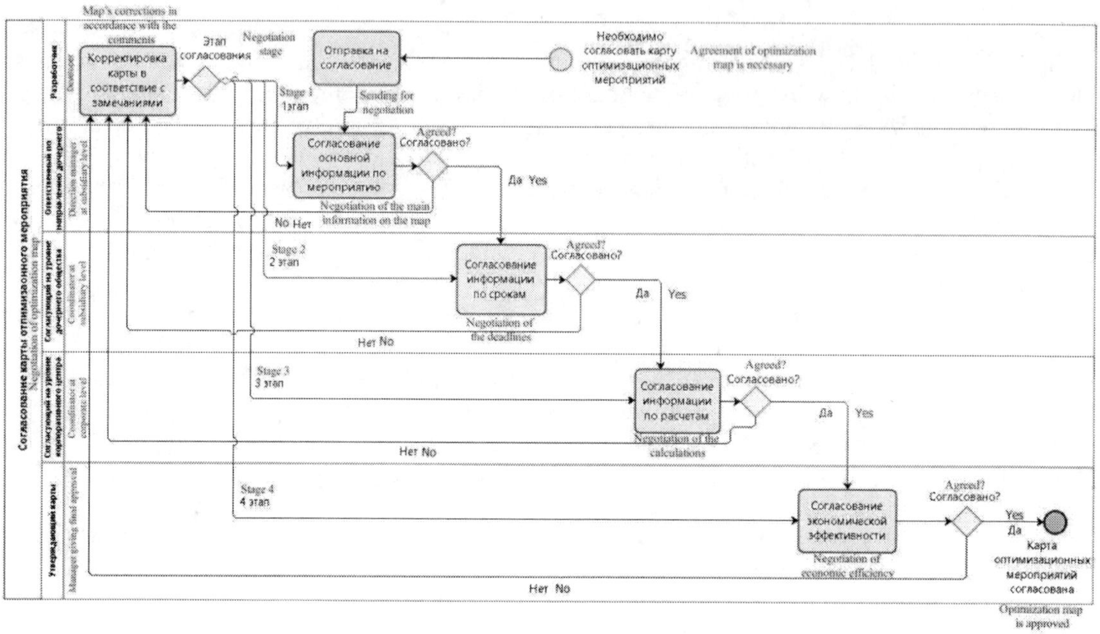

Figure 1. The process of *Optimization Map* approval

To implement the module for creating and verifying a digital signature, the ECDSA algorithm was chosen. The algorithm is based on the complexity of solving the discrete logarithm problem in a group of points of an elliptic curve.

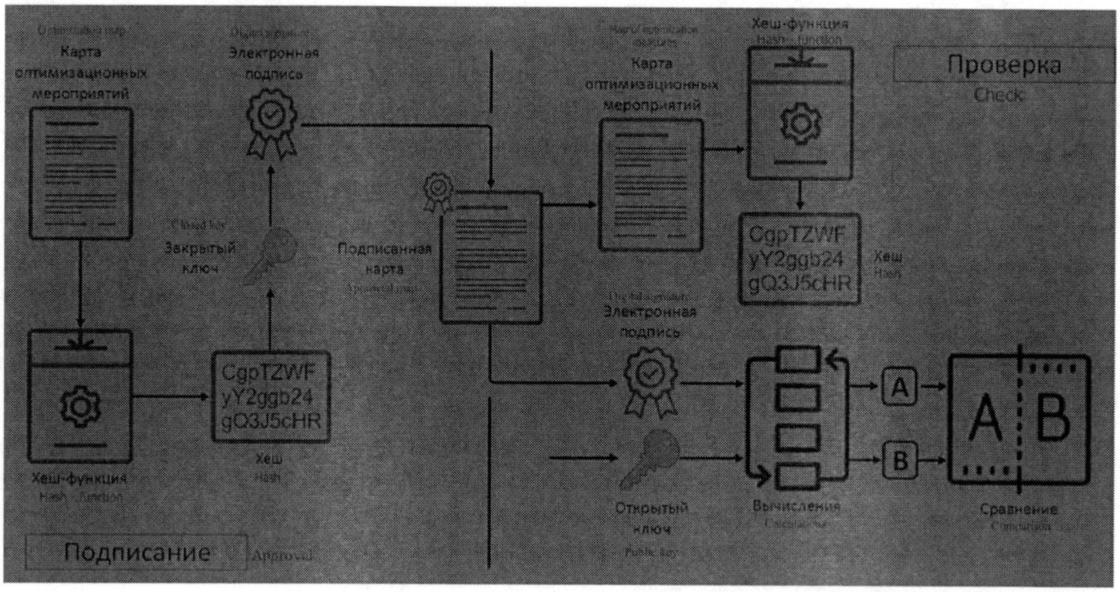

Figure 2. Scheme for signing and verifying a digital signature using the ECDSA algorithm

The following components are involved in the process of signing an electronic document (Figure 3):

IOP Conf. Series: Materials Science and Engineering 1069 (2021) 012027 doi:10.1088/1757-899X/1069/1/012027

- Database that stores records of the document is signed by the user. It includes the hash-sum of the signature, the date and information about the person who signed it. The rest of the system data is also stored in the database.
- Information system of cost optimization for an oil and gas company. In this system, the *Optimization Map* is edited and approved. The information system records information into the database. It also provides users with functionality for working and approving documents.
- Digital Signature Module is a dynamic package connected to the Creatio CRM system, written in the C ++ programming language, which implements the functionality of signing and verifying a digital signature in the process of the *Optimization Map* approval.
- CRM Creatio is a CRM system serving as a basis for the information system of Cost Optimization Program for an Oil and Gas Company development [15].

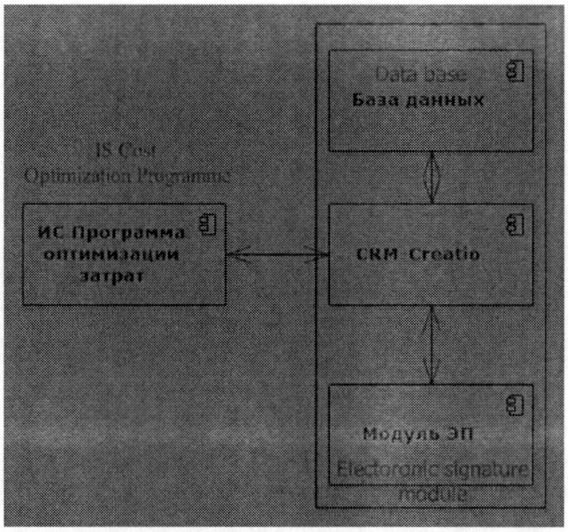

Figure 3. Diagram of *Cost Optimization Program for an Oil and Gas Company* system's components

4. Parallel implementation of the algorithm for creating a digital signature

As part of the development of an information system for cost optimization, the Service Creatio CRM solution is used. To implement the parallel algorithm, the Open Multi-Processing (OpenMP) library in C ++ in the integrated environment Visual Studio 2019 was chosen.

4.1 Description of the module that implements the functions of creating and authenticating electronic signature

For the processes of generating and verifying a digital signature, the signer must have a means of cryptographic protection of information containing public and private keys Q (xq; yp). The private key is stored on a cryptographic security facility, while the public key is known to all participants in the workflow and is stored in the database. As it is described in GOST R 34.10-2018, the variables used in the scheme for verifying and signing an electronic signature are:

V_l - the set of all binary vectors of length l bits;

V^*- the set of all binary vectors of arbitrary finite length;

M - signed initial data, M∈V l;

$h(M)$ - hash-sum of the original data, represented as binary vectors of length l bits

α - coefficient of the binary vector of the hash sum h;

q - the order of the cyclic subgroup of the group of points of an elliptic curve;

P - point of the elliptic curve of order q;

IOP Conf. Series: Materials Science and Engineering 1069 (2021) 012027 doi:10.1088/1757-899X/1069/1/012027

r and s are the values of the digital signature;
ν - electronic signature parameter;
d - the value of the signature private key;
Q - public key of the signature;
$\xi = (\bar{r} \| \bar{s},)$ – a concatenation of two binary vectors $\bar{r}$ and $\bar{s}$, corresponding to the values of the electronic signature r and s.

To obtain the values r and s of the digital signature for the signed initial data on the set of all binary vectors of arbitrary finite length (M $\in$ V ^ *), the algorithm for generating a digital signature is performed (view Figure 4). The input parameters of this process are the private signature key d and the signed initial data M; the calculated result is a digital signature. The method of parallel data processing is used to calculate the point of the elliptic curve and determine the value of the digital signature.

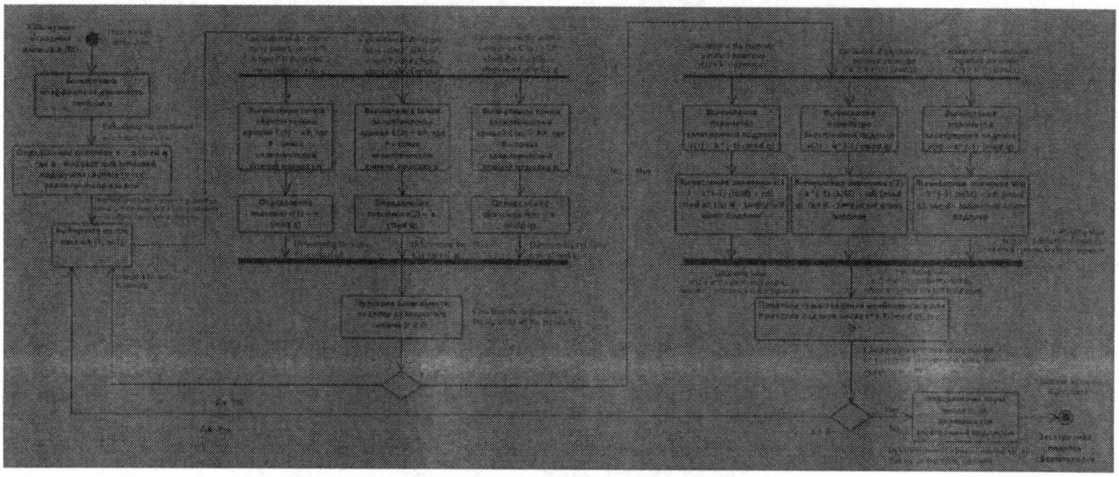

Figure 4. Diagram of the sequence of a digital signature generation

When verifying the digital signature ξ under the data M, the input parameters are the signed data M, the electronic signature ξ and the public key of the signature Q. The calculated result is the validity or unreliability of the signature. In this case, parallel data processing is used to calculate the coefficients determining the point of the elliptic curve. The sequence diagram of digital signature verification is shown in Figure 5.

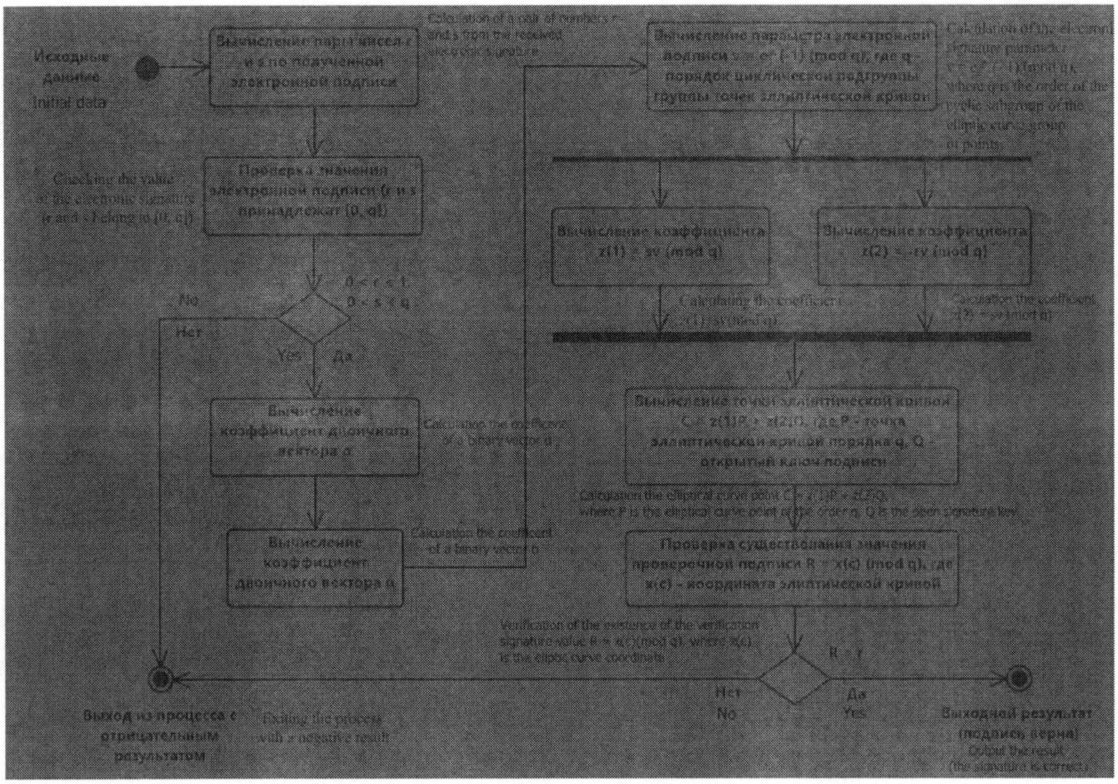

Figure 5. Diagram of the sequence of a digital signature verification

System interface

The section register is a list of entries for the *Optimization Map* block with brief information on each card.

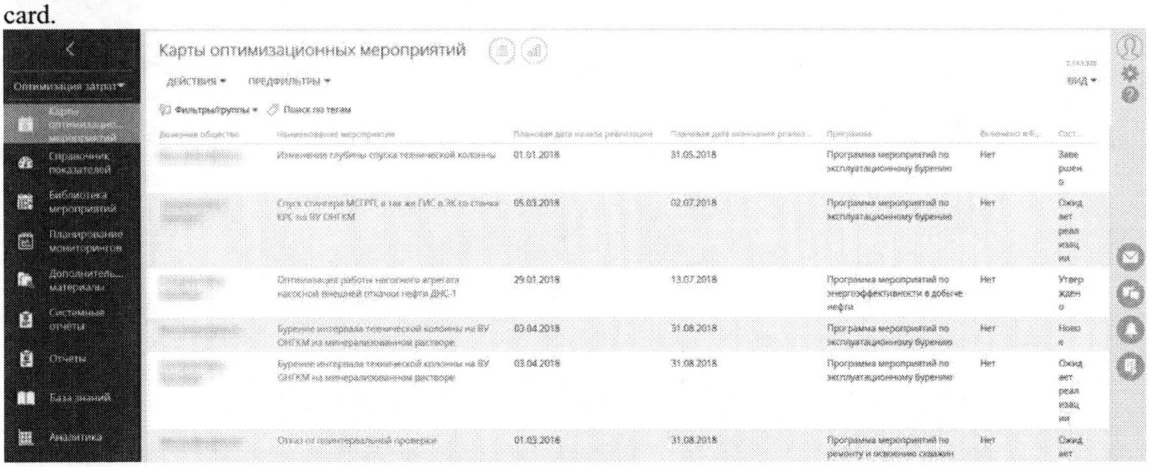

Figure 6. Register of the *Optimization Map* block

To go to the map of a specific event one has to select its title.

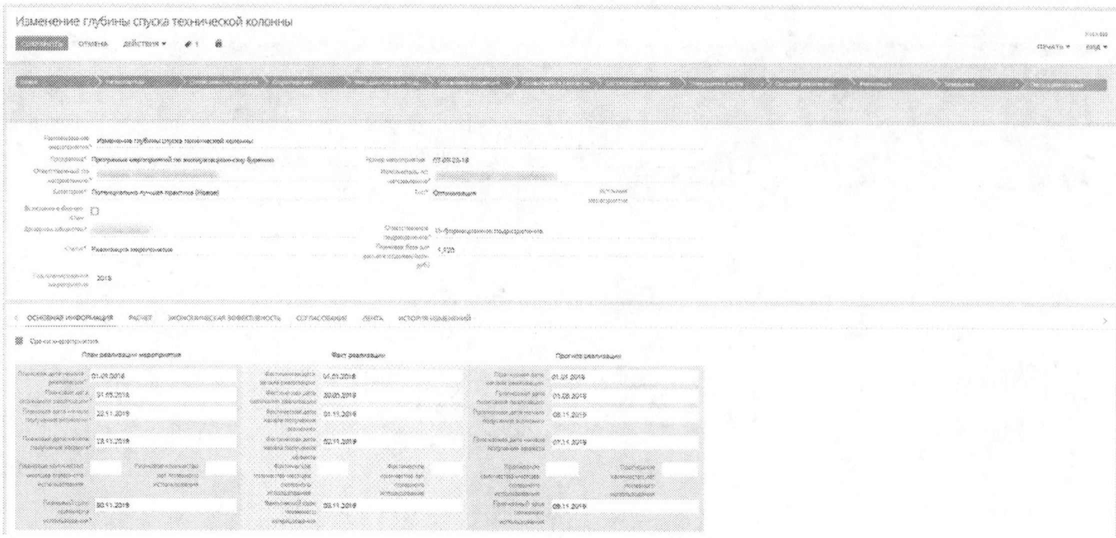

Figure 7. *Optimization Map* example

The *Optimization Map* is divided into several blocks: Basic information, Calculation, Economic efficiency, etc. A definite company manager signs one of the blocks.

Figure 8. *Calculation* tab

Each block is negotiated separately. Users who need to negotiate a certain block of *Optimization Map* have access to the buttons "Agree" and "Reject". These buttons enable them to place their digital signature. After clicking the "Approve" button, a modal window with the confirmation of the *Optimization Map* approval is displayed.

5. Conclusion

In this article, we have presented the results of developing *Optimization Map* section as a part of *The Cost Optimization Program for an Oil and Gas Company* information system. During the development process, we investigated the most efficient way for an oil and gas company's productivity. The developed *Optimization Map* system consists of repeating stages. One of the stages includes

Optimization Map's verification with a digital signature. *Optimization map* builds on the module that enables signing and verifying a digital signature. In this case, digital signature creation has been accessed with the use of the ECDSA algorithm. The chosen algorithm depends on the complexity of solving the discrete logarithm problem in a group of points of an elliptic curve. The method of parallel data processing has been used to calculate the point of the elliptic curve and determine the value of the digital signature.

The improved performance of a company is associated with the distribution of functions and responsibilities among its staff. In the presented scheme, a separate company's manager will take responsibility for a specific stage of the *Optimization Map* information system. The advantages of the present study are connected with the reduced work time of the company's employees and boosted potential of their activities. This has become possible due to the *Optimization Map* information system developed with the use of parallel data processing and the ECDSA algorithm for digital signature creation.

References

[1] Davidson A 2018 *Electronic signatures.* 10.1017/cbo9781316182796.016.

[2] Cano M D & Cañavate-Sanchez A 2020 Preserving Data Privacy in the Internet of Medical Things Using Dual Signature ECDSA. *Security and Communication Networks.* **1-9.** 10.1155/2020/4960964.

[3] Bohm N & Mason S 2019 Electronic signatures and reliance. *Amicus Curiae.* pp. 2-6. 10.14296/ac.v2017i110.5033.

[4] Aleshnikov S I & Gorbachev A A 2019 Methods of Algebraic Curves over finite fields in Public Key Cryptography and Coding Theory. *Nauka. Society. Defense.* **1 (18).** 10.24411/2311-1763-2019-10179

[5] Komarova A V, Menshchikov A A & Korobeinikov A G 2017 Analysis and comparison of electronic digital signature algorithms. GOST R. 34.10-1994, GOST R 34.10-2001 and GOST R 34.10-2012. In: *Cybersecurity Issues.* **1 (19).** 10.21581/2311-3456-2017-1-51-56

[6] Lamba S & Sharma M 2013 *An Efficient Elliptic Curve Digital Signature Algorithm (ECDSA)* pp. 179-183. 10.1109/ICMIRA.2013.41.

[7] Filat O & Yemelianenko T 2020 *Implementing Elliptic Cyptography to Create a Digital Signature.* 10.36074/05.06.2020.v3.30.

[8] Toradmalle D, Singh R, Shastri H, Naik N & Panchidi V 2018 *Prominence Of ECDSA Over RSA Digital Signature Algorithm.* pp. 253-257. 10.1109/I-SMAC.2018.8653689.

[9] Rybalov A N 2018 *On the Generic Complexity of the Discrete Logarithm Problem in Point Groups of Elliptic Curves Over Finite Fields. Application.* https://doi.org/10.17223/2226308X/11/41

[10] Jin H, He D & Jianhua Ch 2010 An IdentityBbased Digital Signature From ECDSA. *Education Technology and Computer Science, International Workshop* pp. 627-630. 10.1109/ETCS.2010.159.

[11] Mehibel N & Hamadouche M 2019 A New Enhancement of Elliptic Curve Digital Signature Algorithm. *Journal of Discrete Mathematical Sciences and Cryptography* **23.** 1-15. 10.1080/09720529.2019.1615673

[12] Trobec R & Kosec G 2015 *Parallel Implementation.* 10.1007/978-3-319-17073-2_8.

[13] Zhang J & Chen L Q 2009 *An Improved Algorithm For Discrete Logarithm Problem* chapter 2 pp 658-661. 10.1109/ESIAT.2009.457.

[14] Saxena S & Kishore N 2017 PRDSA: Effective Parallel Digital Signature Algorithm for GPUs. *International Journal of Wireless and Microwave Technologies.* **7.** 14-21. 10.5815/ijwmt.2017.05.02.

[15] *CRM-system Creatio.* Accessed online: https://www.terrasoft.ru/page/crm-products/ [14.06.20].

Designing a biometric access control concept

A A Troshkov, S V Bogdanova, A N Ermakova, V E Rachkov

Stavropol State Agrarian University, Stavropol, Zootechticheskiy ln., 12, 355000, Russia

E-mail troshkov1954@mail.ru

Abstract. The current legislative and regulatory database of biometric characteristics for the security of information technologies is not informative and has an imperfect system. The main disadvantages include: the lack of system and methodology for building biometric identification / authentication for managing access to protected information resources; insufficient knowledge of all possible biometric characteristics of a person; an imperfect model of the biometric system; a static approach to assessing access control. Therefore, there is a need to develop a biometric identification/authentication system to restrict access to information resources in various areas. Accordingly, we propose to systematize the biometric system for protecting and managing access to information.

Keywords: Encoding, encryption, access control, biometrics.

1. Introduction

Digitalization of society leads to changing the functioning of systems for storing, processing, receiving and transmitting useful information.

However, the increase of the information flow, storage resources require security of information that is regulated by normative-legal act of the security policy (SP).

The security policy is controlled by the governing bodies of enterprises of all ownership forms. An example of information security form and information protection is:

- order (algorithm) of access to resources;
- monitoring access to resources;
- vesting (expanding) the involvement and authority of management;
- segments distribution of information resources.

2. Model of security access control

Implementation of the system information security policy is the establishment of regulatory requirements for the composition and execution of various security mechanisms. This can be achieved by setting high requirements for software, encoding systems (encryption), and organizational and technical measures. Requirements for the functioning dynamics of the information resources describe various rules for identifying and authenticating users and registering events that occur in this area. The highest part of information security mechanisms is the access control and localization system for various types of illegal access. Currently, the information security policy in the field of user access is implemented by security personnel – access control specialists (Figure 1).

Personalized, account information, configuration parameters and encryption codes belong to the highest levels of trusted objects. The objects must also have the required security. Trusted interfaces and a certain number of officials are used for studying, reading, applying, changing and input / output.

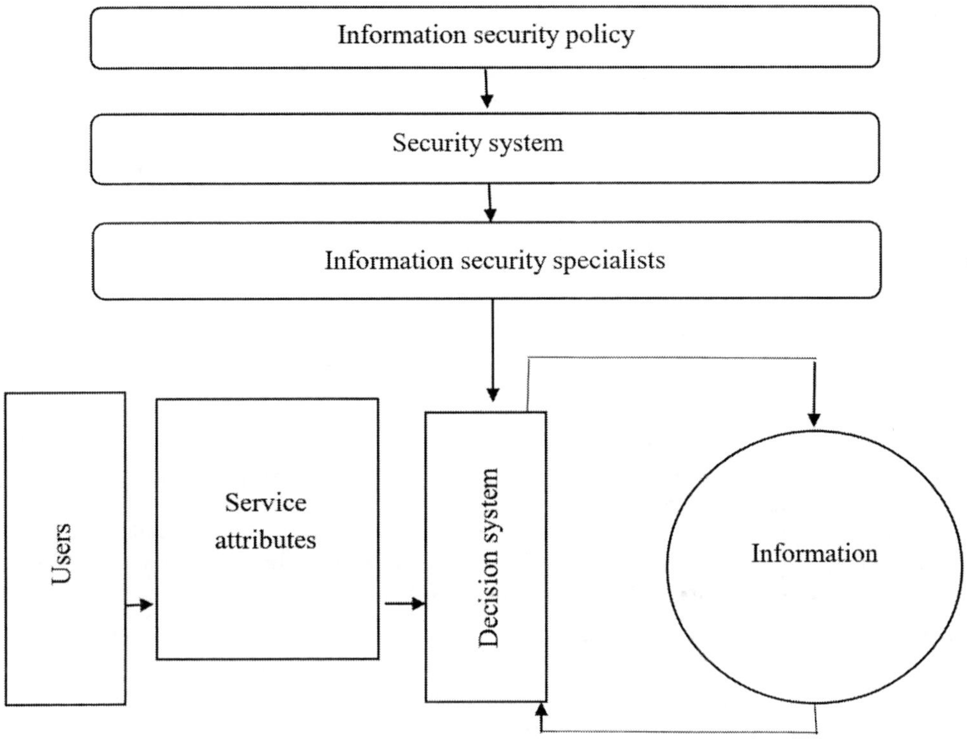

Figure 1. Managing subjects access to information resources

Figure 2 shows the designed access control model for certain special security services.

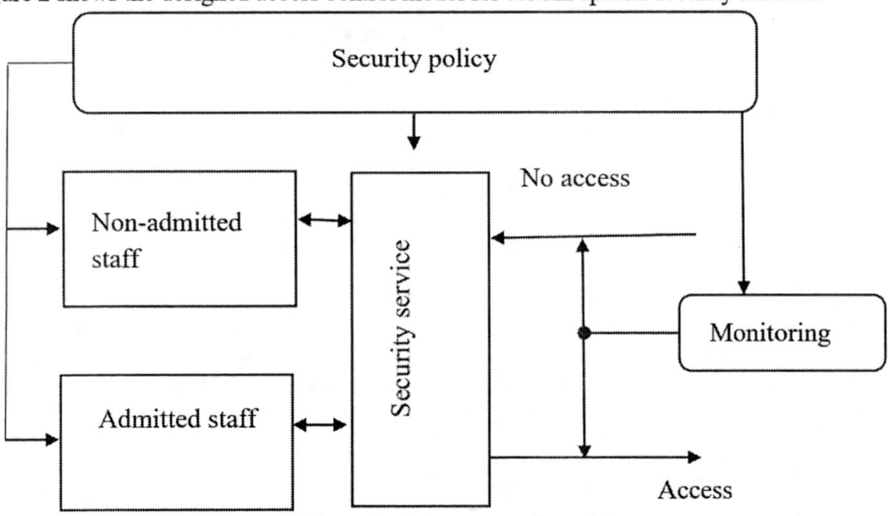

Figure 2. Access control model

The algorithm of model functioning (Figure 2) is based on the rules of access control located in the security policy plan. Since the information resource is decomposed into segments, the segment required for access is set (or personalized) for the users.

For the design of the access matrix, individual or group placement is applied (Figure 3).

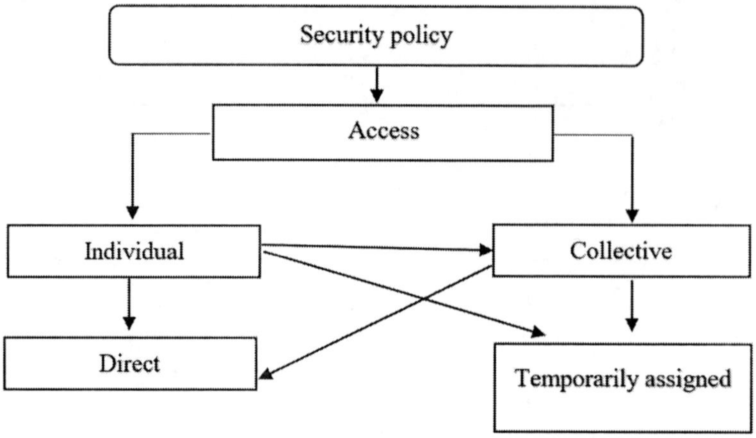

Figure 3. Access matrix

For the formalized algorithm, access control lists approved by the manager are used. Each ACL element contains rules for accessing a segment of an information resource.

However, access decisions in various scientific sources show a lack of resistance to illegal access. To solve this scientific and technical problem, we propose to involve the introduction of biometric characteristics (BC) of the user into the procedure of functioning of the identification system. In particular, to improve the access matrix by adding BC to individual and group permissions (Figure 4).

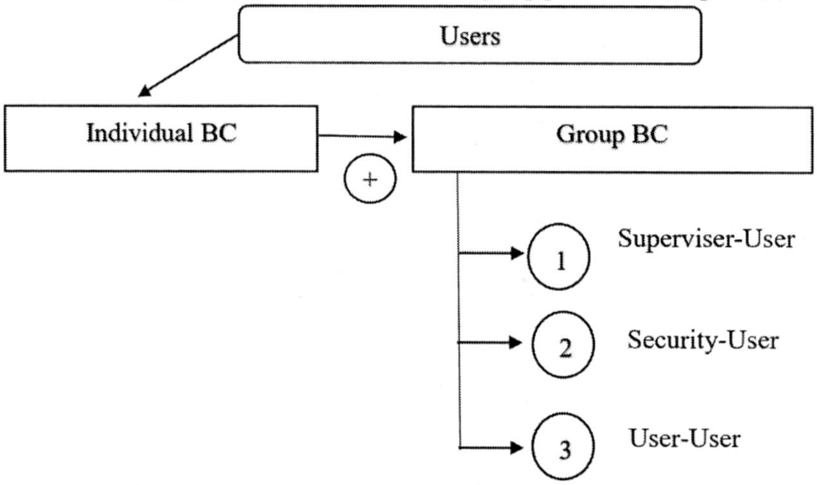

Figure 4. Improved user matrix

3. Functioning of group biometrics system

Figure 4 demonstrates conceptual proposals which are made for the use of the following methods for managing access to information resources in a group BC:

1 Σ BC " Manager-user»

2. Σ BC "Security-user»

3. Σ BC "User-user"

Analysis of work in the field of security mechanisms and services indicates that such proposals were not previously available. Preliminary calculations show that the characteristics of RAR and RAF have a dynamic improvement when using an improved user matrix.

BC grouping for user access allows to create an attribute ΣBC with high distinctiveness of identification by rules:

- user group – with individual BC with assigned priorities;
- creation of groups – with a dedicated list where the personal data of group members are listed, including other groups;
- all groups are part of the administrative information cloud.

Segments of information resources are defined according to the user's mandate. To accommodate collisions, a BC or BC group are allocated, which describe the priority in the cloud (Figure 5).

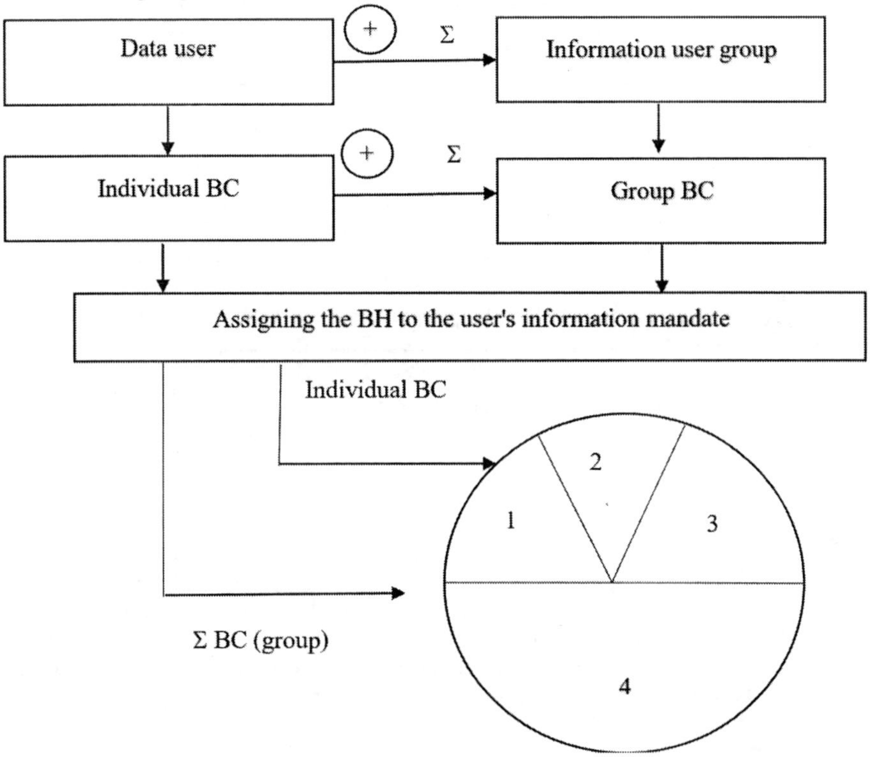

Figure 5. Location of collision and allocation of priorities

Figure 5 highlights the presence of rules: if they contradict each other, priority is given to the decisive rule (according to the user's mandate), which determines more specific access in accordance with the permission to access certain segments of information. For more convincing, the user can provide the subject authentication protocol (if necessary). Using such proposals and models, it is possible to organize work with information of various levels of secrecy, which eliminates the mixing of information of different labels and categories, which was previously known as a lack of access control system operation.

For example, a generalizing algorithm for the representation of group BC in Figure 6 is given. From this figure we form a biometric system:

S-u (Supervisor - user)
S - u (Security -user)
U-u (User-user)
From the system, we design the following algorithm (Figure 6)

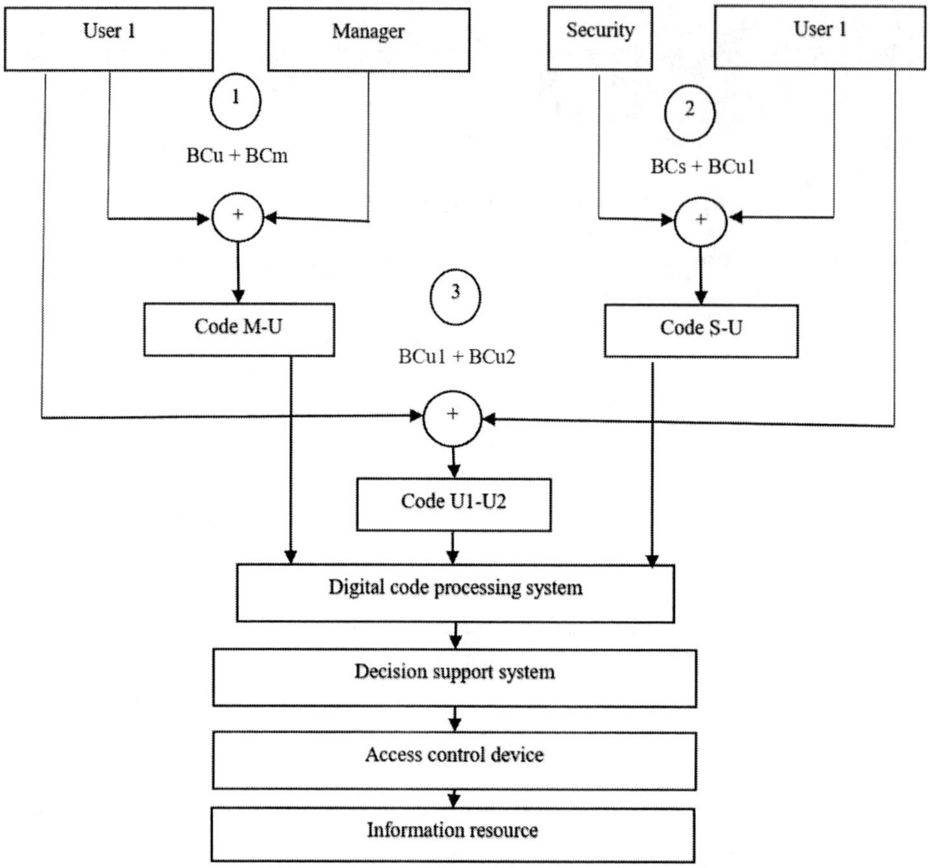

Figure 6. Algorithm of the system functioning

Summation (+) of BC allows you to increase the stability of coding, improve the quality of identification, despite its complexity.

4. Conclusion

The presented biometric concept makes it possible to form a dynamic algorithm for the functioning of access to information resources, to carry out the distribution of powers in the biometric system. In accordance with the code, the distribution of segments of information resources is proposed. The application of the presented concept allows you to form an access matrix in the information security policy of an enterprise of various forms of ownership, correctly distribute the user's mandate for access control.

References

1. Wasserman F 1992 Neurocomputer technology: Theory and practice. Per. s Engl. M.: Mir
2. S Burnet, S Payne 2002 Cryptography. Official guide to RSA Security Binom-Press p 384
3. Khoreyev P 2009 Methods and means of information protection in computer systems: a textbook *Academy* p 256
4. Alferov, A, Zubov A, Kuzmin A, Cheremushkin A 2009 Fundamentals of cryptography: a textbook Moscow: Helios ARV p 245

5. Troshkov M, Troshkov A, Shuvaev A 2016 Protection of business information in the field of agro-industrial complex management *Scientific and practical journal "Bulletin of the Stavropol agro-industrial complex.* **2** 22/1 pp 135-139
6. Troshkov A, Zhuk A, Rachkov V 2016 Modern threats and methods of information protection in information systems. Collection of scientific papers on the materials of the all-Russian NPC: "Institutional transformations of national economic systems" Stavropol, Stgau, pp 78-85.
7. Troshkov A, Troshkov M 2017 Synthesis of a biometric key for the formation of a cryptographic code that controls access to information resources *Collection of scientific articles based on the materials of the International scientific and practical conference* pp 224-228
8. Norman T L 2017 Foundational Security and Access Control Concepts Electronic Access Control pp 21-42
9. Poniszewska-Maranda A 2012 Role Approach in Access Control Development with the Usage Control Concept Proceedings of the 11th IFIP TC 8 international conference on Computer Information Systems and Industrial Management
10. Bhalla A 2020 The latest evolution of biometrics Biometric Technology Today 8 pp 5-8

Application of face recognition technologies in access control and management systems

A N Marenkov, T M Gelagaev, V Yu Kuznetsova

Astrakhan State University, Astrakhan, Tatischeva str., 20A, 414056, Russia

E-mail: arhelia@bk.ru

Abstract. The article shows the relevance of the widespread use of computer vision technology on the example of face recognition as part of the access control and management system. The scheme of access control implementation with face recognition technology is described. The use of this technology makes it possible to increase the level of information security of enterprises and, as a consequence, reduce the possible financial damage from the implementation of attacks on their assets from illegitimate penetration into the protected area through the access control system using passes of legal users.

Keywords: security; access control and management system; face recognition; neural networks; safety.

1. Introduction

The use of computer vision in the field of security provides a huge impetus for the improvement of access control systems (ACS). Representatives of medium and large businesses that use access control systems at their facilities provide information that every fiftieth passage through the turnstile is carried out in violation of security requirements - to enter the territory with limited access, cards of legal users or fake access cards are used. Face recognition technologies are systematically introduced into many technological processes, including security systems. ACS with the use of face recognition technology helps to detect situations when an attacker uses a legal user's card for passage, by comparing the face of the cardholder from the organization's database with the portrait of the one who is trying to enter the protected area. Improving the access control system in this way will increase the level of information security of enterprises and, as a result, reduce the possible financial damage from the implementation of attacks on their assets. In addition, contactless identification of users by recognizing their faces is relevant in an unfavorable epidemiological situation.

2. The principle of operation of a classic ACS

The access control and management system is one of the most popular and effective systems for protecting areas with limited access. ACS restricts the passage to the protected area, while not interfering in any way in the business processes of the organization. In addition, the system monitors the movements of employees within the organization and records their hours worked, which contributes to violations of the work schedule.

Main functions:

- restriction of access to the premises of the protected object;

- keeping time records of working hours for each employee;
- fixing the time of arrival and departure of visitors;
- personal and temporary control over the opening of internal premises;
- control over the movement of employees around the facility;
- registration and notification of cases of attempts to enter the protected premises;
- integration and interaction between video monitoring and security and fire alarm systems.

The classic principle of ACS operation is as follows: it is an access system based on key cards with RFID tags.

The abbreviation RFID is derived from the term Radio Frequency Identification, which means radio frequency identification in Russian. An RFID tag consists of three components:

- a chip that stores identification information and is responsible for communication with the reader;
- Antenna that allows information to be transmitted between the tag and the reader;
- shell or body.

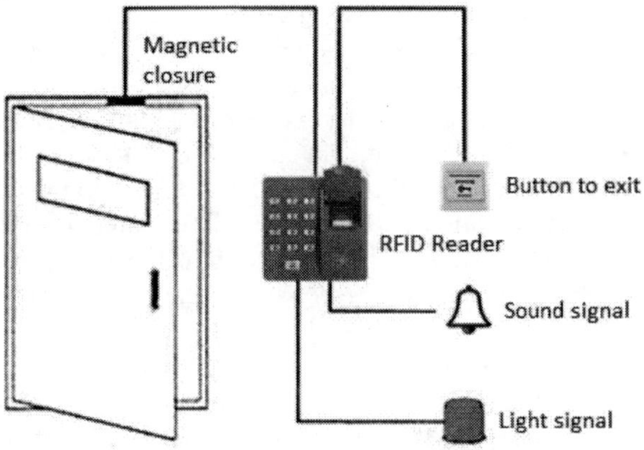

Figure 1. The classic scheme of access control at the enterprise

Currently, RFID systems can be found absolutely everywhere, starting with public transport (paying for travel by means of a card tied to a user's account), staying at a hotel (a smart card is a key to enter the room), ending with a foreign passport (new generation), which can be obtained with an electronic contactless RFID chip [3].

Practice shows that this type of access control has a number of disadvantages that can provoke vulnerabilities associated with unauthorized access to the information resources of the protected object. The most striking example of such a drawback is the use of someone else's cards to enter a protected area, for example, this is possible if an employee has forgotten the card of the house and asked his colleague to "lend" him a card for the passage, or, worse, an attacker used a lost or stolen card to pass through.

Also, a significant problem for ACS is the widespread cloning of keys and access cards. In the press and on ad sites, you can often see information about services for cloning access keys for anyone, for example, door phone users often turn to such a service. However, the use of such keys in an access control system significantly reduces the effectiveness of ensuring security in an organization. Many access control systems do not prevent the simultaneous use of several identical keys or access cards. The cloned key not only allows access to the enterprise but also opens all internal doors that were allowed for the authentic key.

Most often, this problem is solved by installing biometric scanners (fingerprints, retina, etc.), however, in an unfavorable epidemiological situation, these systems are becoming less in demand. At the same time, contactless fingerprint reading systems show poor performance. According to a report by the American National Institute of Standards and Technology (NIST), the accuracy of contactless devices when recognizing one finger was relatively low - only 60-70%.

IOP Conf. Series: Materials Science and Engineering 1069 (2021) 012029 doi:10.1088/1757-899X/1069/1/012029

3. Face recognition for security purposes

Face recognition is one of the most promising methods of biometric contactless identification of a person by face. According to analysts from Future Market Insights, the global market for contactless biometric technologies will grow at an average annual rate of 17.4% from 2020 to 2030 and will reach $ 70 billion by 2030. It is expected that contactless technology will be more in demand due to the coronavirus pandemic in the world and surface hygiene problems such as contact fingerprint scanners [4].

Face recognition is the process of matching the facial images of people caught in the camera lens with photographs from a database of previously saved facial images of reference standards, for example, employees of an organization. According to the structural implementation of the face recognition system, there are 3 schemes:

Analysis of the video stream on the server - the IP camera sends the entire video stream to the server for processing and analysis. On the server, specialized software searches for a face in the video sequence and compares the images of faces obtained from the video stream with the base of reference faces.

The disadvantages of such a scheme will be high network load, high cost of the server, even the most powerful server can be connected to a limited number of IP cameras, i.e. the larger the system, the more servers. The advantage is the ability to use an existing video surveillance system.

Analysis of the video stream on the IP camera - the images will be produced on the camera itself, and the processed metadata will be transmitted to the server.

Disadvantages - special cameras are needed, the choice of which is extremely small at the moment, the cost of cameras is higher than that of conventional ones. Also, in systems of different manufacturers, the issue of storage and size of the database of recognized faces of standards will be solved in different ways, as well as issues of interaction between the software on the camera and software on the server.

Advantages - connection of an almost unlimited number of cameras to one server.

Analysis of the video stream on the access control device - the camera is built into the access control device, which, in addition to face recognition, which occurs on the device, performs the functions of access control, usually through a turnstile or an electric lock installed on the door. The database of reference faces is stored on the device, and usually no longer in the form of photographic images.

Disadvantages - as a rule, all such devices are produced for indoor use. Advantages - Low cost of systems compared to video surveillance systems used for face recognition.

As a rule, all the proposed methods of face recognition are mainly implemented using 2D images, since Despite the development of three-dimensional models, the base of such standards is still rather scarce, and the equipment for organizing this kind of recognition is expensive.

Analysis of face recognition approaches. Despite the variety of existing approaches, it is possible to define a general algorithm for face recognition.

The main stage of the described process is the face recognition itself, which is ensured by calculating features and identifying the similarity or dissimilarity of photographs. Let's consider several methods that can be applied in the face recognition algorithm.

Flexible comparison method on graphs. The essence of this method is reduced to graph matching. Faces are represented as graphs with weighted vertices and edges. At the stage of recognition, one of the graphs - the reference one - remains unchanged, while the other changes in order to best fit the first. In such recognition systems, graphs can represent both a rectangular lattice and a structure formed by characteristic (anthropometric) points of the face. The difference (distance) between two graphs is calculated using a certain price deformation function that takes into account both the difference between the values of the features calculated at the vertices and the degree of deformation of the edges of the graph [5].

Neural networks. Currently, there are about a dozen types of neural networks (NN). Neural networks are trained using a set of training examples. The essence of training comes down to adjusting the weights of intraneuronal connections in the process of solving an optimization problem using the gradient descent method. In the process of learning the neural network, key features are automatically extracted, their importance is determined and relationships between them are built. It is assumed that a trained NN will be able to apply the experience gained in the learning process to unknown images due to generalizing abilities. The best results in the field of face recognition (according to the results of the analysis of publications) were shown by the convolutional neural network [7].

Hidden Markov models. One of the statistical methods of face recognition is hidden Markov models (HMM) with discrete time. HMM uses the statistical properties of signals and takes into account their spatial characteristics directly. The elements of the model are a set of hidden states, a set of observed states, a matrix of transition probabilities, an initial probability of states. Each has its own Markov model. When recognizing an object, the Markov models generated for a given base of objects are checked and the maximum observable probability is sought that the sequence of observations for a given object is generated by the corresponding model.

4. Prototype development

In the developed prototype, it is proposed to use a neural network, which will be trained with a teacher. The neural network will be trained on two images, where the result of comparison will be true or false (the person in the photographs is the same person or not). In addition, face recognition can be carried out using standard libraries [9].

The teacher creates a dataset containing at least two photographs of one person from many others [9].

Reasons for choosing this approach:

- the neural network is retrainable;
- the neural network can be "retrained" by adding new data sets;
- a trained neural network gives quick answers (within the framework of the problem being solved "similar / not similar");
- you can always increase or decrease the similarity threshold.

The following structure of an ACS prototype with face recognition technology is proposed:

- Arduino Mega to control RFID readers and turnstiles;
- RFID readers;
- IP cameras for capturing images;
- a computer with the ACS application installed;
- server with database.

Arduino Mega. Arduino electronic devices have long established themselves in the programmable electrical engineering market as a high-quality, multifunctional and inexpensive product. Therefore, in this project, this electronic device of the Mega model will be used, which in turn has 256 KB of flash memory and 8 KB of RAM in its arsenal.

RFID reader. These readers are needed to read UID (User identifier) from smart cards.

IP camera. For better system operation, you will need IP cameras with certain characteristics:

1. The presence of WDR (Wide Dynamic Range) - this parameter that affects the illumination.
2. The number of frames per second - the more frames per second the camera takes, the higher the probability that the camera will take the desired picture. This task requires a minimum of 20 frames per second
3. Resolution of the video surveillance camera - higher resolution will provide higher detail. High resolution will have a positive effect on the process of identifying markers by the neural network in the resulting image.
4. Varifocal lens - a lens where it is possible to change the focal length.

A computer. Minimum requirements for the normal operation of the ACS application:

- processor 2 cores, frequency 1.1 GHz;
- Windows 7 and newer;
- RAM 2 GB;
- installed .NET Framework 4.8;
- 150 MB of free space.

Server with a database. The server must have SMSS from Microsoft installed.

The conceptual model of the ACS operation is shown in Figure 2.

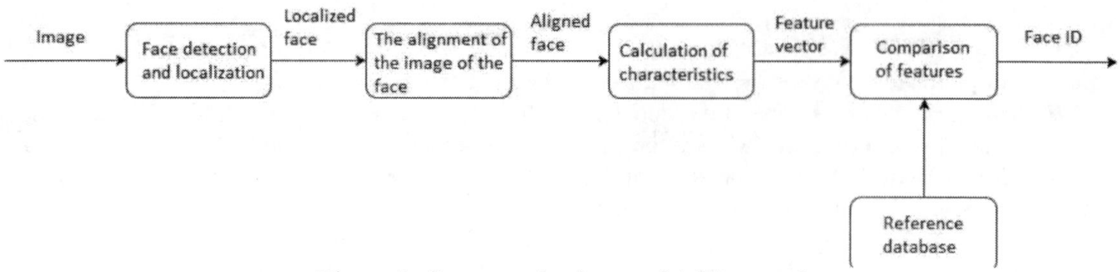

Figure 2. Conceptual scheme of ACS operation

5. The scenario of using ACS with face recognition technology

The owner of the smart card must bring the smart card to the reader. The reader receives the UID of the presented card and sends it (the UID of the smart card) to the Arduino electronic device. The Arduino, in turn, sends the UID of the read smart card and the ID of the reader from which the UID was received to the computer (with the ACS application installed) via the COM port. On the computer, the ACS application sends a request to the database in order to check the presence of this smart card in the database, if the UID of the smart card does not exist, then we have an attacker. When a card is detected, an image bound to the UID of the read smart card is unloaded from the database. The ACS application analyzes the coincidence of the face image obtained from the database with the image obtained from the IP camera. If the probability of a match is below a certain predetermined threshold, it means that we are facing an attacker and the system will deny him access. If the probability of image coincidence is higher than the threshold value, the ACS application will send the turnstile ID and the "open" command to the Arduino electronic device, and the Arduino, in turn, will send the "open" command to the required turnstile.

The block diagram of the proposed solution is shown in Figure 3.

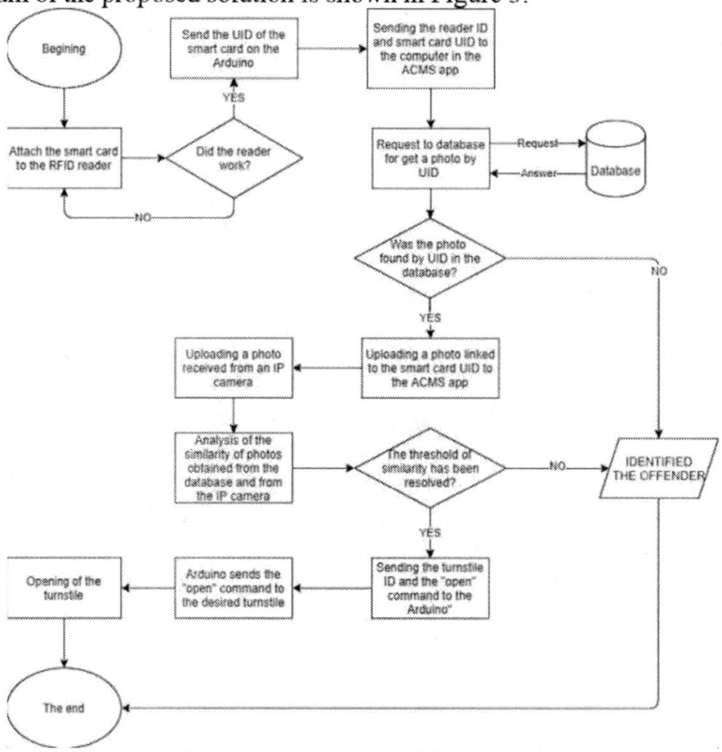

Figure 3. Flowchart of the process

6. Conclusion

- Face recognition technologies are systematically introduced into many technological processes, including security systems. ACS with the use of face recognition technology helps to detect situations when an attacker uses a legal user's card for passage by comparing the face of the cardholder from the organization's database with the image of a person trying to enter the protected area.
- The paper proposes a general scheme of the access control and management process using face recognition technology and demonstrates a conceptual diagram of the proposed hardware solution.
- Improving the access control system will improve the level of information security of enterprises and, as a result, reduce the possible financial damage from attacks on their assets.
- Contactless identification of users by recognizing their faces is relevant in an unfavorable epidemiological situation.
- Potential users of this system can be both government and commercial organizations interested in ensuring the confidentiality regime on their own territory.

References

[1] Vorozheikin M R, Chernova S V 2019 *General information about the face recognition system* (Publishing house "Sputnik +" № 47) p 35-37.

[2] Rudinskaya E A, Paringer R A 2019 *Development of a face detection algorithm using combinations of Haar cascades* (Proceedings of ITNT-2019 Publishing house: Novaya tekhnika) p 6-12.

[3] RFID tags - the ultimate guide to choose [site]: URL: https://securityrussia.com/blog/rfid-metki.html

[4] Site of the news service "Kiosksoft" [site] URL: https://kiosksoft.ru/news/2020/05/28/tochnost-beskontaktnoj-identifikacii-otpechatkov-palcev-poka-ostaetsya-nizkoj-71797

[5] Vlasenko A V, Dzoban P I, Zhuk R V 2020 *Overview of machine learning tools and their use in the field of cyber security* (Caspian Journal: Management and High Technologies №1) p 144-155.

[6] Marenkov A N, Prikhodko A A, Kuznetsova E O 2019 *Intelligent video surveillance systems: problems and functionality* (Problems of increasing the efficiency of scientific work in the military-industrial complex of Russia, 2nd All-Russian scientific practical conference)

[7] Marenkov A N, Demin K S 2020 *Methods for recognizing human movements using technologies of neural networks* (Mathematical methods in engineering and technology – MMTT).

[8] Viola P, Jones M J 2004 *Robust real_time face detection* (International Journal of Computer Vision vol. 57, № 2) pp. 137-154.

[9] Bradski G, Kaebler A 2008 *Learning OpenCV* (Published by O'Reilly Media), p 495-512.

[10] Adelson E H and Bergen J R 1986 *The Extraction of Spatio-Temporal Energy in Human and Machine Vision* (Proceedings of Workshop on Motion: Representation and Analysis) p. 151-155.

IOP Conf. Series: Materials Science and Engineering 1069 (2021) 012030 doi:10.1088/1757-899X/1069/1/012030

Methods for Modeling Sociopolitical Processes to Address Issues of Integrated Security in Caspian Region

K A Markelov, I M Azhmukhamedov, V Yu Kuznetsova

Astrakhan State University, 20a Tatischev str., 414056 Astrakhan, Russia

E-mail: arhelia@bk.ru

Abstract. The article dwells on the issue of ensuring integrated security in the Caspian region. It was noted that this region, due to its geographical location, is strategically important for the implementation of Russia's internal and foreign policy, so any changes in it should be carefully thought over and analyzed for possible consequences. To address the issues of ensuring integrated security, the authors considered the main methods for modeling social and political processes that can be applied to simulate the consequences of specific managerial decisions in the Caspian region.

Keywords: Caspian Sea; integrated security; mathematical modeling.

1. Introduction

The Caspian region has always been an important issue in the foreign and internal policy for Russia, and in light of the consequences of the economic crisis, the emphasis in the policy and economy was shifted from Europe to the regions, as a result of which the Caspian region began to attract the attention of not only Eurasian researchers but also the world political elite. Its peculiarity lies in the fact that the Caspian region, being the cradle of the Eurasian civilization, is primordially a kind of a balanced cocktail, characterized by mixing nationalities, religions and cultures, and any interference in this balance can lead to serious consequences both at the intercultural and economic levels. On this basis, any possible changes in this area cannot but affect the national interests of Russia, just as they cannot isolate it from participation in them. Russia's close attention to this geopolitical direction is also caused by the fact that it is a strategically important region in terms of its location, rich biological and hydrocarbon reserves, energy, and economic potential, as well as other parameters. Therefore, ensuring comprehensive security in the Caspian region is an urgent task of state importance.

In this regard, it is necessary to take into account all aspects, factors, interests and goals of actors of sociopolitical processes, and special attention should be paid to interethnic conflicts when making decisions, since the Caspian region is primordially multi-religious and multi-ethnic. In addition, modern society is entering the post-industrial informational period of its formation [4]. Along with expanding creative opportunities, it also leads to increasing threats to national security, associated with violation of the established modes of using information and communication systems, infringement of the constitutional rights of citizens, the spread of malicious programs, as well as the use of modern information technologies for hostile, terrorist and other criminal action.

All these parameters require in-depth system analysis. And the impossibility to test hypotheses and assumptions experimentally limits a decision-maker in making decisions about social or political processes. This leads to the need for modeling real conditions. Subsequent models, reflecting complex sociopolitical objects, are so complex and multifaceted that it is hard to imagine their effective use beyond mathematical descriptions and the use of computer technology.

2. Methods of modeling

Modeling is a research method based on construction and examination of models of the final objects of research. Studying a model gives a fairly reliable idea of the modeled phenomenon or object to the extent that the model adequately reproduces this phenomenon. Of course, modeling in the social sphere, including the political one, differs significantly from modeling physical, technical and similar objects. However, all social and political processes have both qualitative and quantitative characteristics. This makes it possible to build models of all types to study political objects, including mathematical and computer models.

In practice, the modeling process appears as a complex sequence of interrelated stages (Figure 1). The first of them is a clear statement of the problem at hand. Then the researcher must determine the object of modeling and the object or angle in which he/she wants to consider this object. Then comes the turn of collecting information, a primary study required to form a conceptual model. Next, a conceptual model is formed. If mathematical formalization is necessary, a respective task is selected and a mathematical (or computer) model is adapted to it (or a new one is created), a study is carried out to collect empirical information, mathematical (or computer) data processing is carried out, resulting in an information model that replaces the object under study. An experiment is carried out on this model - mathematical or logical, which is followed by the final stage of modeling – interpretation of the result.

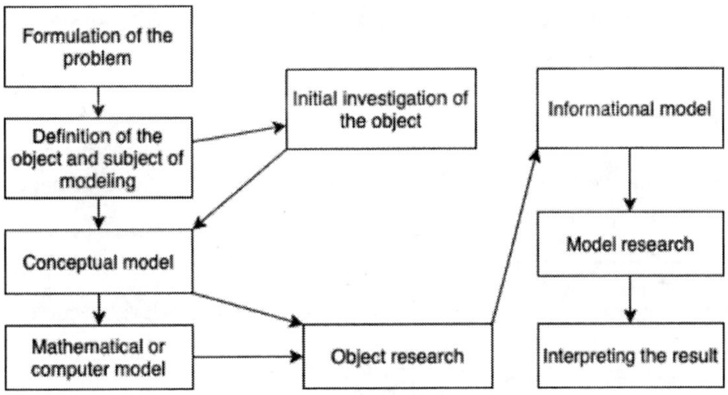

Figure 1. The modeling steps

As a subject area of scientific knowledge, the modeling of social and political processes started to form quite recently. Papers on the modeling of social and political processes by Russian and U.S. mathematicians began to appear in the mid-70s of the 20th century.

There is a class of modeling methods that can be considered for addressing the issues of ensuring integrated security in the Caspian region.

V.A. Shvedovsky [5] developed probabilistic mathematical models using the theory of nonlinear dynamical systems. On the basis of statistical information of sociological research data, the author created mathematical models for changing components of the socio-psychological potential of specific ethnopolitical centers. This method can be used to identify regional tensions in the Caspian states in order to make a timely decision to localize hotbeds of unrest and prevent revolutionary sentiments.

A relatively similar problem can be solved through analysis of hierarchies. This approach acts as a general methodology aimed at solving a wide range of problems in the field of making and evaluating the effectiveness of managerial decisions. In other words, the method of analyzing hierarchies allows to evaluate, structure and synthesize various factors, which allows to choose the optimal alternative when several criteria are important. Besides, this approach is actively used by scientists to analyze interethnic and interreligious contradictions. Assessment can be carried out both electronically and orally. Papers by T. Saaty and K. Kearns give specific examples of the application of the hierarchy analysis method. This assessment is relevant for the Caspian region, which is a multicultural area [1].

A.A. Samarsky and A.P. Mikhaylov dedicated a number of papers to mathematical modeling using analogies with technical systems, which made it possible to model power systems and build models of corrupt structures and models of a globalizing society [7]. This approach can be used to analyze political alliances and assess the degree of globalization of the Caspian states' population.

The use of the agent-based method to model processes in social and political systems has become popular since the second half of the 20th century. First of all, this is due to the fact that the methods of computer modeling have found their successful application in the Humanities and natural sciences. Since the 1990s of the last century, the agent-based method has been presented as a powerful tool to model social systems. Besides, the ability to carry out computation on supercomputers makes it possible not only to enable the creation of models of interaction between several people or small communities, but also to imitate the behavior of global social systems and processes within them. This provides a technical opportunity in an artificial environment to simulate processes in different regions, states and throughout the world in general.

The first mentions of the agent-based model of political processes in the aspect of analyzing international interaction and changes in sociocultural norms appeared in papers by T. Schelling and R. Axelrod [8,10]. The contribution of Schelling and Axelrod's research to the scientific area of modeling social and political processes is that the described models and obtained results showed that agent-based modeling has broad prospects for use, so it began to be used to analyze the processes of social, cultural and political norms, views and values at the international level.

It is also worth noting that agent technologies allow to reproduce complex interactions of "political subjects" almost with realistic accuracy in a virtual environment that imitates the structure of sociopolitical systems and the conditions for their functioning, as well as to test multivariate hypothetical scenarios in the framework of political analysis.

Therefore, analysis of the available approaches to modeling social and political processes showed that each of the modeling methods considered above can cover only one of the aspects of the integrated security of the Caspian region (Figure 2).

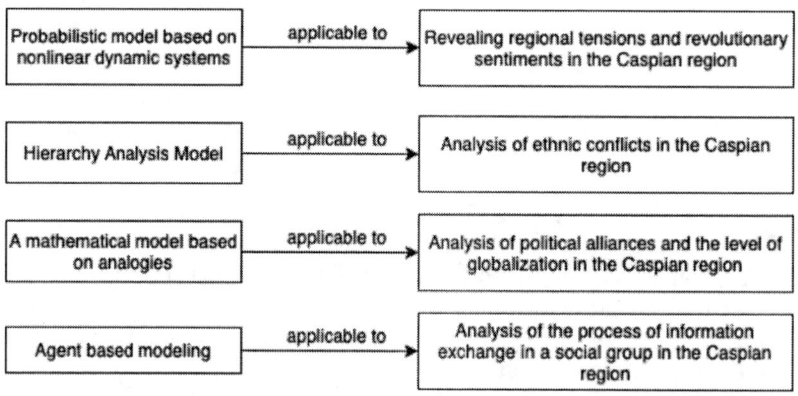

Figure 2. The methods and their applications to problems in the Caspian region

In order to combine the results of each simulation, it is advisable to use a cognitive approach. Cognitive modeling of political processes can combine each of the stages and, at the same time, reflect the specifics of the cultural and ethnic diversity of the Caspian region to the fullest extent possible.

3. Cognitive modeling

Cognitive modeling is actively used to solve problems in the management of various systems. Its use is mainly aimed at creating formal models and methods that provide an intellectual process of solving problems by taking into account human cognitive capabilities (such as perception, cognition, understanding, representation, etc.) when solving control problems. At the same time, tasks that can be attributed to the class of modeling and control problems, where the use of a fuzzy cognitive model for their solution seems to be effective and appropriate, have the following characteristics:

- have some degree of uncertainty of an objective or subjective nature;
- can be assessed both in quantitative and qualitative measures;
- do not have an analytical description of most of the interconnections between parameters.

A cognitive map is a sign or weighted graph over many factors, i.e. directed graph, the vertices of which are associated with factors. In the case of a signed graph, the edges are matched with signs (+ or -), and in the case of a weighted graph – weights in one or another scale. The weights are searched either by means of statistical processing of information or by expert means.

Changes in factors are carried out step by step until the system's response is determined, after which, using a multicriteria choice, many favorable scenarios are determined and ranked [57].

Different interpretations of vertices, edges and weights on edges, as well as different functions that determine the influence of the connections on factors, lead to different modifications of cognitive maps and means of their study. At the same time, interpretations may differ both in terms of content and mathematics. Due to the presence of many modifications of cognitive maps, we can talk about different types of models, based on these maps.

There are five types of cognitive maps according to the type of relations used:

- evaluating the focus of attention, associations and the importance of concepts (concepts);
- showing the dimension of categories and cognitive taxonomies;
- representing influence, causality and system dynamics (causal cognitive maps);
- reflecting the structure of arguments and conclusions;
- illustrating frames and codes of perception [2].

The control problem in cognitive maps is stated as follows. The situation factors include control factors (factors exposed to direct influence of the decision maker) and target factors, the change or stabilization of which is the goal of management. A specific control decision (strategy) is a choice of a set of control factors.

In a number of interpretations of the sign graph, the possibilities for comparing and ranking solutions are very limited. They can be compared only by the set of target factors on which they have the desired (positive or negative) influence.

The influence analysis method proposed in T. Sawaragi's paper solves this problem, since it is based on the following assumptions [9]:

1. The strength of the influence of one factor on another along a given path depends on the length of this path (that is, on the number of edges in it).
2. The more parallel influences (along different paths) exist between factors, the stronger the influence between them.

Let P_{ij}^m and N_{ij}^m be the number of positive and negative paths of length m going from the factor x_i to the factor x_j, respectively. Then the total positive and negative influences of the factor x_i on the factor x_j are determined as follows:

- positive influence

$$\bar{P}_{ij} = \sum_{m=1}^{\infty} f(m) P_{ij}^m,$$

- negative influence

$$\bar{N}_{ij} = \sum_{m=1}^{\infty} f(m) N_{ij}^m,$$

where $f(m)$ is a monotone non-decreasing function of the path length m which determines the degree of weakening of the influence on the path from x_i to x_j.

A monotonically non-decreasing and differentiable function is usually chosen as $f(m)$

$$f(m) = \alpha m \ ,$$

where $\alpha \in [0; 1]$ is the coefficient that determines the degree of attenuation [10].

With a decrease in α, the influence of long paths on the final result decreases, therefore, by changing α, one can analyze the influence of paths of different lengths.

To compare various strategies, various variants of the evaluation function $V(s_{ij}, c_{ij})$, are considered, where s_{ij} is the total influence of factor i on factor j and s_{ij} is the consonance of the effect of factor i on factor j, which are determined from the following relations:

$$s_{ij} = \bar{P}_{ij} + \bar{N}_{ij};$$

$$c_{ij} = (\bar{P}_{ij} - \bar{N}_{ij})/(\bar{P}_{ij} + \bar{N}_{ij}).$$

The consonance c_{ij} is a measure of the difference between positive and negative influences. The larger it is, the more definite the nature of the influence. In particular, the function $V(s_{ij}, c_{ij})$ must meet the following requirements.

Let strategy i be characterized by a pair (s_{ij}, c_{ij}), and strategy i' - by a pair $(s_{i'j}, c_{i'j})$. Then, if $V(s_{ij}, c_{ij}) > V(s_{i'j}, c_{i'j})$, i is preferable to i'.

If $c_{ij} = 0$, $V(s_{ij}, c_{ij}) = 0$ for any s_{ij}.

If $c_{ij} > 0$, V $V(s_{ij}, c_{ij})$ increases monotonically in both variables; if $c_{ij} < 0$, $V(s_{ij}, c_{ij})$ decreases monotonically in both variables.

Under some reasonable assumptions, it is advisable to choose the evaluation function in the form $V(s,c) = VS(s) \cdot VC(c)$.

More detailed characteristics of the interaction of factors can be identified using fuzzy cognitive maps.

The most common approach to calculating fuzzy influences is as follows.

Let there be m paths between f_i and f_j, and $I_r(f_i, f_j)$ denotes the influence of f_i on f_j along the r-th path, and $T(f_i, f_j)$ denotes the total influence of f_i on f_j along all m-th paths. Then

$$I_r(f_i, f_j) = \min_p w_{p,p+1},$$

$$T(f_i, f_j) = \max_{1 \le r \le m} I_r(f_i, f_j),$$

where $w_{p,p+1}$ is the weight of the directed edge from f_p to f_{p+1} on the r-th path.

Thus, the operation $I_r(f_i, f_j)$ selects the weakest link in the r-th path, and the operation $T(f_i, f_j)$ selects the strongest link $I_r(f_i, f_j)$.

In problems of dynamic analysis, fuzzy values are attributed not only to connections, but also to factors. In this case, in contrast to the weights of the links, which are considered constant during the analysis, the value assigned to the factor v_i is the value of some function $y_i(t)$, which depends on the weights of the incoming edges and the values of the factors input to v_i, and changes with time.

The vector $Y(t) = (y_1(t), y_2(t), \ldots, y_n(t))$ of the values of all factors of the situation at the moment t forms the state of the situation at the moment t: $X(t)$. The set of edge weights w_{ij} is given by the adjacency matrix of the graph $W = \{w_{ij}\}$. The presence of a value for a factor allows not only to assess the strength of the influence on the factor, but also to express the result of total influences in the form of a specific value.

The concept of the state of the situation allows us to talk about the development of a situation in time under the influence of various external influences, expressed in changes in the values of factors, that is, to set a forecast problem (direct problem), as well as to explore the possibilities of managing the situation, that is, to look for influences leading to the desired (target) state (inverse problem).

Let us consider a simpler case that is quite often encountered in many applied problems, when these functions, firstly, are the same for all factors, and secondly, they depend not on the values of the input factors but on their increments (i.e., the increment value has the same effect in any state) [53].

Let us assume that the range of values of each factor v_i is a linearly ordered set (scale) of linguistic values:

$$Z_i = \{z_{i1}, z_{i2}, \ldots\ldots, z_{iq}(i)\},$$

where z_{i1} and $z_{iq}(i)$ are the minimum and maximum elements of the set, respectively, and $k < l$ implies $z_{ik} < z_{il}$.

The powers of the scales $q(i)$ are different for different factors. The increment of the factor value is characterized by the sign of the increment and its magnitude. For the current value $y_i(t) = z_{im}$, passing to the elements Z_i: $\{z_{i(m+1)}, z_{i(m+2)}, \ldots, z_{iq(i)}\}$ gives a positive increment $\tilde{P}^{+}$; negative increment $\tilde{P}^{-}$ is obtained by passing to $\{z_{i(m-1)}, z_{i(m-2)}, \ldots, z_{i1}\}$.

In the general case, when passing from z_{im} to z_{il}, the factor gains an increment $\tilde{P}^{sign(l - m)}$. Obviously, for $m < l$, the increment will be positive, and for $m > l$ it will be negative.

For computational convenience, when analyzing situations, we define the mapping E: $Z_i \rightarrow [0, 1]$ of the discrete linguistic scale $Z_i = \{z_{i1}, z_{i2}, \ldots, z_{ir}\}$ onto the segment [0, 1] as follows. We split the segment [0, 1] into r equal segments, the boundaries of which are denoted in the ascending order $b_0 = 0, b_1, \ldots, b_{r-1}, b_r = 1$. Let us put $(z_{ik}) = (b_{k-1} - b_k) / 2$ (z_{ik} is mapped to the center of the k-th line segment). This mapping allows to make the algorithms of the fuzzy cognitive model numerical. The inverse mapping E^{-1}: $[0, 1] \rightarrow Z_i$ is a homomorphism: all points lying in the interval (b_{k-1}, b_k) are mapped to one point z_{ik}.

Using direct mapping, the state of the situation is represented in a numerical form, and further calculations are performed with a numerical representation of the state of the situation $X(t)$. Reverse mapping is used only for qualitative interpretations of analysis results.

Thus, cognitive maps allow to combine various modeling methods, which makes it possible to create an effective tool for studying complex systemic processes occurring in sociopolitical institutions through the modeling of interrelated and mutually influencing relations between many social and political actors, which allows to track evolution and dynamics of sociopolitical and economic relations in the Caspian region.

4. Conclusion

- The Caspian region is characterized by mixing nationalities, religions and cultures, and any interference with this balance can lead to serious consequences both at the intercultural and global economic and political levels.
- The world community is interested in ensuring comprehensive security, while modern realities require the adoption of managerial decisions that can upset the balance of a particular geopolitical region. In this regard, modeling is the optimal method for predicting the consequences of certain managerial decisions in relation to the Caspian region.
- The use of various modeling methods makes it possible to create an effective tool for studying complex systemic processes occurring in sociopolitical institutions through the modeling of interrelated and mutually influencing relations between many social and political actors, which allows to track the evolution and dynamics of sociopolitical and economic relations in the Caspian region.
- The use of fuzzy cognitive modeling allows to reflect the specifics of decision-making in the Caspian region to the fullest extent possible.

References

[1] Saaty T. L., Kearns K. 1991 *Analytical Planning; The Organization of Systems* (Pergamon Press, Oxford, Translated to Russian. Paperback edition, RWS Publications, Pittsburgh).
[2] Azhmukhamedov I. M., Protalinskiy O. M. 2013 *Methodology for modeling poor formalized weakly structured sociotechnical systems* (Astrakhan: Astrakhan State Technical University

Press) pp. 144-154
[3] Kuznetsov O.P. 2006 *Cognitive modeling of poorly structured situations* (Moscow Polytechnical Museum).
[4] Brumshteyn Yu. M., Markelov K. A., Martynov I. A. 2020 *Images of political leaders of leading foreign countries in russian citizens' consciousness: general analysis of management questions* (Astrakhan: Caspian region: politics, economic, culture №3) pp. 147-156.
[5] Shvedovsky V. A. 2000 *Dynamic model of electoral behavior* (Mathematical modeling № 8) pp. 46–56.
[6] Shabrov O. Ph. 1996 *Systems approach and computer modeling in political science research* (Social sciences and modernity № 2) pp. 100-110.
[7] Samarsky A. A., Mikhailov A. P. 1997 *Mathematical Modeling: Ideas, Methods and Examples* (Moscow)
[8] Axelrod R. 1976 *The Structure of Decision: Cognitive Maps of Political Elites* (Princeton University Press)
[9] Sawaragi T., Iwai S., Katai O. 1986 *An integration of qualitative causal knowledge for user-oriented decision support* (Control Theory and Advanced Technology v.2) pp. 451-482.
[10] Shelling T. 2016 *Micromotives and Macrobehavior* (Moscow Institute Gaydara Press).

Poisoning attack of the training dataset for on-line signature authentication using the perceptron

Evgeny Kostyuchenko[1], Balgit Lodonova[1], Yakov Usoltsev[1], Artem Konovalov[1]

[1]Tomsk State University of Control Systems and Radioelectronics, 40 Lenina St., Tomsk, 634050, Russia Federation

E-mail key@fb.tusur.ru

Abstract. The study of the dependences of the effectiveness of a poisoning attack for the task of on-line signature authentication using a perceptron to make a decision. The dependences on the structure of the neural network, the volume of poisoning sets and the noise of the training sample were studied. Conclusions are drawn on the lack of the ability to detect such attacks by changing the overall accuracy of the system.

Keywords: Training sample poisoning, Biometric authentication, On-line signature, Neural network

1. Introduction

Attacks on machine learning algorithms, and in particular on neural networks, can lead to various vulnerabilities in systems using these algorithms, in particular, an attacker can deceive the recognition algorithms of certain images [1, 2] or system developers there will remain the possibility of unauthorized access to the system [3]. Considering how important the component of the modern information space is machine learning [4], the tasks associated with the study of attacks on machine learning algorithms are now more than ever an urgent topic. Watching over development trends of modern information technologies we can observe significant increase areas of application to machine learning systems. Now system of speech recognition [5] or complex system of security systems [6] are created on neural networks basis.

As examples of such attacks, adversarial attacks [7, 8] and attacks at the training stage [9] can be distinguished.

The process of preparing a training sample is one of the key points of machine learning that determines its outcome. For this reason, providing counteraction to attacks while ensuring the security of the system based on machine learning methods is relevant and counteracting poisoning attacks of a training dataset [10] is important.

The poisoning attack to dataset is analyzed [11]. The essence of this attack is that when training a neural network, an attacker injects his data into the training set. Within the framework of the system under study, an attacker can embed his data sets in the data of legal users in order to increase the percentage of statistical errors of the second kind and gain potential access to the system under the guise of a legal user [12].

IOP Conf. Series: Materials Science and Engineering 1069 (2021) 012031 doi:10.1088/1757-899X/1069/1/012031

As part of a review aimed at identifying typical set poisoning attacks and evaluating their effectiveness, we can distinguish such works as [13-18]. Consider them, the methods used within the system and the results obtained in more detail.

In [13], a study is proposed of counteracting the methods of selecting features for training. Selection methods such as LASSO, ridge regression and the elastic net are considered. The general methods of machine learning are analyzed, it is shown that poisoning of 5% of the training dataset of the class leads to a complete disruption of the system.

The work [14] explores the stability of federated learning systems against training set poisoning attacks. The possibility of an attacker generating new training examples and using them to poison a training sample is shown. It is shown that this approach to attack allows achieving accuracy up to 80% for the attacker.

The vulnerability of the least squares method for the algorithm uLSIF, which is used to calculate the importance ratio for the adaptation of a covariant shear region to poisoning attacks, was shown in [15]. The results allow us to talk about the applicability of training dataset poisoning for the method under consideration.

The work [16] shows the possibility of the combined effect of the poisoning attack of the training sample regarding the integrity and accessibility of information. Attack efficiency for logistic and linear regression methods is shown.

In [17], the effectiveness of an attack for an intrusion detection system is considered. The Edge Pattern Detection (EPD) algorithm is analyzed to develop a new poisoning method that attacks several machine learning algorithms used in IDS.

In [18], the effectiveness of pollution attacks of training sets for intrusion detection systems, in particular, for mobile applications, is shown. A solution has been proposed to increase the detection efficiency of such attacks by at least 15%.

The analysis of the work allows us to talk about the relevance of the study of poisoning attacks of the training set based on new research in this direction over the past 2 years. It is seen that these attacks can be applied to various systems, in particular, intrusion detection systems and identification systems. Various characteristics, such as voice [2, 19], keyboard handwriting [20], off-line [21] and on-line [22] signature can be used as input for authentication systems using biometric characteristics. This work is devoted to the study of the effect of a poisoning attack on a training dataset on one of the systems not represented in existing works, namely, an identification system based on signature dynamics.

2. Description of the existing on-line signature authentication system

The system without attack is a script on a mathematical package MATLAB [23] that includes:

1) extraction of data sets of signatures from the database;

2) creation of a neural network with specified parameters;

3) splitting the signature datasets into three parts – for neural networks training, for selection decision threshold and for primary network testing with retraining;

4) random generation of new sets based on existing sets with a deviation of 1% for increasing the number of input data for training;

5) neural networks training;

6) selection decision threshold;

7) testing a trained neural network;

8) obtaining and correct presentation of various statistical errors [24].

Input data for neural networks training is signature dataset, several dozen for each user. Signature data sets are presented in the form of 144 signature parameters [22] that provide complete information about the dynamics of signature writing.

The study was conducted on dataset of 2445 parameterized signatures from 22 users. User is identified in system by a unique number for each user from 1 to 22. 13 users are legitimate on the basis of which the network will be trained. The rest 9 users are not registered in the system, and they will be used for testing system.

For all users, a process is provided for a tenfold increase in the number of signatures by generating nine signatures based on one with making a random change for each of the 144 signature parameters, not exceeding 1% of the parameter value.

By default, created neural network takes 144 parameters to input layer, have 30 neurons in intermediate layer and return vector of result. Number of elements in this vector is equal to the number of legitimate users in the system and displays to whom the entered signature belongs (Figure 1).

Signatures from legitimate users are divided into three sets: training set, set for selection decision threshold and set for primary training, in the ratio of 60%, 20% and 20%.

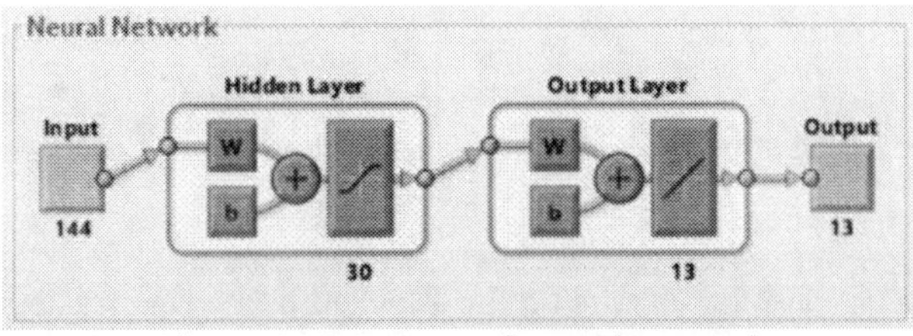

Figure 1. Structure of neural network

Based on the results of testing the trained neural network, data is formed on the sum of the errors of the first and second kind, the system error and the total error of the first and second kind with a tenfold priority of the error of the second kind.

3. Variants of the investigated attack parameters

As a part of a research on machine learning with biometric authentication and attacks, we conducted studies with gradual increase of various parameters of the neural network learning.

Parameters were increased within predetermined intervals.

In particular, we studied the effect on the output of the following parameters:

- number of neurons in intermediate layer;
- the percentage of inclusion of alien set under the guise of a legitimate user;
- number of users in system;
- coefficient of noise.

3.1. Dependence on the number of neurons in the intermediate layer

In result, we changed the number of neurons in intermediate layer and received plots of dependences of the total error of the first and second kind at tenfold priority when attacking the system (Figure 3) and without attack (Figure 2).

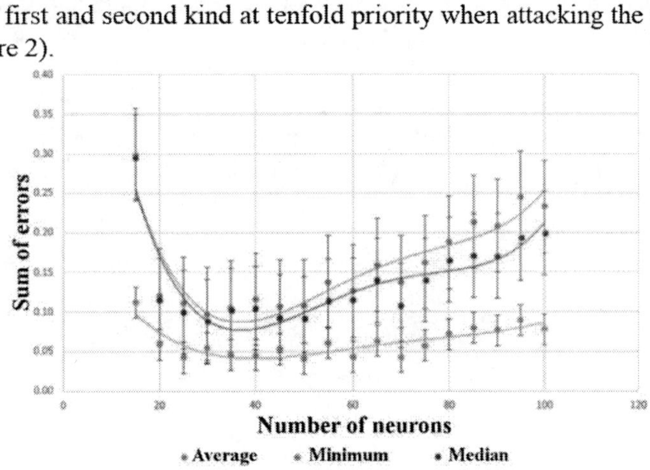

Figure 2. Dependence of the sum of errors sum at tenfold priority of second kind error to number of neurons in intermediate layer

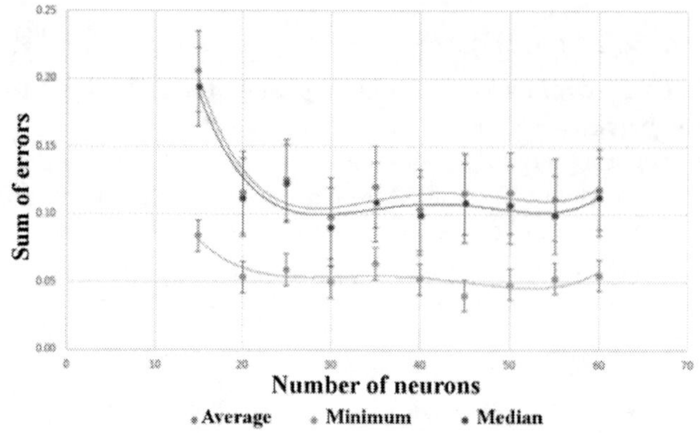

Figure 3. Dependence of the sum of errors sum at tenfold priority of second kind error to number of neurons in intermediate layer

Also error of the second kind that occurs on the attacked user value was received. The value reflects the probability of an intruder penetration in system as legal user.

The following graph shows dependence statistical error of second kind value of the number of neurons in intermediate layer (figure 4).

As seen from the graphs, gradual increase of number of neurons in intermediate layer doesn't influence poisoning attack efficiency. The extremum position is blurred for legal user, but extremums value is similar in all studies.

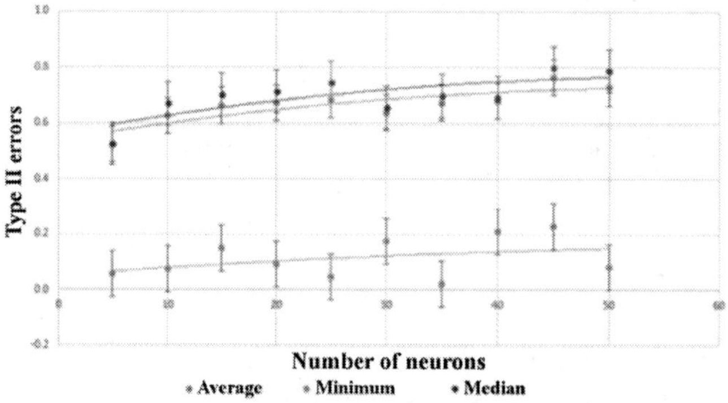

Figure 4. Dependence of changes in the statistical error of the second kind to number of neurons in intermediate layer

3.2. Dependence on the percentage of inclusion

Similar in methodology study was conducted for estimates of the effect of the percentage of introduced by the attacker sets on system functioning and on the probability of an intruder penetration in system as legal user.

Probabilities are shown on figure 5.

Percentage values introduced sets located on the abscissa axis, value of errors located on ordinate axis.

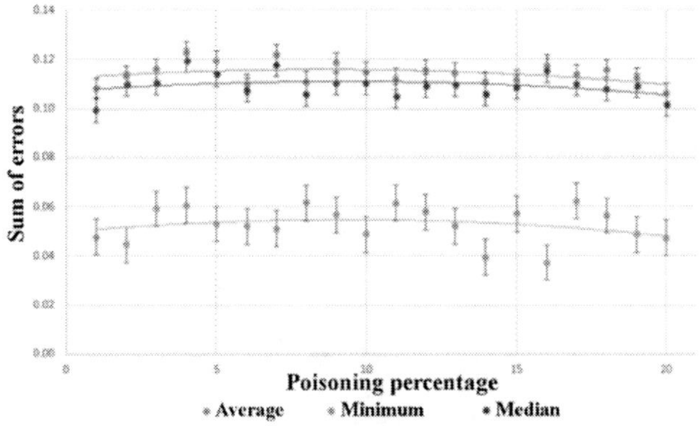

Figure 5. Dependence of the sum of errors sum at tenfold priority of second kind error to percentage of include

As graphs show the percentage of inclusion doesn't influence on systems efficiency. It means poisoning attack is difficult to detect by system performance analysis and by user data. We received value of dependence of the error on attacked user. The error value reflects the probability of an intruder penetration in system as legal user.

The dependence of the effectiveness of the attack on the percentage of poisoning is shown in the figure 6.

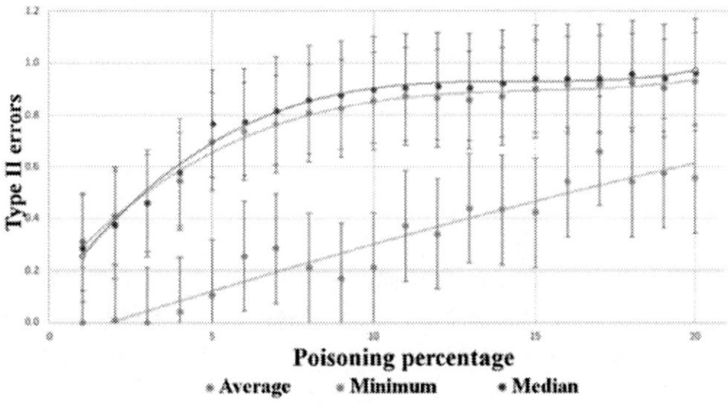

Figure 6. Dependence of changes in the statistical error of the second kind to percentage of include

As seen from the dependency graphs percentage of poisoning sets influences user errors. Therefore, by using a set poisoning attack, an attacker is theoretically able to gain access with minimal risk of being detected.

3.3. Influence of the number of users on the system operation

We changed the number of users in training set for attacked system and received graphs of dependence of errors sum at tenfold priority of second kind error to the probability of a successful attack.

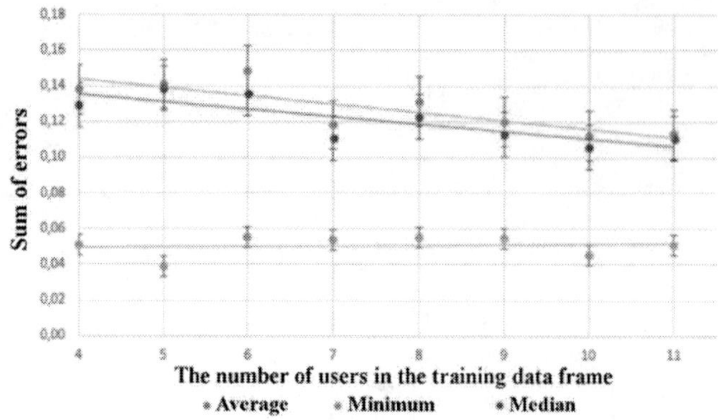

Figure 7. Dependence of the sum of errors sum at tenfold priority of second kind error to percentage of include

As seen from the error graphs increase in the number of users with a fixed number of examples from the user doesn't change errors value. Also it becomes clear attack on system is more effective with a small number of users.

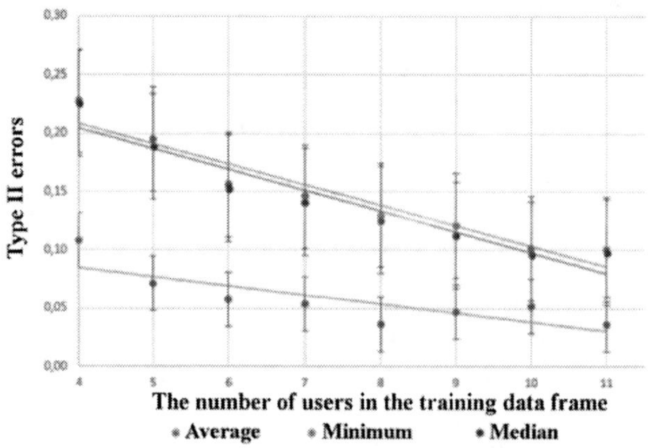

Figure 8. Dependence of errors for attacked user

3.4. Influence of the percentage of noise on the operation of the system

Increase the training set by generation of new sets with a noise on the basis of available is a typical method for enhance learning sample representativeness and the quality of the system [25].

We conducted a study of influence of multiplicity of noising on the system operation with attack on it and without attack.

Noise is generated based on training set with introduction of entropy of 1%.

We changed number of automatically generated sets for the neural network and received graphs of dependence of errors sum at tenfold priority of second kind error and attack efficiency on multiplicity of noising. Experiments were carried out for system without and for attacked system.

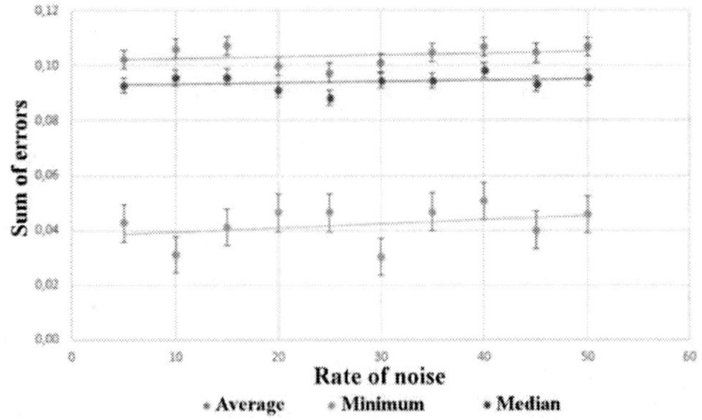

Figure 9. Dependence of error to multiplicity of noising, system without attack and tenfold priority of second kind error

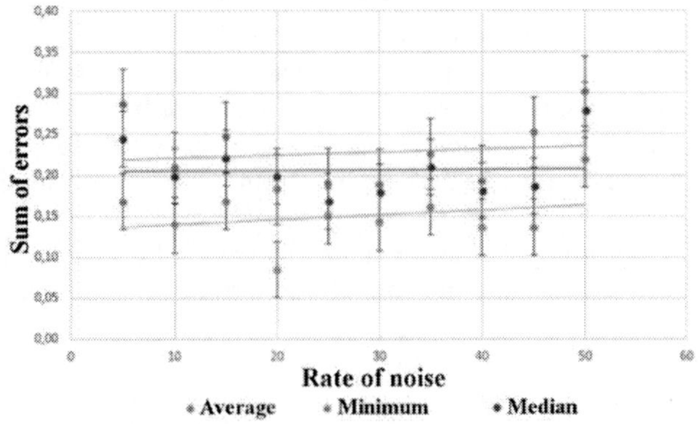

Figure 10. Dependence of error to multiplicity of noising, system with attack and tenfold priority of second kind error

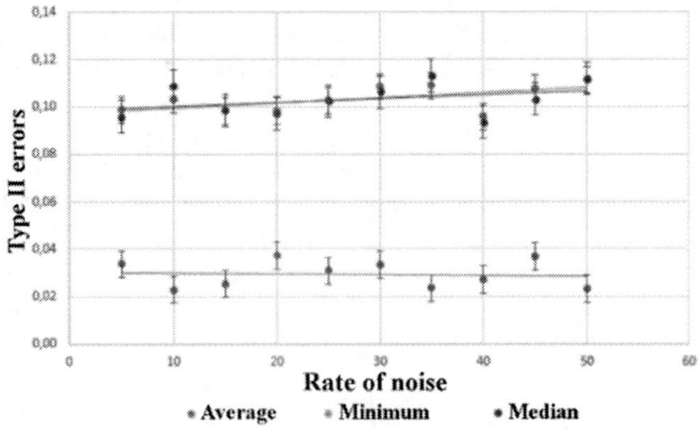

Figure 11. Dependence of attack success to multiplicity of noising, percentage of include is fixed

It is seen, increase multiplicity of noising over 5 doesn't affect error rates which are also near some constant indicator

4. Discussion of results

The obtained dependences were analyzed and an attempt was made to explain them.

Summing up the study we inferred optimal number of neurons, significant different from which leads to a decrease of system efficiency. Change in the number of neurons doesn't affect system efficiency for legal users that in the presence of attack in the absence of. And extremum has blurred position boundaries, although compared to the previously recorded level error value and rate don't change. This shows the problematic nature of attack detection for system. It's impossible to detect using methods of changing system parameters. Also efficiency of attack doesn't depend number of neurons.

Also it`s seen, system efficiency doesn't change under the influence of introduced sets. We can see increased attack efficiency from percentage inclusion. And minimum level of inclusion (1%) increase attack efficiency to 30 % which is significant for system efficiency. Inclusion of 10-20% of sets make access to the system for attacker.

Study of depend errors to numbers of neurons shows that general index doesn't change in investigated range, and probability of successful attack with fixed level of poisoning set decreases for a single user. This can be explained by two factors: firstly, increase of number of users is an increase of total number of sets, as a result, total share of included sets is decrease. Secondly, total number of classes is increase, and all of them can output for system. Attacker needs to be identified only as attacked user for a successful attack. Therefore, increase of total number of classes decrease probability of this match. However, corresponding to attack users output is stay more possible.

Similar in general concept attacks to the neural network were carried out by Jiang, Wenbo, et al.[26]. This neural network was created to recognize road signs. Jiang, Wenbo, et al modeled situation in which attacker included own set in training set. Inclusive set is similar set of attacked user. Inclusive set is modificated to visually practically do not differ to legal set. This inclusion doesn't allow the neural network to correctly identify inclusive set. Varying of inclusion degree was make possible to strike dependence of system accuracy to varatios of number of inclusive set to total set. As a result, system accuracy decrease from 95 to 33 for one user set and from 97 to 63 for another user when varying of poison rate from 0% to 10%.

Similar on approach to poisoning, but different in poisoning set generation method study was created by Biggio, Battista, et al [27]. Varying of poison rate from 0% to 10% was in their publication too. Biggio, Battista, et al used different from the proposed Jiang algoritm to generate poisoning set and received increase of error from 0-4 to 20-28 to different trainig sets.

Unfortunately for us, systems in the presented publications don't have a calculating statistical errors. But we can compare received by us values of statistical second kind errors on a specific user with system efficiency and percentage of test error, which is received in the presented publications.

Increase of percentage of inclusive set rate from 1% to 10% changes value of statistical second kind error from 30% to 87% in our system. We can make a conclusion that attack in this study has a high impact on the system. But the attack can be easier to detect because it involves more aggressive intervention in the system.

Another aspect of neural network operation namely, the effect of noise on network training, was studied Y. Hayakawa, A. Marumoto, and Y. Sawada [28]. They studied method of "white" noise, which consist in no correlation of the generated random variable with training set. This method is comparable to used in our system method.

5. Conclusion

The study was presented on the effect of changing various parameters of neural network training to final classification accuracy.

The study was carried out for the task of on-line signature authentication using a perceptron to make a decision.

Results of the study confirm the inherent for similar attacks on machine learning methods sensitivity to poisoning inclusion.

Even 1% of include is enough to multiple reduction in quality of system operation in relation to the detection of an attacking user.

Further studies in countering to attack on machine learning methods is relevant for such areas of information security as a authentication and detecting invasion [24].

6. Acknowledgment
This research was funded by the Ministry of Science and Higher Education of Russia, Government Order for 2020–2022, project no. FEWM-2020-0037 (TUSUR).

References
[1] Goodfellow I J, Shlens J and Szegedy C 2015 Explaining and Harnessing Adversarial Examples *arXiv:1412.6572 [cs, stat]*

[2] Rakhmanenko I, Shelupanov A and Kostyuchenko E 2019 Fusion of biLSTM and GMM-UBM systems for audio spoofing detection *International Journal of Advanced Trends in Computer Science and Engineering* **8**

[3] Chen B, Carvalho W, Baracaldo N, Ludwig H, Edwards B, Lee T, Molloy I and Srivastava B 2018 Detecting Backdoor Attacks on Deep Neural Networks by Activation Clustering *arXiv:1811.03728 [cs, stat]*

[4] Paudice A, Muñoz-González L, Gyorgy A and Lupu E C 2018 Detection of Adversarial Training Examples in Poisoning Attacks through Anomaly Detection *arXiv:1802.03041 [cs, stat]*

[5] Kipyatkova I S and Karpov A A 2016 Variants of Deep Artificial Neural Networks for Speech Recognition Systems *Тр. СПИИРАН* **6** 80

[6] Iskhakov S Y, Shelupanov A A and Meshcheryakov R V 2017 Simulation modelling as a tool to diagnose the complex networks of security systems *J. Phys.: Conf. Ser.* **803** 012057

[7] Tahir R, Cheng K, Eric L and Khokhar M S 2019 Multi-domain Cross-dataset and Camera Style Transfer for Person Re-Identification *International Journal of Advanced Trends in Computer Science and Engineering* **8** 2035–41

[8] Roli F, Biggio B and Fumera G 2013 Pattern Recognition Systems under Attack *Progress in Pattern Recognition, Image Analysis, Computer Vision, and Applications* Lecture Notes in Computer Science ed J Ruiz-Shulcloper and G Sanniti di Baja (Berlin, Heidelberg: Springer) pp 1–8

[9] Kloft M and Laskov P Online Anomaly Detection under Adversarial Impact 8

[10] Zhao W, Li P, Zhu C, Liu D and Liu X 2019 Defense Against Poisoning Attack via Evaluating Training Samples Using Multiple Spectral Clustering Aggregation Method *Computers, Materials & Continua* **59** 817–32

[11] Steinhardt J, Koh P W and Liang P 2017 Certified Defenses for Data Poisoning Attacks *arXiv:1706.03691 [cs]*

[12] Kwon H, Yoon H and Park K-W 2019 Selective Poisoning Attack on Deep Neural Networks † *Symmetry* **11** 892

[13] Xiao H, Biggio B, Brown G, Fumera G, Eckert C and Roli F 2018 Is feature selection secure against training data poisoning? *arXiv:1804.07933 [cs, stat]*

[14] Zhang J, Chen J, Wu D, Chen B and Yu S 2019 *Poisoning Attack in Federated Learning using Generative Adversarial Nets*

[15] Umer M, Frederickson C and Polikar R 2019 Vulnerability of Covariate Shift Adaptation Against Malicious Poisoning Attacks *2019 International Joint Conference on Neural Networks (IJCNN)* 2019 International Joint Conference on Neural Networks (IJCNN) pp 1–8

[16] Jiang W, Li H, Liu S, Ren Y and He M 2019 A Flexible Poisoning Attack Against Machine Learning *ICC 2019 - 2019 IEEE International Conference on Communications (ICC)* ICC 2019 - 2019 IEEE International Conference on Communications (ICC) pp 1–6

[17] Li P, Liu Q, Zhao W, Wang D and Wang S 2018 Chronic Poisoning against Machine Learning Based IDSs Using Edge Pattern Detection *2018 IEEE International Conference on Communications (ICC)* 2018 IEEE International Conference on Communications (ICC) pp 1–7

[18] Chen S, Xue M, Fan L, Hao S, Xu L, Zhu H and Li B 2018 Automated poisoning attacks and defenses in malware detection systems: An adversarial machine learning approach *Computers & Security* **73** 326–44

[19] Matveev Y 2013 The Problem of Voice Template Aging in Speaker Recognition Systems *Speech and Computer* Lecture Notes in Computer Science ed M Železný, I Habernal and A Ronzhin (Cham: Springer International Publishing) pp 345–53
[20] Kostyuchenko E Y, Viktorovich I, Renko B and Shelupanov A A 2018 User Identification by the Free-Text Keystroke Dynamics *2018 3rd Russian-Pacific Conference on Computer Technology and Applications (RPC)* 2018 3rd Russian-Pacific Conference on Computer Technology and Applications (RPC) pp 1–4
[21] Even S, Goldreich O and Micali S 1990 On-Line/Off-Line Digital Signatures *Advances in Cryptology — CRYPTO' 89 Proceedings* Lecture Notes in Computer Science ed G Brassard (New York, NY: Springer) pp 263–75
[22] Tomsk State University of Control Systems and Radioelectronics, Tomsk, Russia, Samsonov S S, Filimonenko I V and Tomsk State University of Control Systems and Radioelectronics, Tomsk, Russia 2018 Dynamic-signature-based user authentication using a fuzzy classifier *CO* **42** 657–66
[23] MATLAB - MathWorks - MATLAB & Simulink https://www.mathworks.com/products/matlab.html
[24] Kostyuchenko E, Gurakov M, Krivonosov E, Tomyshev M, Mescheryakov R and Hodashinskiy I 2016 Integration of Bayesian Classifier and Perceptron for Problem Identification on Dynamics Signature Using a Genetic Algorithm for the Identification Threshold Selection *Advances in Neural Networks – ISNN 2016* Lecture Notes in Computer Science ed L Cheng, Q Liu and A Ronzhin (Cham: Springer International Publishing) pp 620–7
[25] Reed R and II R J M 1999 *Neural Smithing: Supervised Learning in Feedforward Artificial Neural Networks* (Cambridge, Mass.: Bradford Books)
[26] Jiang W, Li H, Liu S, Luo X and Lu R 2020 Poisoning and Evasion Attacks Against Deep Learning Algorithms in Autonomous Vehicles *IEEE Transactions on Vehicular Technology* **69** 4439–49
[27] Biggio B, Nelson B and Laskov P 2013 Poisoning Attacks against Support Vector Machines *arXiv:1206.6389 [cs, stat]*
[28] Hayakawa Y, Marumoto A and Sawada Y 1995 Effects of the chaotic noise on the performance of a neural network model for optimization problems *Phys. Rev. E* **51** R2693–6
[29] Proshkin N, Barannik E, Lapina M, Alla K and Lapin V 2020 Developing models of IoT infrastructures to identify vulnerabilities and analyse threats *IOP Conference Series: Materials Science and Engineering* **873** 012018

Comparison of recognition using Google and Kaldi to solve the problem of assessing intelligibility

E Y Kostyuchenko[1], L N Balatskaya[1], S S Kharchenko[1], M A Lapina[2]

[1]Biomedical Research Laboratory Tomsk State University of Control Systems and Radioelectronics Tomsk, Russia

[2] Institute of Mathematics and Information Technologies named after Professor N.I. Chervyakova North Caucasus Federal University Stavropol, Russia

E-mail key@fb.tusur.ru

Abstract. The paper is devoted to improving the functionality of the speech intelligibility assessment solution. A previously implemented approach based on the Google Cloud Speech-to-Text solution has shown controversial quality indicators due to its attachment to the global news context. The paper proposes and evaluates the use of the Kaldi system for recognition. This solution not only made it possible to assess the quality of speech without using the Internet, but also to improve the quality of recognition in isolated phrases by excluding the global news context.

Keywords: speech intelligibility, speech recognition, verbal intelligibility, Kaldi

1. Introduction

The problem of obtaining automatic objective assessments of speech quality through intelligibility is highly relevant due to its potential use in several critical areas.

The first such direction is the use of intelligibility when assessing the possibility of information leakage through acoustic channels [1]. This task is relevant both for assessing the current acoustic environment in a room and for assessing the effectiveness of using noise to protect information circulating in the room [2].

Another important area is the use of the obtained assessments of the speech quality during speech rehabilitation of patients after the treatment of oncological diseases of the organs of the speech-forming tract [3].

The problem of the incidence of cancer is relevant both in the world and in Russia. So, in Russia alone in 2018, more than 624,000 [4] cases were registered for the first time in their life, diagnosed with malignant neoplasms. At the same time, the number of registered cases is growing from year to year.

One of the localizations encountered are tumors of the organs of the speech-forming tract, in particular the oral cavity and oropharynx.

Another negative side of the statistics given in the above work is the susceptibility to diseases of the population of working age. Loss of working capacity and getting a disability is a colossal problem from both an economic and psychological point of view.

One of the ways to improve the effectiveness of rehabilitation is to form objective assessments of the quality of pronunciation without the participation of experts, which will allow individual training sessions to be carried out independently. This will increase the intensity of the rehabilitation process through independent training and work with a speech therapist within the framework of staged exercises and checkpoints.

Based on this, the task of automating the receipt of objective assessments of speech quality is urgent. Within the framework of this work, we compare the use of two speech recognition systems: Google Cloud Speech [5] and Kaldi [6] in solving this problem, and the possibility of obtaining such estimates without using access to the Internet is carried out.

2. Review of related works

Consider the classification of existing methods for assessing speech quality, shown in Figure 1 [7].

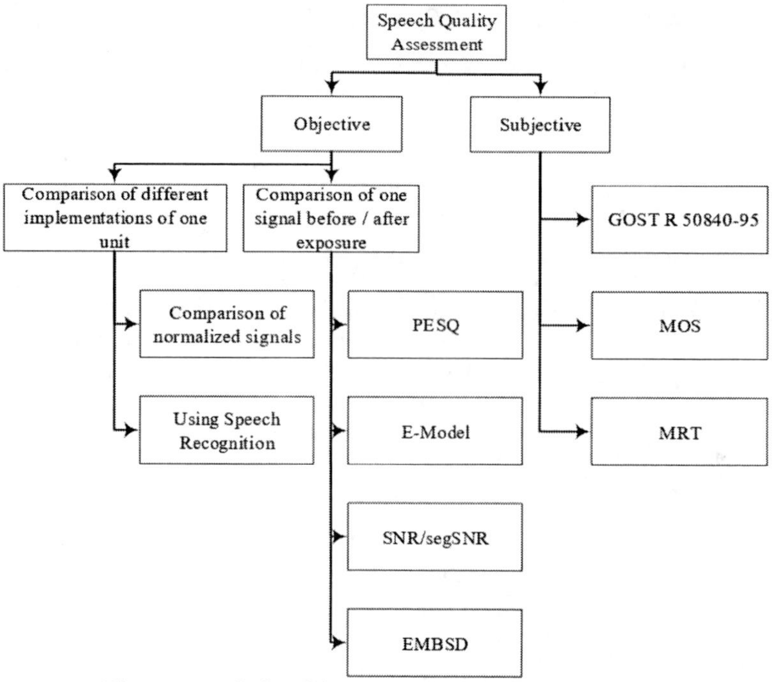

Figure 1. Relationships and relationship Collections

It should be borne in mind that this classification is rather arbitrary. So, the methods referred to it as subjective (GOST R 50840-95 [8], Mean opinion score [9], Modified Rhyme Test [10]) because they use direct expert judgment. Within the framework of the methods themselves, such a method of increasing objectivity as using several experts and increasing objectivity by increasing the number of experts is used. However, the number of such experts varies, as a rule, from 3 to 5, it is impossible to exclude them, therefore the category is designated as Subjective, not only and not so much from the point of view of the subjectivity of the estimates obtained, but also from the point of view of the impossibility of directly excluding the participation of auditors-subjects.

Objective methods are divided into 2 categories: comparison of signals before / after transmission / processing (Perceptual evaluation of speech quality [11], E-model [12], Signal-to-noise ratio [13], Enhanced Modified Bark Spectral Distortion [14]). This approach is popular when assessing the quality of communication channels, however, its use for assessing the intelligibility of a single speaker in its pure form seems to be poorly possible, since even if by analogy we compare the signals before / after the operation, then from the point of view of the methods of this category, these are two fundamentally different signals and their comparison using such methods will not be correct.

The second category is for comparing differing signals. There are two approaches to assessment:

1.Use various methods of signal normalization in order to bring them to such a form that it would be possible to apply methods from the previous category or use difference metrics for signals of identical duration

2.Use an assessment approach using speech recognition methods, in which subjective expert methods can be used, in which the expert is replaced by a recognition system

Within the framework of this work, it is assumed that the latter category of methods will be used, which is used to obtain estimates of intelligibility according to GOST R 50840-95 [8] and compares the application of Google Cloud Speech and Kaldi speech recognition systems.

3. Research methods and data used

3.1. Research methods

As part of the work, the quality of recognition was assessed using three variations of recognition systems:

1.Google Cloud Speech systems in free implementation using the web interface on the site;

2. a commercial version of the Google Cloud Speech system integrated into the software for assessing speech quality using the Google Cloud Speech API;

3. Kaldi speech recognition libraries.

The fundamental difference between Google Cloud Speech and Kaldi systems is that the former uses Internet resources to transmit speech information for processing to Google servers. This approach has the obvious advantage of being able to use Google's vast computing power to recognize our information. This allows using more time-consuming but accurate recognition algorithms or simply increasing the search depth, improving the quality of the selection of the final recognized phrase from the many analyzed variants. However, the use of external resources has three drawbacks, the first of which, the most obvious, is the need for a continuous connection to the Internet for the system to work. In addition, you need to pay for access to Google Cloud Speech resources when implementing recognition of many phrases from an application for assessing speech quality. The third drawback was revealed at the stage of regular operation of the speech quality assessment system, an effect was revealed, the influence of which was imperceptible at the stage of choosing a recognition system within the framework of work []. The operation of the system has shown that the use of Internet resources and the computing power of Google has a drawback in the form of attachment of the used recognition model to global news content. As part of the operation of the system, this was expressed in the regular appearance in the recognized phrase of words that dare to be directly related to what is happening in the world (terrorist attacks, elections, the coronavirus pandemic), but have nothing to do with the words actually used. This is due to the fact that, based on operating experience, the recognition model built by Google depends on the frequency of words used in search queries. This approach is relevant when recognizing search queries, especially related to news. But for our task related to the recognition of a fixed set of phrases, this is a disadvantage. In addition, it is not obvious whether information about the address of the requestor and his previous recognition requests is used in making a decision. If so, this can have a positive effect on the quality of recognition, but negatively on the quality of the estimates obtained (we recognize the same sets of phrases). During the preliminary selection, these effects were not noticed, in this regard, when building the research plan, 2 questions were formulated:

1. Is there a difference between the recognition accuracy on a fixed set of phrases for the Google Cloud Speech system in the free implementation using the web interface on the website and the commercial version of the Google Cloud Speech system? Is the discovery of the above effects related to the fact that the free version does not have access to the additional world context or for some other reason, or was this effect simply missed during preliminary testing?

2. When using the offline version of the Kaldi recognition system, which of the effects will be greater: lack of access to more powerful computing resources, which reduces the quality of recognition, or a

reliable lack of access to the global context and previous recognized phrases, which in our case is an advantage.

To answer these questions, it was decided to conduct an experiment to assess phrasal and verbal intelligibility as part of a comparison of approaches 1-3, as well as to compare the recognition results between the implemented versions of the 2-3 system in terms of recognition of individual characters using the Levenshtein distance.

Verbal and phrasal intelligibility is understood as the proportion of correctly recognized words or phrases of the reference signal in the recognized [8].

The Levenshtein distance is defined as the minimum number of one-character operations (namely, insert, delete, replace) required to convert one sequence of characters into another [15].

The measurements are taken on the records made before the operation, before the beginning of rehabilitation and during its process. At the same time, it is not the absolute values of the indicators that are of decisive importance, but the ratio between them (after the surgical treatment, the speech quality decreases, but then increases in the process of rehabilitation).

Thus, during the study, the aforementioned indicators were measured in at least three marked control points both for individual users (the assessment is speaker-dependent, since the comparison occurs with the speech of a particular speaker before the operation), and by the averaging method after normalization according to the indicator before the operation.

3.2. Description of the used speech base

For the experiment, we used patient records made at the Oncology Research Institute of the Tomsk National Research Medical Center during treatment and speech rehabilitation procedures. For the study, records of patients with at least 3 sessions were selected. The number of such patients is 24.

The recordings were made in mono with a sampling rate of 11025 Hz. The number of recordings within one session is 25. The spoken phrases and their transcription are presented in Table 1.

Table 1. Phrases used in the assessment of speech

№	No. Phrase and transcription		
1.	A white veil lay in the fields bʲɛłaja pʲɛlʲɛna lʲɛʐała na pɔlʲax	9.	The path ran into a clay ledge trɔpʲinka upʲɛrłasʲ v ɣlʲinʲistij ustup
2.	Heroes of the front came to school v ʂkɔłu prʲijɛzʐalʲi ɣʲɛrɔji frɔnta	10.	The sun rose over the forest sɔłnt͡sʲɛ pɔdnʲałɔsʲ nał lʲɛsɔm
3.	White steam spreads over puddles bʲɛłij par rasstʲiłajɛtsʲa nad łuʐamʲi	11.	There were nurses at the entrance v pɔdjɛzdjɛ stɔjalʲi sanʲitarkʲi
4.	The tank crew understood the task ɛkʲipaʐ tanka pɔnʲał zadat͡ʂu	12.	The meeting place became known stałɔ izʲvʲɛstnɔ mʲɛstɔ vstrɛt͡ʂʲi
5.	This unit works well ɛtɔt błɔk rabɔtajɛt xɔrɔʂɔ	13.	The site is under surveillance na ut͡ʂastkʲɛ vʲɛdjɛtsʲa nablʲudjɛnʲijɛ
6.	Steppe fires begin nat͡ʂʲinajutsʲa stjɛpnijɛ pɔʐarɨ	14.	The teacher entered the class v kłass vɔʂjɛł prɛpɔdavatjɛlʲ
7.	Students water the garden ut͡ʂʲjɛnʲikʲi pɔlʲivajut ɔɣɔrɔd	15.	The log split in two pɔlʲɛnɔ raskɔłɔłɔsʲ nadvɔjɛ
8.	Heavy pouring is over tjaʐjɛłij pɔlʲjɛm zakɔnt͡ʂʲiłsʲa	16.	Need to load the gun nadɔ zaradʲitʲ ruʐʲjɛ

17.	The cart began to creak tjɛlʲɛɣa nat͡ʂała skrʲipʲɛtʲ
18.	Mother took the child to the garden matʲ ɔtvʲɛła rɛbʲɛnka v sad
19	The guys were sitting on the shore rɛbʲata sʲidjɛlʲi na bʲɛrɛɣu
20.	The store sells apples v maɣazʲinʲɛ prɔdajutsʲa jabłɔkʲi
21.	The director compared income with expense dʲirɛktɔr sravnʲił dɔxɔd s rasxɔdɔm
22.	Tall rye swayed vɨsɔkaja rɔʐʲ kɔłɨxałasʲ
23.	Flowers glittered in the valley t͡sʲvʲɛtɨ pʲɛstrɛlʲi v dɔlʲinʲɛ
24.	The wonderful smell of the forest refreshes t͡ʂudnɨj zapax lʲɛsa ɔsʲvʲɛʐajɛt
25.	The day was amazingly good djɛnʲ bɨł udʲivʲitjɛlʲnɔ xɔrɔʂ

The total number of recording sessions is 79. The total number of records analyzed in the evaluation is 1975. The distribution of the number of users depending on the number of sessions is presented in Table 2.

Table 2. Distribution of the number of users depending on the number of sessions

Number of sessions	Number of users
3	17
4	7

4. Description of experiment results

Table 3 shows examples of the original phrase and the result of its recognition by all three considered methods, as well as the quality metrics used for them.

Table 3. Examples of the original phrase and the result of its recognition by the three considered methods, as well as the quality metrics used for them (1 - phrasal intelligibility, 2 - verbal, 3 - Levenshtein distance normalized by the length of the phrase, subtracted from 1)

Phrase recognizer type	Number of users	1	2	3
The original	The site is being monitored	1	1	1
Google web	Well, let's dream plot.	0	0,25	0,55
Google API	Kirov plant observation	0	0,25	0,45
Kaldi	Observed frequently	0	0,5	0,72

Table 4 shows the average values for the recognition quality of the three considered methods, as well as the quality metrics used for them (1 - phrasal intelligibility, 2 - verbal, 3 - Levenshtein distance normalized by the length of the phrase, subtracted from 1). The study for the Google web interface using Levenshtein distance was not conducted due to the lack of the ability to automate the receipt of estimates.

IOP Conf. Series: Materials Science and Engineering 1069 (2021) 012032 doi:10.1088/1757-899X/1069/1/012032

Table 4. Average values for the recognition quality of the three considered methods, as well as the quality metrics used for them (1 - phrasal intelligibility, 2 - verbal, 3 - Levenshtein distance normalized by the length of the phrase, subtracted from 1). Results before surgery, after and after speech rehabilitation sessions

Phrase Recognizer Type		1	2	3
Google web	Until about.	0,53	0,79	-
	After about.	0,22	0,39	-
	After p.	0,38	0,61	-
Google API	Until about.	0,47	0,70	0,78
	After about.	0,15	0,31	0,68
	After p.	0,30	0,52	0,76
Kaldi	Until about.	0,51	0,73	0,92
	After about.	0,27	0,51	0,76
	After p.	0,44	0,61	0,92

5. Analysis of the obtained results

Consider the results obtained.

1. Speech quality scores obtained using the API are lower than those obtained using the web interface. This fact is significant and within the framework of the conduct can be explained by the influence of two factors.

- Manual estimation using the web-Interface was spread over time (due to manual labor costs) and because of this could be obtained taking into account the changing temporal context. However, this assessment is obtained in a manual processing mode and its effectiveness is achieved through the use of methods that do not correspond to the initial class of assessment methods. This is due to the need to perform manual actions, which contradicts the initial principle of automating the assessment.

- there is no information about the same Google recognition algorithms for data received from the web interface and API. In the conditions of the negative impact of accounting for the global information context, additional capabilities implemented in the API interface can be negative for solving the recognition problem on a limited set of phrases.

2. The tendency for the assessment to fall after surgery and its recovery during rehabilitation is confirmed regardless of the assessment method. Taking into account the verification of this approach for recognition using the web interface, it can be assumed that the results obtained for using an alternative software version based on the Kaldi library do not contradict the results obtained on the basis of expert assessments.

3. Estimates obtained on the basis of more small units of the series (phrase-word-sign) are less variable, which is confirmed by the possibility of accepting a larger number of intermediate values of the interval [0; 1] with a larger number of individual values, and the results of the experiment. An important consequence of this is the great promise of evaluations focused on individual signs than on words and,

especially, phrases. This result has been incorporated into speech quality assessment software as a separate metric based on normalized Levenshtein distance.

6. Conclusion

As a result of the study, answers were obtained to two main questions posed before its conduct:

1.The results obtained using the Google Cloud Speech web interface look more preferable than the API version. This fact can be explained:

- lack of use of the global context,
- various processing algorithms aimed at recognizing arbitrary spontaneous speech, but not individual fixed phrases taken out of context. An additional factor is the atypical nature of these phrases from GOST in comparison with standard search engine queries;
- time-spaced testing of the web interface.

2. Benefits of using the offline version of the Kaldi library showed better results on fixed phrases. This allows us to talk about the modification of the existing program for assessing the quality of speech by replacing the recognition module.

3. In addition to the initial tasks of the study, the transition to elementary recognition units (from phrases / words to signs) makes it possible to increase the sensitivity of the resulting estimates of the quality of pronunciation, regardless of the difference between the concepts of letter-sound.

The results obtained are implemented in the third version of the speech quality assessment based on intelligibility by replacing the recognition module with the Kaldi module and changing the main quality metric to the Levenshtein distance.

Acknowledgments

This research was funded by a grant from the Russian Science Foundation (project 16-15-00038)

References

[1] Likov Y, Morozova G and Kukush V 2017 Influence of features of information leakage channels on intelligibility of eavesdropped voice messages *Technology audit and production reserves* **1** 4–8

[2] Seitkulov Y N, Boranbayev S N, Tashatov N N, Davydau H V and Patapovich A V 2005 Speech information security assessing in case of combined masking signals *Journal of Theoretical and Applied Information Technolog* **12** 3270-3281

[3] Kostuchenko E, Novokhrestova D, Tirskaya M, Shelupanov A, Nemirovich-Danchenko M, Choynzonov E and Balatskaya L 2019 The Evaluation Process Automation of Phrase and Word Intelligibility Using Speech Recognition Systems *Speech and Computer* Lecture Notes in Computer Science ed A A Salah, A Karpov and R Potapova (Cham: Springer International Publishing) **11658** pp 237–246

[4] Zaridze D G, Kaprin A D and Stilidi I S 2018 Dynamics of morbidity and mortality from malignant tumors in Russia *Voprosy Onkologii* **64** 578–591

[5] Speech-to-Text: Automatic Speech Recognition *Google Cloud* https://cloud.google.com/speech-to-text

[6] Kaldi ASR https://kaldi-asr.org/

[7] Kostyuchenko E and Novokhrestova D Speech Rehabilitation After Combined Treatment of Cancer and the Formation of a Set of Syllables for Assessing Speech Quality *CEUR-WS* **2552** 11

[8] GOST R 50840-95 Speech transmission over varies communication channels. Techniques for measurements of speech quality, intelligibility and voice identification 1995

[9] Daengsi T, Wutiwiwatchai C, Preechayasboon A and Sukparungsee S 2014 IP Telephony: Comparison of Subjective Assessment Methods for Voice Quality Evaluation *Walailak Journal of Science and Technology (WJST)* **11** 87–92

[10] Brungart D S, Barrett M E, Cohen J I, Fodor C, Yancey C M and Gordon-Salant S 2020 Objective Assessment of Speech Intelligibility in Crowded Public Spaces *Ear and Hearing* **41** 68S

[11] Rix A W, Beerends J G, Hollier M P and Hekstra A P 2001 Perceptual evaluation of speech quality (PESQ)-a new method for speech quality assessment of telephone networks and codecs *2001 IEEE International Conference on Acoustics, Speech, and Signal Processing. Proceedings (Cat. No.01CH37221)* 2001 IEEE International Conference on Acoustics, Speech, and Signal Processing. Proceedings (Cat. No.01CH37221) **2** pp 749–752
[12] G.107 : The E-model: a computational model for use in transmission planning 2015
[13] Quackenbush S R, Barnwell T P and Clements M A 1988 *Objective Measures Of Speech Quality* (Englewood Cliffs, N.J: Prentice Hall)
[14] Yang W and Yantorno R 1999 Enhanced modified bark spectral distortion (embsd): an objective speech quality measure based on audible distortion and cognition model 163
[15] Wagner R A and Fischer M J 1974 The String-to-String Correction Problem *J. ACM* **21** 168–173

Software tools communication process models for Modbus TCP/RTU for diagnostics using machine learning approaches

Yuri Sosnovskiy[1], Maria Lapina[2], Vitalii Lapin[3], Massimo Mecella[2,4]

[1] Crimean Federal University named after V.I. Vernadsky, Simferopol, Russia

[2] North-Caucasus Federal University, Stavropol, Pushkin str., 1, 355017, Russia

[3] Stavropol Regional Clinical Consulting and Diagnostic Center, Stavropol, Lenina str, 304, 355017, Russia

[4] SAPIENZA Università di Roma, via Ariosto 25, I-00185 Roma, Italy

E-mail sosnovskiy.yv@cfuv.ru

Abstract. The work proposes a software solution to the problem of generating data sets that simulate information control processes in microprocessor control systems and automated process control system (APCS) at the lower and middle levels using the Modbus TCP/RTU protocol. Such datasets (dumps) with controlled characteristics, records of computer attacks and information and technical impact (ITI) of various types are advisable to use for training spacecraft recognition systems based on artificial intelligence and machine learning. Also, due to the possibility of introducing records of any types of attacks, it is possible to test intrusion detection systems, intelligent sensors of the spacecraft for their efficiency. The analysis of the existing software, the possibility of its combination to solve the problem is given. The structure of a data transmission system (DTS) fragment of a real VCE ACS and its representation as a set of device identifiers, addresses, registers and data types. The model of communication processes Modbus TCP/RTU was built, the results of modeling are presented. The features and limitations of the software used are discussed, as well as the prospects for improving the model.

Keywords: Software tools, Machine learning, Communication process models, Generating data sets

1. Introduction

Industrial network data transfer protocols are an integral part of systems that demonstrate a positive trend in the number of implementations - automated process control system (APCS), microprocessor control system (MCS), as well as intelligent devices included in the popular category of Internet of Things (IoT), Smart City, etc. [1]

It is worth noting that the set of options for the used data transfer protocols in a wide area of "smart" devices is quite large and, by no means, is not limited to the options for Modbus implementations. Even more "orthodox" automation systems allow the use of a variety of their own data transfer protocols. At the same time, most of these protocols have well-defined structures of packets or frames and sets of commands, which to some extent makes them similar to the category under consideration.

In connection with the massive use of such "intelligent" devices, the problem of monitoring traffic for potential errors in the operation of devices and various ITVs arises [2, 3]. In the case of using standard TCP/IP protocols, a wide range of monitoring, control, intrusion prevention, etc. programs are available [4]. However, when using the protocols of the lower (field) level of the automated process control system, standard solutions have low efficiency or are not applicable at all.

In this regard, the problem of developing intelligent sensors for the protocols of the lower level of APCS, MCS arises sharply [5, 6]. Accordingly, traffic sets are required, the so-called. "Dumps" that have predefined properties that allow, taking into account a priori information, to configure, test or evaluate the effectiveness of the corresponding solution. In addition, when using approaches based on artificial intelligence (AI) or machine learning (ML), large sets of real or model data are required to train the sensor system and test its effectiveness [7, 8, 9].

2. Setting goals, objectives

The aim of the work is to develop a solution that allows generating sets of traffic protocols Modbus RTU (serial interface, most often based on RS422 / 428 or 232) and Modbus TCP (based on Ethernet) with specified properties, such as the number of active devices, address ranges, sets of commands used and transmitted values, errors in data packets.

Additionally, it is required to generate a "man-in-the-middle" type ITV and record data exchange processes in a dump.

To achieve a given goal, it is required to solve a number of logically and functionally related tasks:

1. Creation of the structure of the simulated network of devices, selection of the master, endowing the slaves with the specified properties (number of inputs, ranges of values, options for changing them, number of outputs, options for controlling outputs);
2. Development of requirements for traffic dumps and, accordingly, for software components;
3. Analysis of existing software solutions;
4. Implementation of clauses 1 and 2 in the software environment of existing or own components;

In addition to purely practical and scientific tasks, the implementation of the corresponding solution will raise the educational process in the field of automation to a higher level by providing students with a tool for modeling and visualizing processes in the local self-government network and the possibility of studying the influence of certain types of ITV on the communication subsystem.

3. Selecting a fragment of the simulated system

As a real prototype for creating a model of Modbus TCP/RTU communication processes, domestic equipment for automation was chosen, manufactured by OWEN, consisting of:

- measuring-regulator OWEN TRM200. Contains two universal inputs for connecting a wide range of sensors, built-in RS-485 interface (OWEN protocol, Modbus ASCII / RTU) - 2 units;
- frequency converter of general industrial application PChV101-K18-A. Contains a built-in logic controller and a built-in RS-485 interface (OWEN protocol, Modbus ASCII / RTU) - 1 unit;
- programmable logic relay PR110 and interface module PR-MI485 for organizing work under the Modbus ASCII / RTU protocol - 1 set;
- touch panel controller OWEN SPK107, which also includes the Modbus ASCII / RTU protocol - 1 unit;

On the basis of the specified equipment, the simplest stand was assembled, in which, by means of Modbus, data are read from the meter-regulators, the frequency converter and the programmable relay are controlled. The structure of the simulated network fragment (slave devices) is revealed by the list of protocol identifiers. Table 1 provides a description of the address space and sets of transmitted parameters of the Modbus RTU data transmission network of the lower (field) level - the level of microprocessor control systems (MCS).

Table 1. Description of the logical structure of the DTS MCS

Device, id	Registers	Functions	The values

TRM1, 1	ch1 0x0000 ch2 0x0001	0x03	signed int16, static, 200 signed int16, sinusoidal, [140,260], T=2 hours
TRM2, 2	ch1 0x0000 ch2 0x0001	0x03	signed int16, sinusoidal, [100,300], T=4 hours signed int16, accidental, [180,190]
PChV1, 3	CTW 0xC34F	0x10	word (register command word)
	REF 0xC359	0x10	word (job register RS-485) Int16 (job register RS-485, frequency)
	STW 0xC417	0x03	word, static, 49392 (status word)
	MAV, 0xC421	0x03	word, static, 51061 (current values)
	FREQ, 0x3F01	0x03	word, static Int16. = REF * 10
	POW, 0x3EE3	0x03	word, static
	CURR, 0x3F0B	0x03	word, static
PR1, 7	var1, 0x200	0x03	signed int16, sawtooth, [0,8640], T=100c
	var2, 0x202	0x03	signed int16, sinusoidal, [10,400], T=1 hour
	var3, 0x204	0x03	signed int16, static, [200,500]
	var4, 0x206	0x03	signed int16, static, 120, fluctuation 5%
	var5, 0x208	0x03	signed int16, static, 255
	var6, 0x20A	0x03	signed int16, static, 111, fluctuation 10%
	var7, 0x20C	0x06	signed int16
	var8, 0x20E	0x06	signed int16

The address space of devices, registers and variables described in Table 2 is the basis for modeling. It reproduces some variants of the address spaces of real devices and can be verified in a real model using physical devices. In addition, it also makes it possible to simulate various protocol and ITV errors for the communication subsystem of the MCS. As a result, it allows to obtain test sequences of data transmitted to the data transition system (DTS) MCS with predetermined and known errors and interventions.

Table 2. Description of the simplified logical structure of the DTS MCS model, taking into account the restrictions

Device, id	Registers	Function	The values
TRM1, 1	ch1 0x0000 ch2 0x0001	0x03	word, static, 200 word, sinusoidal, [140,260], T=0,5
TRM2, 2	ch1 0x0000 ch2 0x0001	0x03	word, sinusoidal, [100,300], T=1 час word, accidental, [180,190]
PChV1, 3	CTW 0xC34F	0x06	word (register command word)
	REF 0xC359	0x06	unsigned int16 (frequency recorded by the client)
	STW 0xC417	0x03	word (status word)
	FREQ, 0x3F01	0x03	word, frequency (calculated)
	POW, 0x3EE3	0x03	word, power (calculation)
	CURR, 0x3F0B	0x03	word, current (calculation)
PR1, 7	var1, 0x200	0x03	word, sawtooth, [0,8640], T=100c
	var2, 0x202	0x03	word, sinusoidal, [10,400], T=0,5 часа
	var3, 0x204	0x03	word, accidental, [200,500]
	var7, 0x20C	0x06	unsigned int16 (255)
	var8, 0x20E	0x06	unsigned int16 (100)

The most difficult, from the point of view of communication interaction, is the PChV device, which has more than 40 registers available for reading and writing through the communication interface. Key registers and their relationship in the model are shown in Table 3.

Table 3. Key registers and their relationship in the PSC model

Register, modbus address	Description and set of register values (discrete or interrelation of output and input)
CTW, 0xC34F	Control word register. Command set: stop 2C 63, start straight 2E 63, start reverse: 2E 62.
REF, 0xC359	RS-485 reference register - setpoint. Value from 0 to 1000 (operational from 0 to 400).
FREQ, 0x3F01	If the commands start forward / reverse and (REF> 2), then FREQ = REF-2, otherwise FREQ = 0.
STW, 0xC417	Status output word. Enabled / disabled is determined by command from CTW. Set of values - based on REF setpoint value: enabled, no errors: 0F 03 (FREQ <= 100); on, overheating: 8F 03; on, current limit: 4F 03; on, overheating, current limit: CF 03; on, out of frequency range: 0B 03 (FREQ> 102); on, out of frequency range, overheating: 8B 03; on, out of frequency range, current limit: 4B 03 (FREQ> 200) AND (CURR> 5); on, out of frequency range, overheating, current limit: CB 03 (FREQ> 200) AND (CURR> 6); off, no errors: 07 03; off, overheating: 87 03 (CTW = 11363) AND (REF> 400).

CURR, 0x3F0B	If (FREQ <76) then CURR = (FREQ * 0.1) + Random (0..5 * 0.2);
	If (FREQ> 75) AND (FREQ <201)) then CURR = (FREQ * 0.06) + Random (0..5) * 0.2);
	If (FREQ> 75) AND (FREQ <201) then CURR (FREQ * 0.06) + Random (0..5) * 0.2);
	If (FREQ> 400) then CURR = 0 (residual current circuit simulation)

POW, 0x3EE3	POW= CURR * 380

The operation algorithm of this device (PChV1,3) in a simplified form, sufficient for work tasks, is as follows.

1. Turning on the drive, as well as changing the direction of its rotation is carried out by writing the CTW control word into the register 0xC34F of the PCHV with a given combination of bits, according to table 3;
2. Changing the speed setpoint - writing REF to register 0xC359. This command is transmitted multiple times with different values of the drive rotation speed setpoint (within the specified range), at random times when the drive is running;
3. Reading the current state of the STW drive - register 0xC417;
4. Reading the current values of the drive: speed FREQ (register 0x3F01), power POW (register 0x3EE3), drive current CURR (register 0x3F0B).
5. P3 and P4 are performed repeatedly, at random times during the operation of the drive;
6. Turning off the drive - writing the CTW control word to register 0xC34F with the corresponding combination of bits;
7. After some, random time - repeating cycle 1-6.

The rest of the devices (TPM1, TPM2, PR1) were not deliberately complicated. TPMs have, by analogy with their real prototype, two registers available for reading, and in the case of PR1, 5 variables are introduced into the model, three of which are available for reading with the corresponding range of values indicated in Table 2, and two variables available for records. However, the values in them do not affect the values in the var1-var3 variables.

4. Overview of existing software components

Specialists in the field of automation use various software solutions to configure and control the operation of the lower (field) level of the APCS. A practically relevant overview of the most famous of them as of 2016 is given in the material "Software for testing and setting up devices and networks based on MODBUS" at the link [10]. However, below is a number of software tools, with which it is advisable to supplement the material [10].

To simulate the master in the Modbus network, with some restrictions, primarily concerning the processes of writing to the registers of the slave device, you can use the OPC-server program (manufactured by NPF Krug). Terms of use - "freeware" limited to 30 tags. It provides a convenient interface for configuring and monitoring the status of tags, provides statistics on the use of the interface, and allows operation both in the case of a serial interface and via the TCP protocol.

The Modbus TCP/RTU emulator from ardSoft is extremely convenient for simulating the communication component of the slave devices. Simulates groups of slave devices, makes it possible to operate with an unlimited number of devices and tags. Provides access to groups of slave devices via a built-in proxy server via a serial port or Ethernet (including using localhost). Has a convenient organization of slave groups and tags. A convenient control tool is the built-in data stream sniffer.

Weprex is a program of a domestic author that simulates a Modbus master device within the RTU and TCP interfaces. Has the ability to implement all basic Modbus functions, graphing of the recorded values is available. It can read both one register and a group in one request.

Termite (Modbus RTU terminal) - fully functional terminal, free. Transfer function, templates for all Modbus functions. There is a possibility of automatic scanning according to a predefined template. Supports only serial ports.

The "EAT Console" program is designed to control remote devices from a personal computer via the RS-232 / RS-485 serial interface (USB – RS-485 converter) based on the Modbus RTU protocol. Requires Eclipse.

Multiway (Omron) - Modbus TCP, RTU terminal, Modbus functions. Auto poll according to a pre-written template. Shows response delay. Has a built-in protocol sniffer, but it is only available for serial connection.

Conclusions on the use of existing software. In general, for the task at hand – obtaining a dump of Modbus TCP/RTU protocol traffic (namely, part of Modbus traffic, excluding the TCP protocol overhead), existing software is sufficient. In particular, the program emulator Modbus devices version 2.34.63.950 has a free unlimited license and allows you to create virtual servers for connecting slave devices, select the type of protocol (Modbus RTU or TCP) and specify the port through which you can use third-party software to interact with slave devices. For each slave device, the creation of so-called. "Tags" – in fact, a description of the Modbus device registers, their addresses, valid Modbus function codes, and values. It is worth noting the possibility of programmatically generating values in such registers both based on typical laws (random, harmonic, sawtooth), and by writing scripts that can calculate the value to the current register based on other values, including from other virtual registers slave device.

It is also advisable to use software of domestic authors – Weprex as a model of the master device. It allows you to connect both serial and TCP to real hardware and, in this case, to a model slave server. Provides a basic set of Modbus functions for reading and writing registers. At the same time, the number of readable registers within one configuration set does not exceed 12. Thus, for large models it is required to create several configuration sets and manually control the reading of registers.

5. A Modbus TCP/RTU Communication Process Model for Diagnostics Using Machine Learning Approaches

To train a model with the subsequent forecasting of its characteristics, data are needed; they can be obtained by conducting a simulation experiment with the registration (recording) of dumps of communication activity. These dumps will be the basis for building a training dataframe.

To build the model, we used the “ardSoft” program-emulator of Modbus devices. The registers of the emulated Modbus network devices are interrogated using the Werpex software, the transmitted data via the TCP interface is recorded by the sniffer built into the ardSoft software. The structure of the network model of communication processes is shown in Figure 1.

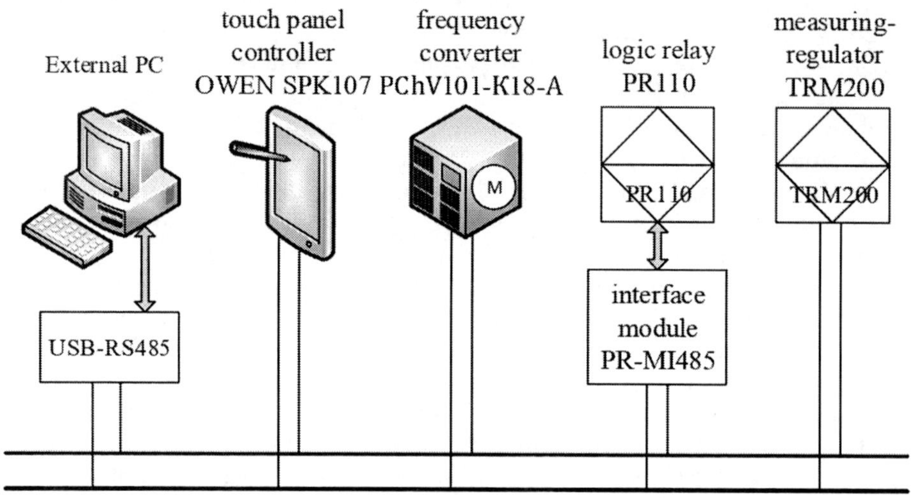

Figure 1. Network structures of the communication process model

The devices included in the model are described above in the "Selecting a Fragment of the Simulated System" section. The formation of a training sample is performed by setting a data change model, the basis for which is Table 2. Table 3 shows data describing the types, values of registers and their relationship for modeling a Modbus device with an integrated PLC - a vector frequency converter (PChVV1). Using the built-in scripting language in the Modbus TCP/RTU emulator program, it is possible to describe such behavioral features. An example of a script for describing the Modbus STW parameter is shown in Figure 2.

```
begin
  if ((CTW = 11874) OR (CTW = 11875)) then
  begin
    self := 3843;
    if (FREQ > 100) then self := 2819;
    if ((FREQ > 200) AND (CURR > 5)) then self := 19203;
    if ((FREQ > 200) AND (CURR > 6)) then self := 51971;
    if (FREQ > 400) then self := 34563;
  end;
  if ((CTW = 11363) AND (REF > 400)) then
  begin
    self := 34563;
  end;
  if ((CTW = 11363) AND (REF < 400)) then
  begin
    self := 1795;
  end;
end;
```

Figure 2. The script of the Modbus TCP / RTU emulator program

Screenshots of an example of a given configuration, polling and data fragments are shown in Figures 3 and 4.

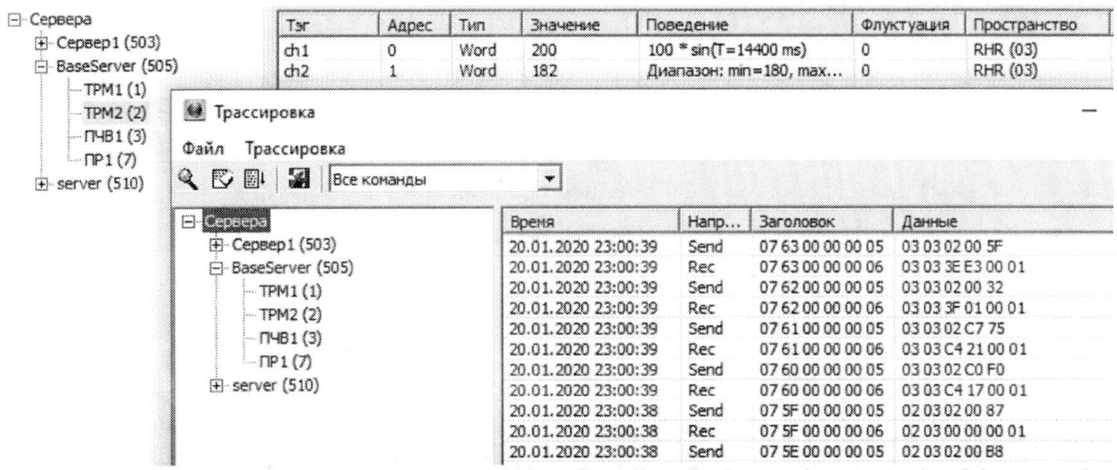

Figure 3. Setting the configuration in the emulator of Modbus devices and an example of data transfer

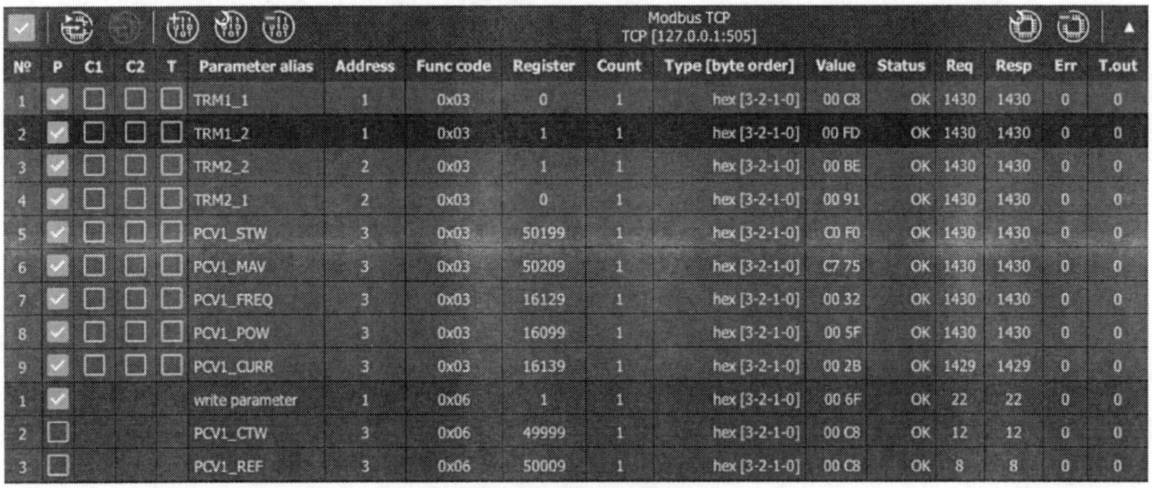

Figure 4. Polling registers

Figure 4 shows the interface of the Weprex program, which simulates the master, which polls the Modbus registers and performs read operations (code 0x03) and write operations on registers (0x06). In this case, the read functions are set at a specified interval, and for writing, manual activation of the register write mode is required.

6. Conclusion

The result of the work is a set of software tools and configuration files for slaves and masters of the Modbus TCP/RTU communication process model. The created model takes into account the specifics of the data transmission system of the simulated equipment. It can be used to generate training and test sequences for diagnosing information security problems using approaches based on machine learning and artificial intelligence.

The model is verified by comparing data sets obtained from the developed model and obtained by registering data from a real system with the specified components. In this case, the service part of the TCP packets was discarded, only Modbus frames were left. The obtained data, with some simplifications, correspond to the data circulating in a real MCS with the specified equipment.

The presented model of Modbus TCP/RTU communication processes for diagnostics using approaches based on machine learning allows:

- supplement the model with new devices, describe their registers and ranges of variable values (with some peculiarities due to Weprex restrictions, described above);
- simulate complex Modbus devices and registers, calculate values in them taking into account the values of other registers of the virtual device;
- generate training datasets for AI systems with pre-introduced distortions, variants of various computer attacks and facilitate the process of marking up such data for later use;
- generate test data sets with specified deviations and spacecraft parameters.

References

[1] "Smart" environments, "smart" systems, "smart" production: a series of reports (a series of green books) within the project "Industrial and technological foresight of the Russian Federation" 2012 / Team of authors; *Center for Strategic Research North-West Foundation* SPb Issue 4 62 p

[2] Tsapko G P, Verigo A A, Katashev A S 2016 Analysis of security risks of automated control systems for technological processes // *Internet-journal "Science"* Volume 8 No.5 http://naukovedenie.ru/PDF/55TVN516.pdf (free access). Title from the screen. Yaz. rus., eng.

[3] Khraisat A, Gondal I, Vamplew P et al. 2019 Survey of intrusion detection systems: techniques, datasets and challenges. *Cybersecur* 2, 20 https://doi.org/10.1186/s42400-019-0038-7

[3] Tiwari Mohit, Kumar Raj, Bharti Akash, Kishan Jai. 2017 INTRUSION DETECTION SYSTEM. *International Journal of Technical Research and Applications* 5 pp 2320-8163

[5] Makeenko N I, Novozhilov I O, Korabeynikov D N 2017 Intrusion detection system // *Modern scientific research and innovations* No 5 [Electronic resource]. URL: http://web.snauka.ru/issues/2017/05/82889 (date accessed: 09/14/2020)

[6] Straub Jeremy 2017 Testing automation for an intrusion detection system. 1-6. https://doi.org/10.1109/AUTEST.2017.8080473

[7] Kang M J, Kang J W 2016 Intrusion Detection System Using Deep Neural Network for In-Vehicle Network Security PLOS ONE 11(6): e0155781. https://doi.org/10.1371/journal.pone.0155781

[8] Tsaia C, Hsub Y, Linc C, Lin W 2009Intrusion detection by machine learning: A review. *Expert Systems with Applications*

[9] Zhuk A, Orel D 2019 Improved Method for Estimating Noise Immunity of Global Navigation Satellite Systems *Proceedings of the 2nd international conference on modelling, simulation and applied mathematics (MSAM2017)* https://doi.org/10.2991/ITIDS-19.2019.54

[10] Software for testing and adjusting devices and networks based on MODBUS [Electronic resource]. URL: https://habr.com/ru/post/281430/ (date accessed: 09/10/2020)

ANALYSIS OF METHODS AND SYSTEMS FOR DETECTING AND COUNTERING UAVS

Elena Basan[1], Vyacheslav Pshikhopov[1], Maria Lapina[2], Massimo Mecella[2,3]

[1] Southern Federal University, Chekov st., 2, 347922, Taganrog, Russia

[2] North-Caucasus Federal University, Stavropol, Pushkin str., 1, 355017, Russia

[3] SAPIENZA Università di Roma, via Ariosto 25, I-00185 Roma, Italy

E-mail ele-barannik@yandex.ru

Abstract. This article discusses methods for detecting UAVs in the radio frequency range. The analysis of the existing UAV detection systems is carried out. An overview of ways to counter UAVs is presented, in particular, by carrying out attacks on navigation and communication systems. A prototype of the UAV detection system, which is based on the UAV, is presented. The UAV detection function has been implemented based on the analysis of the radio frequency spectrum.

Keywords: Unmanned aerial vehicles; detection; opposition; convolutional neural network; training.

1. Introduction

Today, the problem of counteracting small-size unmanned aerial vehicles (SUAV) a copter type, is becoming quite acute. Such devices can damage critical infrastructure. In particular, this is evidenced by the latest incidents with oil bases and airports, when SUAV either blocked the operation of the facility, or their appearance led to significant economic consequences and disruption of the technological process. Counteraction to SUAV may include disruption of accessibility, disruption of functional capabilities of SUAV, disruption of the ability to transmit data through communication channels, interference and manipulation of the communication channel, substitution of transmitted in-formation. Most of the vulnerabilities that can be exploited to compromise the security of SUAV are related to threats to communication channels. If we consider counteraction to the adversary from the point of view of implementing attack scenarios, then we can say that it is necessary to implement an external attack on SUAV. Any external attack scenario begins with reconnaissance. To begin with, it is necessary to find an SUAV to obtain information on the structural and functional characteristics. Information about SUAV can be obtained by several unmasking features:

- noise (acoustic);
- optical (visible);
- infrared (thermal);
- radar (radio);

– radio frequency.

Interfering with the operation of wireless communication channels, intercepting information over radio channels and radar signals, as well as finding transceivers and repeaters of communication is an important tactical task in countering SUAV. Such methods of impact on individual SUAV, such as: interception, jamming, changing location by using a more powerful signal - are fundamental in the implementation of attacks on SUAV groups. The battle space, which previously had only four dimensions (latitude, longitude, altitude and time), now has a fifth dimension: frequency [1].

The author of [1] pays great attention to the issues of stable data transmission, in particular, the connectivity of the information flow or information operations. The concept of information operations has a key meaning in the organization of military operations. In many respects, the successful implementation by the UAV of such actions as reconnaissance, observation, delivery of payload, weapons depend on the success of information operations.

The purpose of this study is to develop a method and system for counteracting small-sized unmanned aerial vehicles (SUAV) through active attacks on communication channels.

2. Analysis of the problem of detecting enemy UAVs

Let's consider some methods of UAV detection. The main means of detecting SUAV are radio location stations. In some cases, SUAV are a difficult target for existing radars. These devices have a small effective scattering area (ESR), which makes their detection difficult enough. In particular, the maximum detection range decreases [2]. In the literature, various methods for detecting unmanned aerial vehicles have been proposed with various approaches, such as: approaches based on the analysis of audio information [3] [4] [5], approaches based on video image analysis using cameras [6], [7] and approaches based on radio frequency sensing [8], [9]. However, each of these approaches has advantages and limitations. Sound techniques can be confused by other sounds in noisy environments, have a limited range and cannot detect UAVs using noise reduction techniques. Whereas camera-based approaches require good lighting conditions, high-quality lenses and ultra-high-resolution cameras to detect SUAV s at long distances, this is certainly much more ex-pensive and more difficult to implement. Radio frequency techniques based on the use of active radar are vulnerable to radio frequency interference [10]. However, the use of deep learning methods offers a great advantage in detecting and classifying SUAV using deep neural networks (DNN), which is also known as multilayer perceptron (MLP). Newer deep learning architectures such as Convolutional Neural Networks (CNNs) are used to detect UAVs. CNNs are used to detect UAVs using CCTV cameras [11] from surveillance images in [12] and Doppler signatures in [13]. In addition, deep learning has been used to solve various time series classification problems using a one-dimensional convolutional neural network (1 DCNN) architecture. This architecture has shown high performance and reliability in the classification of electrocardiogram (ECG) and phonocardiogram (PCG) signals [14], [15]. Multichannel 1DCNN is an extended version of basic 1DCNN with the only difference – the number of input channels. Multichannel 1 DCNN has been used to classify multivariate time series, although it can handle more than one input simultaneously [16]. However, these architectures were only time-dependent and highly dependent on the characteristics of the time domain.

When implementing the UAV detection task, it is also necessary to analyze the spectrum, determine the type of signal modulation, determine at which radio frequencies the highest activity is recorded, and distinguish between noise and a useful signal. Having studied the research on the topic, it can be summarized that the trends in the development of cognitive radio have reached the possibility of solving more complex problems [17]. These tasks include: Classification of new types of modulation as certain types of signals; Classification of unknown signals; Identification of fake signals or signal replay attacks, as well as noise-reduction; Identification of cases of overlapping signals, highlighting each individual signal from the user.

To date, there is a large amount of work devoted to the application of deep learning methods for the classification of radio frequency signals. Examples include spectrum detection [18], MIMO detection [19], channel estimation and signal detection [20], physical layer communication [21], interference detection, [22], stealth suppression [23], [24], power control [25], signal alteration detection [26], and transmitter-receiver scheduling [27]. The classification of radio frequency signals can be used for various applications, for example, radio capture [28], which ultimately can be used in cognitive radio systems [29] subject to dynamic and nondeterministic interference [30]. The classification of modulation using deep neural networks is considered in [31], [32], [33], [34], where the goal is to classify this signal according to the known type of modulation. Various types of datasets have been used to train a deep neural network for modulation classification.

A review of the literature on the concept of UAV detection through object detection and classification using machine learning technologies showed that the use of machine learning facilitates the detection of UAVs in a binary classification model as "UAV" or "not UAV". However, some research in the literature goes beyond the traditional classification for a multiclass classification that identifies UAV types. The first part of the article discusses various tasks such as UAV detection, verification, UAV classification, characterization, and an approach to multiple UAV detection based on radar signals using machine learning methods. The use of high-quality 3D holographic radar along with machine learning in the time domain has been identified as an important research in UAV detection.

The authors of [35] proposed an improved method for UAV detection based on radio frequencies. The proposed methods allowed data to be obtained through a series of experiments in real-world environments using the Phantom 4 drone. Background filtering removes noise to obtain a better dataset. In addition, a frequency estimation method is used using the statistical characteristics of radio frequency signals from the UAV, which helps to identify the presence of the UAV in the controlled area. During the processing phase, UAV detection is programmed through two frequency estimates obtained from statistical fingerprint analysis and spectrum accumulation. The next step is to determine the presence of the drone using a comparison with the bandwidth, which is 9.8 MHz. To analyze and determine the performance of the current algorithm, the obtained data were compared with three well-known methods, such as ANN, CFAR and HOC. The application of the proposed method, along with the elimination of interference, led to the suppression of interference and an in-crease in the detection efficiency. In addition, the proposed methods provide an estimate of radio frequency at distances from 500 to 2800 m.

Focusing on the problem of unwanted drones flying over critical airspace to perform malicious actions, the authors of [36] propose a solution based on the analysis of RF signals. The authors investigated the ability to distinguish radio frequency signals received from UAVs from signals transmitted by other wireless devices. In addition, the detection process used a detector developed using a fusion algorithm that combines analysis of vibration from the UAV structure, detected by monitoring peak frequencies in a specific range from 20 to 50 Hz, as well as to register the shifts of the UAV structure using a detector based on wavelets into a unified classifier for UAV detection. Matthan's proposed solution has been tested in 3 different locations using 7 different UAVs from different manufacturers to ensure the reliability of the solution. The paper shows the effectiveness of using Matthan to distinguish UAV signals from signals generated by other moving devices, and the accuracy of detection, recall and accuracy exceeded 90% in scenarios where the UAV was at a distance of up to 50 meters. Even though the results dropped to 80-85% at a detection distance of 600 meters, the system performance is considered acceptable. However, Matthan's weak point is that he cannot distinguish between different UAVs.

Another study by the authors [37] was aimed at developing an autonomous unmanned aerial vehicle detection system for the classification of unmanned aerial vehicles based on radio frequency signals generated through direct communication between the unmanned aerial vehicle and the PND. The proposed system was developed using two methods. The first method was to implement an active detection and tracking system, in which the system sends a signal and continuously listens to the reflected signals from the UAV propellers. The second method used a passive detection system that listens for signals sequentially. Once the signal is received, it is extracted and analyzed further. According to the authors, the active system was able to distinguish changes between different UAV modes; for ex-ample, between stationary and flying, by significantly increasing the reflection strength of UAV rotor signals at a distance of 3 meters from a frequency of less than 100 Hz. In addition, the authors concluded that the passive eavesdropping detection system was able to distinguish the UAV signals from other signals transmitted or reflected by other objects, such as the UAV camera, engines, propellers, which could affect the communication between the UAV and the PND. In addition, the authors claim that the communication channel be-tween the UAV and the controller was detected at a frequency of 30 Hz in the frequency domain. However, the disadvantage of the system was the fact that changing the distance between the UAV and the detection system from 5 to 50 meters negatively affected the performance of the proposed detection system.

Likewise, a recent study aimed at creating an open-source UAV radio signal dataset and a system that is not only capable of detecting, but also identifying a UAV based on RF communications between the UAV and its PND was published in [38]. The main distinguishing feature between what was discussed above, and this work is that the authors chose to use deep learning algorithms as the main methodology for detecting and classifying UAVs. Three deep neural networks (DNNs) were designed in such a way that the system could achieve three main goals; detection that there is a UAV with an accuracy of 99.7% and fI_score 99.5%, detection and identification of its type with 84.5% and fI_score 78.8%, as well as detection, identification, and determination of flight modes with an accuracy of 46.8% and fI_score 43%. The authors also reported that the algorithm was unable to identify the two UAVs that were manufactured by the same company and their modes. Hence, a solution with different deep learning architectures like CNN can contribute to solving this problem.

3. Development of a system for detecting and countering unwanted SUAV

The analysis of the UAV detection methods showed that this problem can be solved using deep learning methods. At the same time, in most works it is noted that the most suitable tool for solving these problems is the use of ultra-precise neural networks. At the same time, there are even ready-made ground-based systems for detecting UAVs in the radio frequency range, which declare high detection accuracy and range. One of the tasks of the study is not only the detection of the UAV, but also the collection of detailed information about it for the purpose of further countermeasures. At the same time, the methods of countering UAVs, as a rule, are reduced to attacks on the navigation system, noise of the radio channel, suppression using electronic warfare or physical impact, which is confirmed by patent research. The existing set of data for detecting different types of UAVs, as well as the type of UAV activity, is rather limited. It is targeted only at civilian, consumer UAVs. The creation of such a base for military UAVs could be a promising task. In addition, when transmitting data via wireless communication channels, different frequencies, different types of modulation and coding of the transmitted information can be used. Therefore, it is necessary to provide not only activity detection, but also determination of other characteristics of the wireless channel. Thus, for further research, it is proposed to develop a test bench for collecting data on the UAV. The test bench layout is shown in Figure 1. In this diagram, it is proposed to use HackRF as a means of generating and receiving a radio signal. The HackRF One from Great Scott Gadgets is a software-defined radio device capable of transmitting or receiving radio signals from 1 MHz to 6 GHz. HackRF One is an open source hardware platform that can be used as a USB peripheral or programmed to run autonomously. It is designed to test and develop modern radio communication technologies. Device characteristics: Frequency range: 1 MHz - 6 GHz; Bandwidth: 20 MHz. Thus, we can say that this device is not inferior in characteristics to what was used in the previously reviewed works of other researchers. The most important characteristics are frequency range and bandwidth. ONNX (Open Neural Network Exchange) is an open software library for building deep learning neural networks. With ONNX, AI developers can exchange models between different tools and choose the best combination of these tools [39].

Intel Neural Compute Stick 2 is a computing module designed to perform tasks related to artificial intelligence, machine learning, etc. The device is based on Intel Movidius Myriad X Vision Processing Unit (VPU) - a specialized SoC containing 16 general-purpose computing cores, as well as hardware components for acceleration [40].

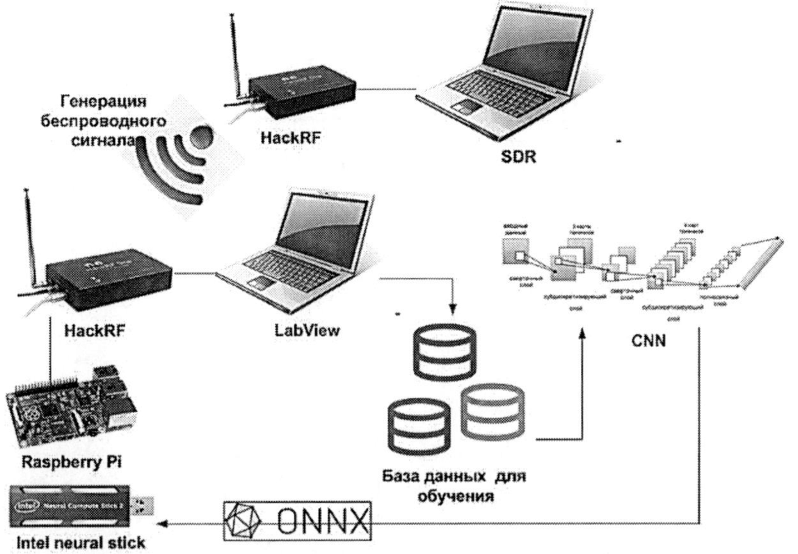

Figure 1. Generalized structural and functional diagram of the stand for preparing UAV detectors

With the help of HackRF, which is connected to a computer with SDR (Software-defined radio, SDR) installed, an RF signal is generated. The signals will be generated according to the developed algorithms, which will include emulation of retail modulation types, and network activity of the UAV. HackRF, on the other hand, will intercept this signal and convert it using the module being developed for the program of the LabView system. The resulting datasets will be fed to the input of the convolutional neural network for training, and then the generated ONNX file will be sent to the Intel Neural Compute Stick 2. Thus, the final device for detecting the UAV is Intel Neural Compute Stick 2, which has quite powerful computing resources, with compactness and small dimensions. After training the neural network and testing the result, it is necessary to integrate the detected UAV module into the UAV adaptive defense system, as well as into the enemy UAV countermeasure system. The architecture was chosen 1DCNN (one-dimensional convolutional neural network), which is shown in Figure 2.
Figure 2 shows that the architecture consists of several blocks:
1) conv - layer of one-dimensional convolution. Convolution is a linear transformation of a special kind of input. If x1 is a feature map in layer l, then the result of a two-dimensional convolution with a kernel of size 2d + 1 and a weight matrix W of size (2d + 1) × (2d + 1) on the next layer will be as follows:

$$y_{i,j}^{l} = func\Big(\sum_{-d \le a, b \ge d} W_{a,b} x_{i+a,j+b}^{l} \Big), \tag{1}$$

where func is the activation function; $\mathbf{y}^{l}_{i,j}$ is the result of convolution at level l; $\mathbf{x}^{l}_{i,j}$ is it input, that is, the output of the entire previous layer.

2) batchnormalization – batch normalization layer. Batch normalization is a technique that solves the problem of changing the distribution of each layer during training relative to the changes in the previous layers. This problem significantly slowed down the learning process and set serious requirements for the initialization of model parameters.

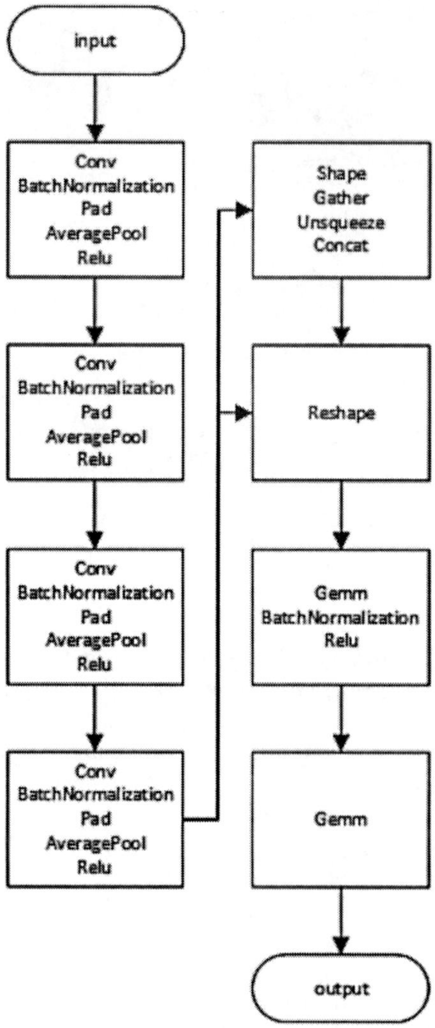

Figure 2. Block diagram of the UAV detector architecture

The idea of the method is to normalize all the values of the features of a given batch independently, by putting a checkmate expectation at 0, and standard deviation at 1. All calculations are performed according to the formulas:

$$\mu_B \leftarrow \frac{1}{m}\sum_{i=q}^{m} x_i \,, \tag{2}$$

$$\sigma_B^2 \leftarrow \frac{1}{m}\sum_{i=q}^{m} (x_i - \mu_B)^2 \,, \tag{3}$$

$$\hat{x}_i \leftarrow \frac{x_i - \mu_B}{\sqrt{\sigma_B^2 + \varepsilon}} \,, \tag{4}$$

$$y_i \leftarrow \gamma \hat{x}_i + \beta \,, \tag{5}$$

where μB – math batch waiting; σ2B – batch variance; $\hat{x}_i$ - intermediate normalization value; y_i – total normalization value; γ, β – trained parameters.

3) pad – an auxiliary layer that adds zero values to the left and right of the input matrix,

4) averagepool – average subsampling layer. Downsampling is a transformation that reduces the dimension of the input data with minimal loss of information. For example, in face recognition with a convolutional network, it is much more important to understand whether there is a face in a photo than to find out which specific pixel it begins and where it ends. Therefore, it is possible to "generalize" the distinguished features, having lost some of the information about their location, but at the same time reducing the dimension.

There are several downsampling operations:

1) operation of taking the maximum:

$$y_{i,j}^{l} = {}_{-d\leq a\geq d,}m_{-d\leq b\geq d}\ x_{i+a,j+b}^{l}\,. \quad (6)$$

2) average operation:

$$y_{i,j}^{l} = \frac{1}{d^2}\sum_{a=0,b=0}^{a=d,b=d} x_{i+a,j+b}^{l}\,, \quad (7)$$

where $y_{i,j}$ is the result of downsampling at the l level; $x_{i,j}$ is its input; d is the size of the downsampling window.

5) relu - a layer using the ReLU activation function. ReLU is one of the most popular non-linear activation functions today:

$$f(x) = \max(0, x), \quad (8)$$

where x is the input value of the function.

Basically, this function is used in the input and hidden layers of the neural network.

6) gather, Unsqueeze, Shape, Reshape – auxiliary layers that change the shape of a multidimensional array into a one-dimensional array,

7) gemm – multilayer perceptron layer,

8) input and output blocks are input and output layers, respectively.

DroneRF was chosen as the dataset. It consists of records of segments of radio frequency activity of several types of drones under the following conditions: 1) in the absence of drones, 2) in the presence of a drone operating in various modes: a) turned on and connected to the controller, b) automatic hovering in the air without physical intervention and control commands from the controller, c) flight without video recording, d) flight with video recording.

The records contain 10.25 seconds. RF activity without drones and about 5.25 sec. drone radio frequency activity in each mode. There are 227 segments in total, each segment is di-vided into 2 equal parts, each of which contains 107 samples. Each sample represents the amplitude of the signal. Note that, based on the analysis of the sources, the most effective and often implemented are attacks on the navigation system and the physical properties of the UAV.

The drones are assumed to be using a 2.4GHz WiFi network. Specification: number of channels: 2, frequency range: 1.2 GHz – 6 GHz, frequency step: <1, gain range: 0 dB - 37.5 dB, maximum instantaneous bandwidth: 40 MHz, maximum I / Q sampling rate: 200 MS / s, ADC resolution: 14 bit. Since the maximum instantaneous bandwidth is 40 MHz, at least 2 receivers (HackRF) must operate to capture the Wi-Fi spectrum (80 MHz), where the first receiver captures the lower half of the frequency band and the second receiver captures the upper half. For training, each segment was divided into 20 virtual channels (106 samples / channel). Each channel was divided into 2048 parts, from each of which an average value was taken. As a result, the minibunch for training was a two-dimensional array of size (20.2048).

UAV protection uses silencers to block control signals and GPS (optional) so that the UAV no longer receives the signals necessary for orientation. If so, the UAV is forced to land and therefore no longer poses a threat. One of the advantages of this method is that it does not damage the UAV. HackRF allows not only receiving but also sending signals. In addition, HackRF can be used to attack navigation systems and noise.

4. Conclusion

The proposed UAV countermeasure detection architecture is a compact and lightweight hardware solution with a sufficiently high computing power, which is very important for a UAV. In addition, unlike the existing UAV detection systems, this solution will be located on the UAV, which performs reconnaissance functions and, due to the possibility of changing the position, can calculate the unwanted UAV in any place, not the required distance.

5. Acknowledgment

This work was supported by the RFBR grant 18-07-00212 "Development of a method and protocol for decision-making for detecting anomalous node behavior in group control systems for autonomous mobile robots".

References

[1] *Adamy, D O* 2015 EW 104 EW against a New Generation of Threats. Boston: Artech House
[2] *Yabov K How to counteract a drone // Military Review.* – URL: https://mensby.com/technology/guns/5386-how-counteract-drone
[3] *Bernardini A, Mangiatordi F, Pallotti E, Capodiferro L* 2017 Drone detection by acoustic signature identification // IS TInt. Symp. Electron. Imaging Sci. Technol pp 60-64
[4] *Kim J, Park C, Ahn J, Ko Y, Park J, Gallagher J C* 2017 Real-time UAV sound detection and analysis system // SAS 2017 IEEE Sensors Appl. Symp. Proc pp 1-5.
[5] *Nijim M, Mantrawadi N* 2016 Drone classification and identification system by phenome analysis using data mining techniques// IEEE Symp. Technol. Hamel. Secur. HST, 2016. – pp. 1-5.
[6] *Aker C, Kalkan S* 2017 Using deep networks for drone detection // 14th IEEE Int. Conf. Adv. Video Signal Based Surveillance AVSS-2017
[7] *Saqib M, Daud Khan S, Sharma N, Blumenstein M* 2017 A study on detecting drones using deep convolutional neural networks // 14th IEEE Int. Conf. Adv. Video Signal Based Surveillance AVSS-2017
[8] *Nguyen P, Ravindranathan M, Nguyen A, Han R, Vu T* 2016 Investigating cost-effective RF-based detection of drones // DroNet 2016 - Proc. 2nd Work. Micro Aer. Veh. Networks Syst. Appl. Civ. Use co-located with MobiSys pp 17-22.
[9] *Ezuma M, Erden F, Anjinappa C K, Ozdemir O, Guvenc I* 2019 Micro-UAV Detection and Classification from RF Fingerprints Using Machine Learning Techniques // IEEE Aerosp. Conf. Proc
[10] *Abeywickrama S, Jayasinghe L, Fu H, Nissanka S, Yuen C* 2019 RF-based Direction Finding of UAVs Using DNN // 2018 IEEE Int. Conf. Commun. Syst. ICCS 2018 pp 157-161.
[11] *Aker C, Kalkan S* 2017 Using deep networks for drone detection // 2017 14th IEEE Int. Conf. Adv. Video Signal Based Surveillance AVSS 2017
[12] *Shijith N, Poornachandran P, Sujadevi V G, Dharmana M M* 2018 Breach detection and mitigation of UAVs using deep neural network // 2017 Recent Dev. Control. Autom. Power Eng. RDCAPE 2017 vol 3 pp 360-365
[13] *Kim B K, Kang H S, Park S O* 2017 Drone classification using convolutional neural networks with merged doppler images // IEEE Geosci. Remote Sens Lett vol 14 no1 pp 38-42
[14] *Kiranyaz S, Ince T, Gabbouj M* 2016 Real-Time Patient-Specific ECG Classification by 1-D Convolutional Neural Networks // IEEE Trans. Biomed. Eng vol 63 no3 pp. 664-675
[15] *Kiranyaz S, Zabihi M, Rad A, Tahir B, Ince A, T, Hamila R* 2016 Real-time PCG Anomaly Detection by Adaptive 1D Convolutional Neural Networks
[16] *Hamila Y, Liu Q, Chen E, Ge Y, Zhao J L* 2016 Exploiting multi-channels deep convolutional neural networks for multivariate time series classification // Front Comput Sci vol 10 no1 pp 96-112
[17] *Miroshnikova N E* 2013 *A review of cognitive radio systems. Information Society Technologies* 108-111

[18] *Davaslioglu K, Sagduyu Y E* 2018 Generative adversarial learning for spectrum sensing // IEEE International Conference on Communications (ICC)

[19] *He H, Wen C-K, Jin S, Li G Y* 2018 A model-driven deep learning network for MIMO detection,

[20] *Ye H, Li G Y, Juang B-H* 2018 Power of deep learning for channel estimation and signal detection in OFDM systems // IEEE Wireless Communications Letters

[21] *O'Shea T. J., Hoydis J.* 2017 An introduction to deep learning for the physical layer // IEEE Transactions on Cognitive Communications and Networking (TCCN)

[22] *Shi Y, Sagduyu Y E, Erpek T, Davaslioglu K, Lu Z, Li J* 2018 Adversarial deep learning for cognitive radio security: jamming attack and defense strategies // IEEE ICC 2018 Workshop - Promises and Challenges of Machine Learning in Comm Networks

[23] *Zheng, Q. Liu, E. Chen, Y. Ge, J. L. Zhao* Exploiting multi-channels deep convolutional neural networks for multivariate time series classification // Front. Comput. Sci. / 2016. – vol. 10. – no. 1. – pp. 96-112.

[24] *Kwang C. C., Prasad R.* Cognitive radio networks / 2009. – p.359

[25] E. G. Villegas, E. López-Aguilera, R. Vidal, J. Paradells Effect of adjacent-channel interference in IEEE 802.11 WLANs // Proc. 2nd Int. Conf. Cogn. Radio Oriented Wirel. Networks Commun. Crown Com / 2007. – pp. 118-125.

[26] *Y. LeCun, K. Kavukcuoglu, C. Farabet,* Convolutional networks and applications in vision // ISCAS 2010 – 2010 IEEE Int. Symp. Circuits Syst. Nano-Bio Circuit Fabr. Syst. / 2010. – pp. 253-256.

[27] *S. Al-Emadi, F. Al-Senaid* Drone Detection Approach Based on Radio-Frequency Using Convolutional Neural Network // IEEE International Conference on Informatics, IoT, and Enabling Technologies (ICIoT'20), 2020

[28] . Shi Y, Erpek T, Sagduyu Y E, Li J 2018 Spectrum data poisoning with adversarial deep learning // IEEE Military Communications Conference

[29] . *Sagduyu Y E, Shi Y, Erpek T* 2019 IoT network security from the perspective of adversarial deep learning // IEEE International Conference on Sensing, Communication and Networking (SECON) Workshop on Machine Learning for Communication and Networking in IoT

[30] . *Erpek T, Sagduyu Y E, Shi Y* 2019 Deep learning for launching and mitigating wireless jamming attacks // IEEE Transactions on Cognitive Communications and Networking

[31] *Shi Y., Davaslioglu K., Sagduyu Y. E.* Generative adversarial network for wireless signal spoofing // ACM Conference on Security and Privacy in Wireless and Mobile Networks (WiSec) Workshop on Wireless Security and Machine Learning (WiseML), 2019.

[32] *Abu Zainab N et al.* 2019 QoS and jamming-aware wireless networking using deep reinforcement learning // T. Erpek, K. Davaslioglu, Y. E. Sagduyu, Y. Shi, S. Mackey, M. Patel, F. Panettieri, M. Qureshi, V. Isler, A. Yener, / IEEE Military Communications Conference (MILCOM)

[33] *Restuccia F et al* 2019 DeepRadioID: Real-time channel-resilient optimization of deep learning-based radio fingerprinting algorithms // S. D'Oro, A. Al-Shawabka, M. Belgiovine, L. Angioloni, S. Ioannidis, K. R. Chowdhury, T. Melodia / ACM Intl. Symposium on Mobile Ad Hoc Networking and Computing (MobiHoc)

[34] O'Shea T, Corgan J, Clancy T C 2016 Convolutional radio modulation recognition networks // International Conference on Engineering Appli-cations of Neural Networks,

[35] *O'Shea T J, Roy T, Clancy T C* 2018 Over-the-air deep learning based radio signal classification // IEEE Journal of Selected Topics in Signal Processing

[36] *Nguyen P et al Matthan* 2017 Drone presence detection by identifying physical signatures in the drone's rf communication // H. Truong, M. Ravindranathan, A. Nguyen, R. Han, T. Vu, / Proceedings of the 15th Annual International Conference on Mobile Systems Applications and Services ser. MobiSys'17 ACM

[37] *Nguyen P et al* 2016 Investigating cost-effective rf-based detection of drones // M. Ravindranatha, A. Nguyen, R. Han, T. Vu / Proceedings of the 2Nd Workshop on Micro Aerial Vehicle Networks Systems and Applications for Civilian Use ser. DroNet'16 ACM

[38] *Zhao M et al* 2018 Through-wall human pose estimation using radio signals // T. Li, M. Abu Alsheikh, Y. Tian, H. Zhao, A. Torralba / The IEEE Conference on Computer Vision and Pattern Recognition (CVPR)

[39] Open Neural Network Exchange [Electronic resource]. URL: https://onnx.ai/ (date of accessed 07.11.2020)

[40] Intel Neural Compute Stick. Artificial intelligence on a flash drive - 2. [Electronic resource]. URL: https://habr.com/ru/company/intel/blog/430492/ (date of accessed 07.11.2020).

The logical-probabilistic model for assessing the information security assessing of the critical information infrastructure subject under destructive influences

E A Maksimova[1], M A Lapina[2], V V Baranov[3], O S Lauta[4]

[1] Russian Technological University MIREA (RTU MIREA) Moscow, Russia

[2] North-Caucasus Federal University, Stavropol, Russia

[3] South Russian State Polytechnic University (NPI) named after M.I. Platov, Novocherkassk, Russia

[4] Military Academy of Communications named after Marshal of the Soviet Union S.M. Budyonny, Saint-Petersburg, Russia

E-mail maksimova@mirea.ru

Abstract. Security of the subject of critical information infrastructure (SCII) is one of the key issues of our time at the international level. One of the guarantors of stability and information security of society and the state as a whole is the reliability of the SCII functioning. The currently existing methods and models for assessing reliability, as applied to CII, are considered without taking into account the factor of influence of the offender, which can have a destructive effect on the subject of CII. This leads to significant errors in the analysis of the information security of the subject of CII, therefore, reduces the effectiveness of the information protection means declared for CII objects.

The proposed logical-probabilistic model is associated with the use of the specifics of the regulatory and legal framework in the field of SCII safety; structured detailing of the CII subject, taking into account the specifics of the subject; stages of the life cycle of the SCII information protection system; highlighted destructive malicious influences of an infrastructural nature; interrelationships of the selected destructs with a number of vulnerabilities on CII objects. The model takes into account physical, physicochemical, chemical, biological, operational factors of threat realization as factors affecting the decrease in the reliability of the functioning of the CII subject. The assessment of reliability indicators is carried out taking into account the characteristics of recoverable and non-recoverable objects, backup methods. The model allows you to assess the reliability of the subject of CII as an integrative characteristic, which is an indicator of the assessment of information security.

Keywords: Critical Information Infrastructure, International Level, Security of Society, Destructive Effect, Significant Errors, Logical-Probabilistic Model, Legal Framework, Recoverable and Non-Recoverable Objects, the Assessment of Information Security.

1. Introduction

The legislation of the Russian Federation [1-5] defines the areas of operation of enterprises and organizations that affect the quality of life and health of society, in order to ensure their safety. State bodies, government agencies, Russian legal entities and individual entrepreneurs operating in the selected areas, which are subjects of critical information infrastructure (SCII), are responsible for maintaining the stability of the area to which they belong.

Necessity Assessment of the reliability of the functioning of CII subjects is characterized by the possibility of assessing the probability of failure-free operation of CII facilities and preventing failures in the functioning of CII spheres, which guarantees stability and information security of society and the state as a whole. Thus, the problem of assessing the reliability becomes key in the comprehensive assessment of IS SCII.

2. Models and methods

The "input" research was carried out in three directions: analysis of models and methods for assessing the reliability of the subject of CII, analysis of ways to improve the reliability of the subject of CII, analysis of existing software systems for assessing the reliability of technical systems.

In the course of analyzing the models and methods for assessing the reliability of the subject of CII, the following were considered: the probabilistic method (the model "block-diagram"), the logical-probabilistic method, the model "fault tree", Markov model [6-13]. The following criteria are defined for the analysis: reliability assessment based on the interdependence and interrelation of CII objects; calculation of indicators in a time interval; calculation of indicators at time t; sufficiency for the research of complex systems; relative ease of calculation; reliability assessment when implementing various methods of redundancy; reliability assessment when applying various recovery strategies; high estimation accuracy; simplicity of circuit construction. According to the selected criteria, the probabilistic method was determined as the most effective (the model "block-diagram").

To analyze the ways to improve the reliability of the functioning of the CII subject, the following criteria are highlighted: the accuracy of calculations, the convenience in reserving sub-objects, the relative simplicity of calculations, the involvement of reserve sub-objects in the work, the inclusion of reserve sub-objects as needed, the guarantee of replacing a failed sub-object with a reserve one. The following methods of redundancy are analyzed: loaded redundancy, unloaded redundancy, light redundancy, sliding redundancy. Based on the analysis results, the most effective and appropriate backup methods are loaded and loaded redundancy.

In the research of software systems for assessing the reliability of technical systems, ARBITR (PC ASM SZMA), ASONIKA-K, ASRN (2000, 2002), Isograph (England, USA), RAM Commander (Israel), Windchill Quality Solutions (Relex) (USA) were considered. By criteria: the ability to take into account the specifics of CII; taking into account the likelihood of the implementation of threats leading to a malfunction of the CII; the possibility of an integral assessment for the system; analysis of the interaction scheme of elements in the form of a block diagram (sub-objects, CII objects); system lifecycle accounting; the ability to assess the reliability of systems other than industrial. As a result of the analysis, the WQS software package was determined as satisfying the largest number of selected criteria. However, to solve the problem, it is necessary to fulfill all the criteria, which leads to the need to develop an appropriate model.

3. Description of the simulation system

To implement the modeling process, sub-objects are selected inventory units of the CII subject: $InvU_1$ – PC, $InvU_2$ – servers, $InvU_3$ – ACO, $InvU_4$ – system software, $InvU_5$ – application software, $InvU_6$ – SRZI, $InvU_7$ – communication channels, $InvU_8$ – peripheral devices; $InvU_9$ – CII sub-object specific for each individual CII subject, $InvU_{10}$ –- processed and stored information, including configuration information. The reliability of the CII objects and the CII subject as a whole depends on their reliability.

In order to reduce the error in the assessment of information security, the life cycle of the subject of CII has been determined and analyzed in accordance with GOST 34.601-90 [17]. On the basis of the "Life Cycle of SIS of the CII subject", the types of destructive influences that lead to the emergence of vulnerabilities in the protection system of CII objects at different stages of the life cycle of the CII subject are identified. Vulnerabilities can be used to implement threats, attacks by malefactors on the subject of CII as a whole. Selected types of destructive influences (destructions): Destr 0 – Errors associated with the primary development of the plant's AS; Destr 1 – errors at the categorization stage; Destr 2 – Infrastructure Analysis Errors; Destr 3 – Errors in the formation of requirements for the information security system of the CII subject; Destr 4 – Errors in the development of the concept of information security for a CII subject; Destr 5 – Errors in the development of technical specifications; Destr 7 – Errors in the technical design; Destr 8 – Errors entering; Destr 9 – Errors while accompanying the subject of CII [18-20].

Vulnerabilities of typical systems are analyzed by the method of constructing vulnerability trees. The relationship between the selected destructs and a number of vulnerabilities has been revealed. Destructions Destr 0, Destr 1, Destr 2, Destr 3, Destr 4, Destr 5, Destr 6, Destr 7 are associated with the vulnerability of the lack of compliance with information security requirements in the design of information security solutions. Destruction Destr 8 is associated with a vulnerability in the absence or insufficient level of knowledge in the field of information security among the maintenance personnel. Destr 9 destruct identifies a number of vulnerabilities, such as the problems of delimiting and controlling access for contractors associated with the need to provide temporary access to a limited amount of equipment without the ability to affect the rest of the system, as well as cancellation of such access at the end of work; lack or insufficient level of knowledge in the field of information security among service personnel; long service life of vulnerable components due to the complexity of upgrading equipment and systems.

Physical, physicochemical, chemical, biological, operational, implementation of threats by intruders have been identified as factors affecting the decrease in the reliability of the functioning of the CII subject.

When assessing the reliability of SCII, the indicators of the reliability of restored and non-renewable objects are evaluated. For recoverable ones - mean time between failures, mean time between failures, probability of recovery, mean time to recover, availability factor, technical utilization factor. For non-recoverable – probability of failure, probability of failure-free operation, time of failure-free operation, failure rate.

4. Discussion

The proposed mathematical model for assessing the reliability of the subject of critical information infrastructure is represented by the function of the model for assessing the reliability of the subject of CII:

$$F = F(P_{thr}, P_{rel_i}),$$

where P_{rel_i} – probability of failure-free operation of CII facilities;

P_{thr} – the likelihood of an IS threat.

P_{thr}– can take the following values:

$$P_{thr} = P_{Destr_i},$$

$$P_{thr} = P_{exp}.$$

where P_{Destr_i} – the likelihood of the implementation of threats generated by destructions at different stages of the CII life cycle;

P_{exp} – expert assessment of the likelihood of threats.

In the model for assessing the reliability of the subject of CII P_{Destr_i} is predicted by the least squares extrapolation method based on the existing statistics on the implementation of threats at enterprises and organizations operating in the areas of CII.

If it is necessary to assess the reliability of the CII subject, based on the probability of the implementation of threats received by the expert, the software provides the ability to enter the value P_{exp}, different from the predicted value P_{Destr_i} based on existing statistics.

P_{Destr_i} for destructs related to infrastructure errors, can take on the values:

$$P_{Destr_i} = P_{Destr2}, P_{Destr10}.$$

P_{Destr_i} for destructs associated with the remaining stages of the life cycle of the CII subject, it can take on the values:

$$P_{Destr_i} = P_{Destr1}, P_{Destr3}, P_{Destr4}, P_{Destr5}, P_{Destr6}, P_{Destr7}, P_{Destr8}, P_{Destr9}.$$

The occurrence of errors, destructions, at different stages of the life cycle of the information security system of the CII subject generates vulnerabilities in the CII protection system, which attackers can exploit to implement threats to the CII subject.

To rate P_{exp} an expert can use the value of the frequency probability of the implementation of threats [16]:

$$P_{exp} = \lim_{N\to\infty} \frac{n}{N},$$

where N is the total number of incidents that occurred during the year; n - the number of specific incidents per year.

To get the value P_{Destr_i} – the likelihood of threats being realized is extrapolation by the least squares method [16]:

$$P_{Destr_i} = a * X + b,$$

where X is a symbol of time;

a – coefficient calculated by the formula:

$$a = \frac{\sum_{i=1}^{n}(act*X)-(\sum_{i=1}^{n} X*\sum_{i=1}^{n} act)/n}{\sum_{i=1}^{n} X^2-(\sum_{i=1}^{n} X)^{\frac{2}{n}}}$$

b – coefficient calculated by the formula:

$$b = \frac{\sum_{i=1}^{n} act}{n} - \frac{a * \sum_{i=1}^{n} X}{n}$$

where *act* is the actual values of a number of dynamics (in this case, the values from the threat statistics in the CII spheres);

n – the number of levels in the time series (the number of years (months) for which the statistics of threats in the areas of CII were taken).

Probability of failure-free operation of CII facilities P_{rel_i} depends on the probabilities of failure-free operation of CII subobjects and their interconnections:

$$P_{rel_i} = P(InvU_1, InvU_2, InvU_3, InvU_7, InvU_9)$$

where $InvU_1$ – PC; $InvU_2$ – server; $InvU_3$ – ACO; $InvU_7$ – channels of connection; $InvU_9$ – CII sub-object specific for each individual CII subject.

When calculating the probabilities of failure-free operation of CII objects, we do not take into account such CII sub-objects as system and application software, information security tools, peripheral devices, processed and stored information, since these CII sub-objects in case of failure will not critically affect the CII object in comparison with the output out of order, for example, ASO, a server for collecting technological information or a dispatcher's workstation.

In the course of the research, all indicators of the reliability of recoverable and non-recoverable CII sub-objects were considered. The analysis showed that to assess the reliability of the subject of CII, the indicator "probability of failure-free operation".

To calculate the assessment of the reliability of CII facilities P_{rel_i} three cases are considered, where pi is the probability of failure-free operation of CII sub-objects:

1. Scheme with a series connection of n dependent sub-objects of the CII:

$$P_{rel_i} = \prod_{i=1}^{n} p_i\,(t).$$

2. Scheme with parallel connection of n dependent KII sub-objects:

$$P_{rel_i} = 1 - \prod_{i=i}^{n}(1 - p_i\,(t)).$$

3. Serial-parallel circuit. To assess the reliability of CII facilities, a phased simplification of the circuit and the use of formulas to assess the reliability of circuits with serial and parallel connection are used.

Assessment of the reliability of the subject of CII $P_{subj}$is calculated similarly to the schemes for assessing the reliability of CII objects: using calculations for parallel and serial connection of CII objects. After assessing the reliability of the CII objects, to assess the reliability of the CII subject, the formation of a structural diagram of the interconnection of the CII objects is carried out and, based on the probabilities of the failure-free operation of the CII objects and the probability of the implementation of threats, the reliability assessment of the CII subject is calculated.

$$P_{subj} = P_{rel_i} * (1 - P_{thr}).$$

In the event that the reliability of the CII subject is insufficient (P_{subj} <= 0,5), it is necessary to apply redundancy methods to improve reliability.

To increase the reliability of the CII subject, the following methods and calculations are used to reserve CII sub objects, where p_i is the probability of failure-free operation of the CII sub objects:

1. When using loaded redundancy (for a system with a serial connection of n sub objects with general redundancy with a multiplicity of k), the reliability of the CII subject is calculated as follows:

$$P_{subj} = 1 - (1 - P)^{k+1} = 1 - \left(1 - \prod_{i=1}^{n} P_i(t)\right)^{k+1}.$$

2. When using loaded redundancy (for a system with a serial connection of n objects with separate redundancy with a multiplicity of k), the reliability of the CII subject is calculated as follows:

$$P_{subj} = \prod_{i=1}^{n}(1 - (1 - P_i(t))^{k+1}).$$

3. When using unloaded redundancy (systems with unloaded redundancy of multiplicity k (total sub objects $k + 1$)), the reliability of the CII subject is calculated as follows:

$$P_{subj} = 1 - \frac{1}{(K+1)!}\prod_{i=1}^{k+1}(1 - P_i\,(t)).$$

Based on the results of the assessment of the reliability of the CII subject before and after the reservation, the reliability gain coefficient is calculated:

$$G_p = \frac{P_{before}}{P_{after}},$$

where P_{before} – assessment of the reliability of the subject of CII before reservation;

P_{after} – assessment of the reliability of the subject of CII after reservation

To assess the level of reliability of the subject of CII, the value 0.5 was chosen as the threshold value. In the case of assessing the reliability of the subject of the CII below 0.5, recommendations are proposed to improve the reliability of the CII subject, which include the methods of loaded and unloaded redundancy.

To implement the proposed model, a functional model for assessing the reliability of the subject of critical information infrastructure has been developed in the form of a context diagram in the IDF0 notation (figure 1).

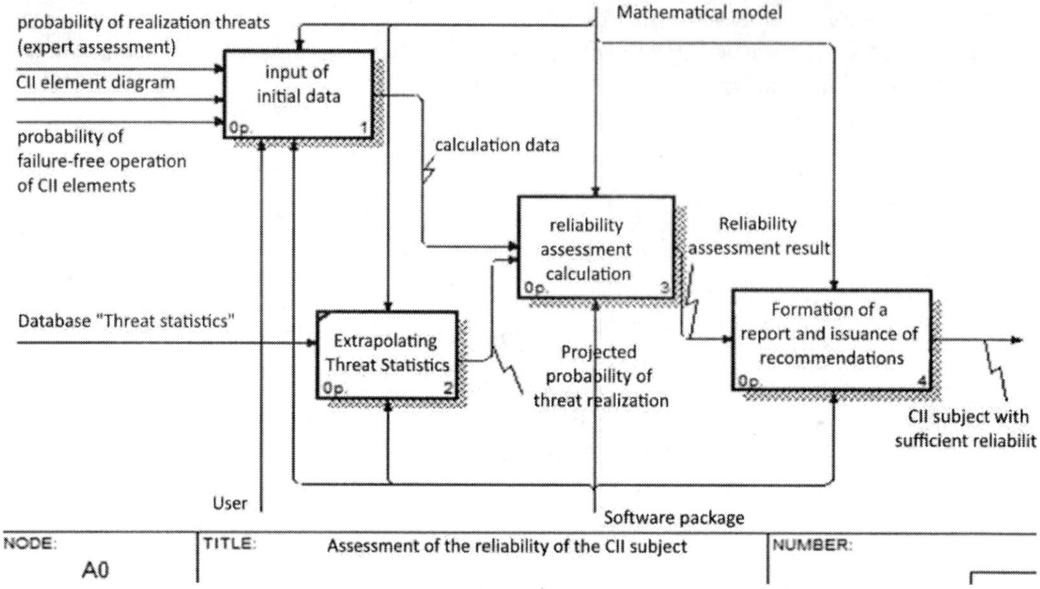

Figure 1. Context diagram for assessing the reliability of the subject of critical information infrastructures in the IDEF0 nutation

The architecture of the software package for the implementation of this model is shown in figure 2.

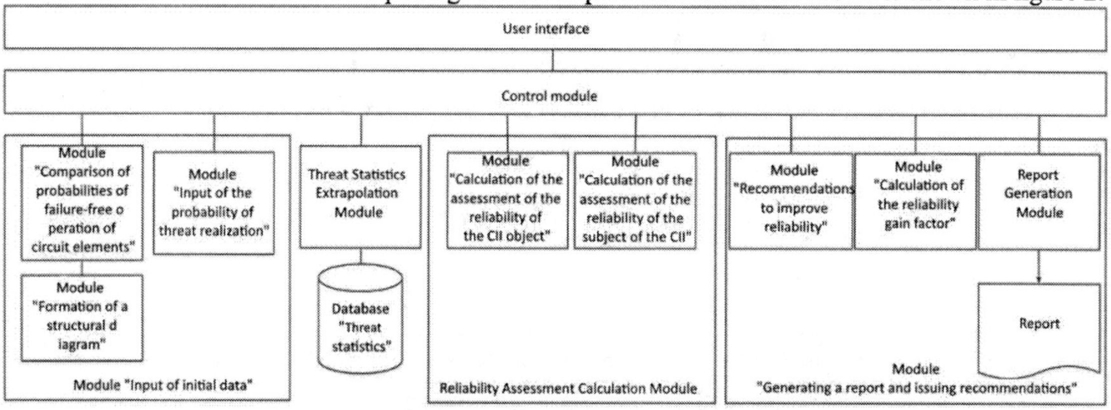

Figure 2. Architecture of the software package for the implementation of the reliability assessment model for the subject of critical information infrastructure

5. Experimental research

In the course of the work, experimental studies were carried out to assess the reliability of CII subjects operating in various industries, which include various types of CII objects. The problem to be solved is to increase the reliability of the subject of the CII if the initial assessment of the reliability of the CII subject is less than 0.5.

Experimental studies involved not only assessing the reliability of the CII subject, based on the presented schemes of CII objects, but also, if necessary, the subsequent application of backup methods to the CII sub objects in order to increase the reliability of the CII subject.

So, for example, the assessment of the reliability of the subject of the CII of the hospital, which includes the object of the CII, which is the personal data information system (PDIS), a structural diagram was taken for the experiment (figure 3).

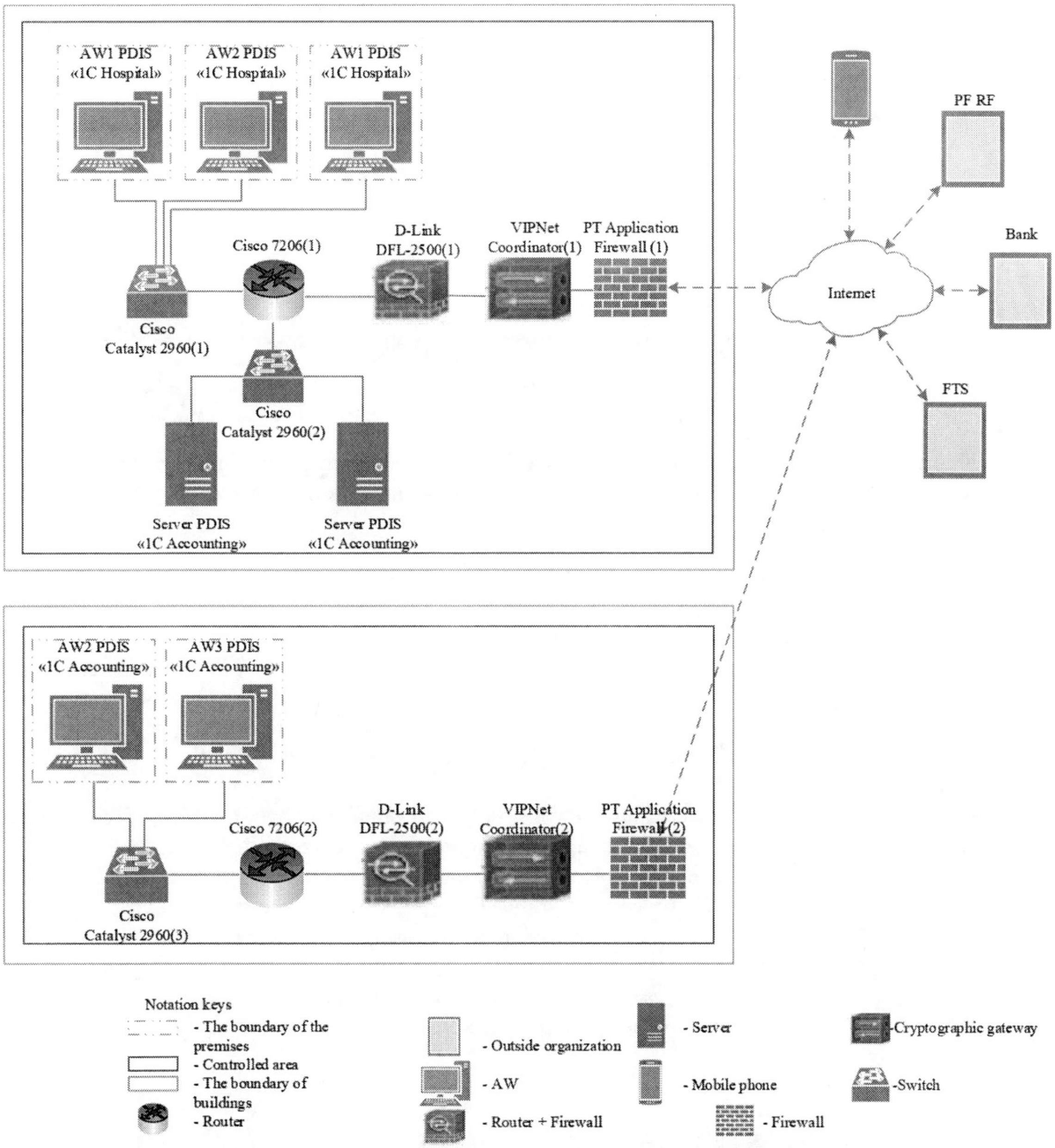

Figure 3. Structural diagram of the object of the critical information infrastructure of the information system of personal data of the subject of the CII – Hospital

To conduct an experiment according to the scheme in Figure 3, a special scheme is built (figure 4), which displays the mutual influence and interaction of CII sub objects. In accordance with the values of the probabilities of no-failure operation of the CII sub objects from the experimental research plan, the reliability of the CII subject – Hospital.

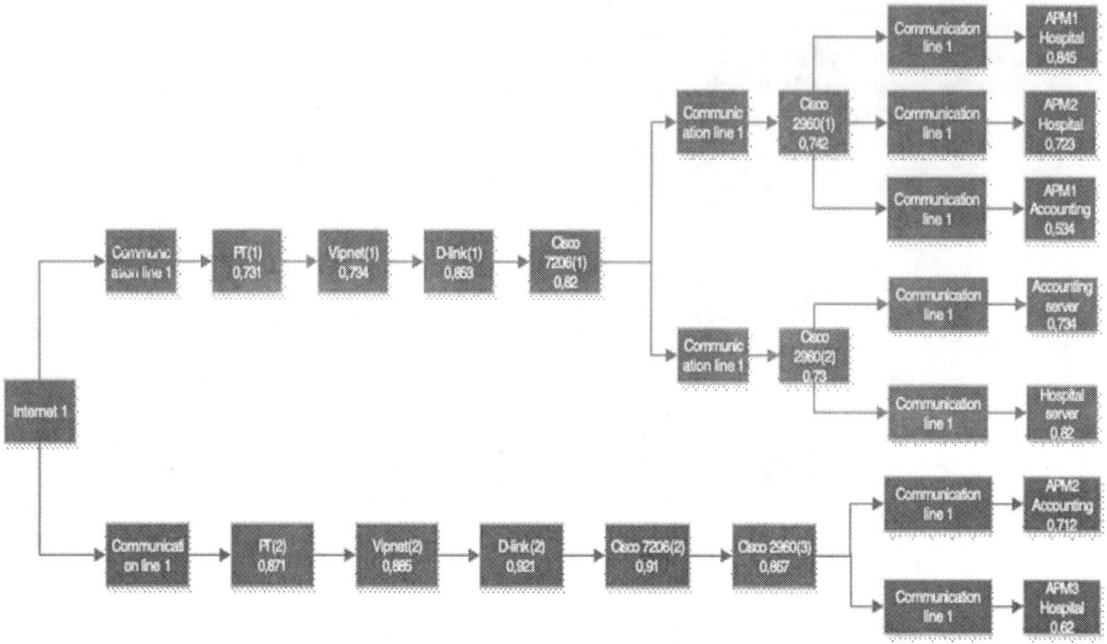

Figure 4. Scheme of interaction between the critical information infrastructure sub objects of the critical information infrastructure object of the personal data information system. Experiment - Hospital (Screen Copy)

To assess the reliability of the subject of CII, the probability of the implementation of threats, predicted on the basis of the existing statistics of threats implemented at enterprises and in organizations operating in the areas of CII, was taken into account.

The assessment of the reliability of the subject of the CII-Hospital showed an insufficient level of reliability. In connection with this, redundancy methods were applied to CII sub-objects. Figure 5 shows a diagram of the CII ISPD object using the methods of loaded and unloaded redundancy.

For the CII ISPD object, loaded and unloaded redundancy was applied for the following KII sub objects:

- For D-link, acting as routers and firewalls, unloaded redundancy was applied, since it is more efficient for the system so that the backup D-links come into operation after the main.
- For routers Cisco 7206 routers, unloaded redundancy was applied, since it is more efficient for the system so that the standby Cisco 7206 comes into operation after the failure of the main.
- For the Cisco 2960 switches, unloaded redundancy was applied, since it is more efficient for the system so that the redundant Cisco 2960s come into operation after the failure of the main.
- For ISPD servers, a loaded reservation was applied, since it is necessary that the backup servers collect, process and store information on a par with the main.

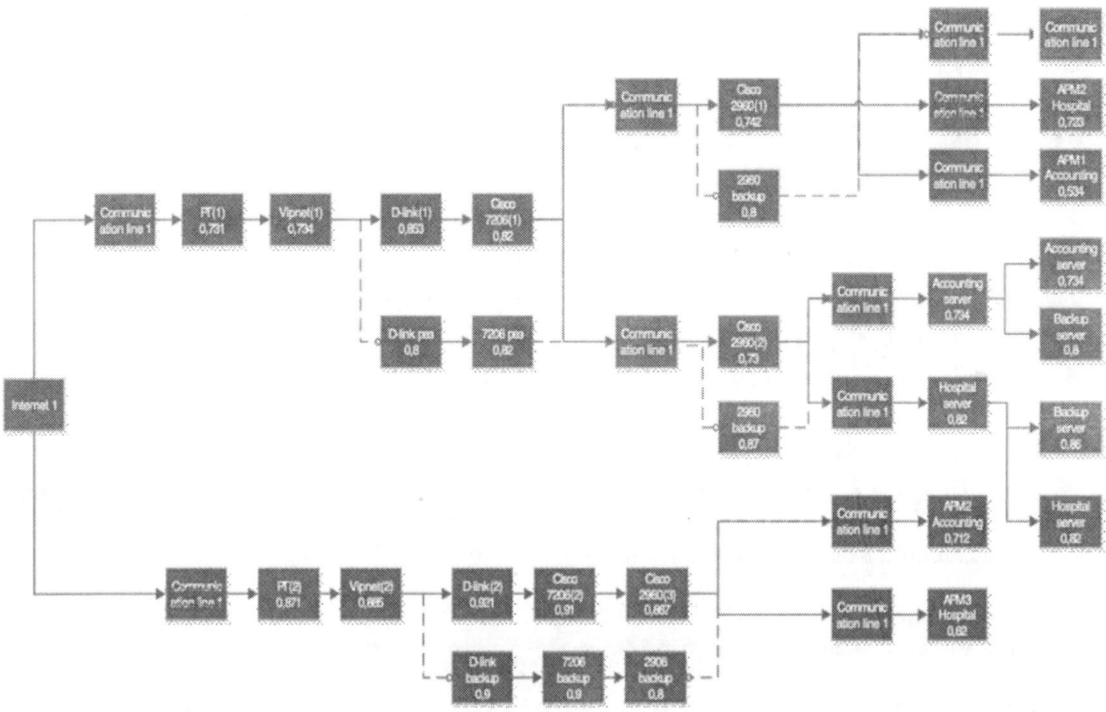

Figure 5. Scheme of interaction of the critical information infrastructure sub-objects of the critical information infrastructure object of the personal data information system. Application of redundancy. Experiment – Hospital (screen copy)

The analysis of the results of experimental studies of the software complex for the implementation of the reliability assessment model of the CII subject is presented in table 1.

Table 1 – Analysis of the results of the experimental research "Subject of CII – Hospital"

Subject CII – Hospital		
Assessment of object reliability		
An object	Reliability assessment before redundancy	Post-backup reliability assessment
PDIS	0,69	0,89
Assessment of the reliability of the subject of CII		
A subject	Reliability assessment before redundancy	Post-backup reliability assessment
Hospital	0,43	0,56
Reliability win rate		
Reliability win rate	0,77	

6. Analysis of experimental research results

The results of experimental studies show an increase in the level of reliability of the subject of CII using the results of the proposed model. So, as a result of the application of backup methods, the level of reliability of the subject of CII – Hospital increased by 0.13 and the reliability gain was 0.77.

7. Conclusion

The proposed logical-probabilistic model is associated with the use of:

- the specifics of the regulatory framework in the field of CII safety,
- structured detailing of the CII subject, taking into account the specifics of the subject,
- stages of the life cycle of the CII information protection system,
- highlighted destructive malicious influences of an infrastructural nature,
- the relationship of the selected destructs with a number of vulnerabilities on CII objects,
- as factors affecting the decrease in the reliability of the functioning of the CII subject – physical, physicochemical, chemical, biological, operational, implementation of threats by intruders,
- evaluating the reliability indicators of recoverable and non-recoverable objects, backup methods and allows you to assess the reliability of the subject of CII as an integrative characteristic, which is an indicator of the assessment of information security.

8.Acknowledgments

This work was supported by the Russian Ministry of Science (Information security, project №3)

References

[1] On the security of the critical information infrastructure of the Russian Federation: Federal Law of July 26 2017 N187-FZ (as amended and supplemented) [Electronic source]. – URL: http://www.consultant.ru/document/cons_doc_LAW_220885/
[2] On approval of requirements for ensuring the security of significant objects of critical information infrastructure of the Russian Federation: Order of the FSTEC of December 25 2017 N239 (as amended and supplemented) [Electronic source] – URL: https://fstec.ru/en/53-normotvo
[3] On the approval of requirements for the creation of security systems for significant objects of the critical information infrastructure of the Russian Federation and ensuring their functioning: Order of the FSTEC of December 21 2017 N235 [Electronic source]. – URL: https://fstec.ru/normotvorcheskaya/akty/53-prikazy/1589-
[4] On approval of the Procedure for informing the FSB of Russia about computer incidents, responding to them, taking measures to eliminate the consequences of computer attacks carried out against significant objects of the critical information infrastructure of the Russian Federation: Order of the FSB of Russia dated June 19 2019 N282 [Electronic source] – URL: https://www.garant.ru/products/ipo/prime/doc/72198410/
[5] On approval of the List of information submitted to the state system for detecting, preventing and eliminating the consequences of computer attacks on information resources of the Russian Federation and the Procedure for submitting information to the state system for detecting, preventing and eliminating the consequences of computer attacks on information resources of the Russian Federation: Order of the FSB of Russia from July 24 2018 No367 [Electronic source] - URL: https://www.garant.ru/products/ipo/prime/doc/71941504/
[6] Shubin R A 2012 Reliability of technical systems and technogenic risk / R.A. Shubin Tambov Publishing and Printing Center FGBOU VPO TSTU P 24-51
[7] Mozhaev A S 1999 General logical-probabilistic method for analyzing the reliability of complex systems Uch allowance L VMA 68 p
[8] Mozhaev A S, Gromov V N 2000 Theoretical foundations of the general logical-probabilistic method of automated system modeling SPb VITU 144 p

[9] Remenyte R, Andrews J D 2006 A simple component connection approach for fault tree conversion to binary decision diagram Availability Reliability and Security ARES 2006 / The First International Conference 20-22 April P 8
[10] Woo Sik Jung Sang Hoon Han and Jaejoo Ha 2015 A fast BDD algorithm for large coherent fault trees analysis Reliability Engineering & System Safety Vol 83 Issue 3 pp 369-374
[11] GOST R IEC 61165-2019 Application of Markov methods 2019 M Standartinform [Electronic source] – URL: https://files.stroyinf.ru/Data2/1/4293727/4293727169.pdf
[12] GOST R 51901.15-2005 (IEC 61165: 1995) Risk management. Application of Markov methods 2005 M Standartinform [Electronic source] – URL: http://docs.cntd.ru/document/1200041677
[13] Zhuk A, Orel D 2019 Improved Method for Estimating Noise Immunity of Global Navigation Satellite Systems Proceedings of the 2nd international conference on modelling, simulation and applied mathematics (MSAM2017) https://doi.org/10.2991/ITIDS-19.2019.54
[14] Kalmykov I, Lapina M, Kalmykov M, Voloshin E, Provornov I 2019 Development of Imitation-resistable Authentication Protocol for Low-orbital Space Satellite Sommunication System: CEUR Workshop Proceedings YSIP3 – Proceedings of the Young Scientist's Third International Workshop on Trends in Information Processing
[15] Kalmykov I, Lapina M, Kononova N, Kalmikov M 2018 Development of the protocol "electronic cash" with inspection correction rules of the electronic e-cash number for e-commerce systems: CEUR Workshop Proceedings REMS 2018 – Proceedings of the 2018 Multidisciplinary Symposium on Computer Science and ICT Pp 147-153
[16] Linets G I, Melnikov S V, Govorova S V, Medenec V V, Lapina M A 2018 Decrease energy consumption of transport telecommunication networks due to the usage of stage-by-stage controlling procedure: CEUR Workshop Proceedings REMS 2018 – Proceedings of the 2018 Multidisciplinary Symposium on Computer Science and ICT Pp 181-190
[17] GOST 34.601-90 2009 Interstate Standard Information Technology. Set of standards for automated systems. Automated Systems. Stages of Creation. M Standartinform [Electronic source] – URL: http://docs.cntd.ru/document/gost-34-601
[18] Maksimova E.A. 2020 Evaluation of information security of the subject of critical information infrastructure under destructive influences [Text] monograph / E. A. Maksimova; Feder. state ed. educated. institution of higher. education "Volgogr. state university" Volgograd: VolGU Publishing House 95 p
[19] Maksimova E A Research of algorithms for secure data transfer between objects of critical information infrastructure // Collection of reports of the XXIII plenum of FUMO VO IS and the All-Russian scientific conference "Fundamental problems of information security in the context of digital transformation" (INFOSE-2019) reports of the XXIII plenum of FUMO VO IB and the All-Russian scientific conference. Resp. editor: V.I. Petrenko 2019 pp 157-163
[20] Maksimova E A, Shikhverdiyeva A Sh 2019 Management of the operation of critical information infrastructure facilities // Materials of the XVI All-Russian School-Conference of Young Scientists "Management of Large Systems" Tambov Publishing Center of the Federal State Budgetary Educational Institution of Higher Education "TSTU" 2019 pp 392-397

Algorithm for predicting the health of an information system after exposure to cyberattacks

A V Dushkin[1], S S Kochedykov[2]

[1]National Research University of Electronic Technology, Moscow, Zelenograd, Shokin square, 1, 124498, Russia

[2]Voronezh Institute of High Technologies, Voronezh, Lenin str., 73A, 394043, Russia

E-mail a_dushkin@mail.ru

Abstract. The presented work proposes a systematic approach to the development of software for cyberattack detection systems and predicting the health of the information system. A formalized descriptive model of an intrusion detection system is presented. A complex of models is described that allow making diagnostic decisions about the state of software performance.

Keywords: Information security; Information system; Information; Cyberattack; Software

1. Introduction

The analysis performed allows us to state the fact that the information system (IS) [1] is an object of destructive influences (DI) aimed at violating information security. An example of such destructive information influences are cyber attacks [2], with the help of which criminals try to penetrate the databases, block or disable electronic objects, disrupt the work of the security service, destabilize the work of the IS. In these conditions, in the course of the structural-parametric synthesis of IS, the issues of comprehensive information security support come to the fore by: a) timely detection, identification and monitoring of destructive influences on IS [3]; b) operational control of the functional efficiency of the IS in conditions of destructive influences [4]; c) assessing the impact of cyberattacks on the functioning of IS and managing protection measures and actively countering cyberattacks [5].

2. Formulation of the problem

We represent the structure of the IS in the form of a classical unit [6-8]. With its help, it is possible to give a unified description of the deterministic and stochastic objects both in continuous and in discrete time, subject to development. For the purpose of a formalized description of information processing processes in an IS and counteraction to destructive influences, we will proceed from the fact that a set of functional components (FC) make up the designed IS. They can be developed independently or purchased from other manufacturers. At the same time, the set of externally system characteristics of these functional components (FC) is set by control commands, performed functions, needs for information and resource support [9-11].

It is necessary to develop an intrusion detection system (IDS) model that can be used to describe the processes and algorithms for diagnosing the operability of IS software.

3. Formalized description of information processing processes and intrusion detection systems

Let's depict in fig. 1 the structure of the IS software, taking into account the parameters controlled by the IDS.

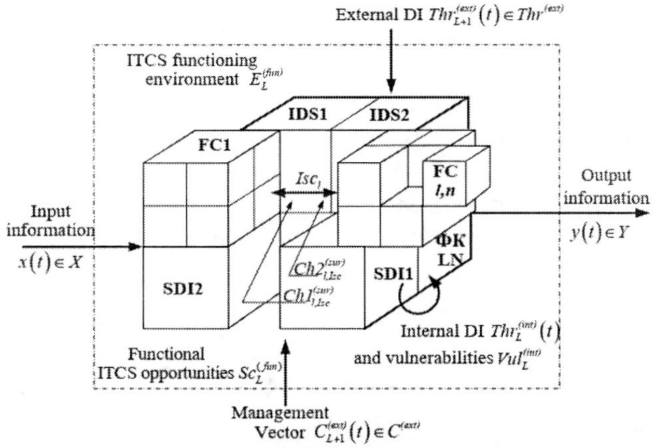

Figure 1. Block diagram of the IS software, taking into account the parameters controlled by IDS

A description of a structured IDS that maintains the required level of security in a cyberattack environment. $C_{l,Isc} = \langle c_{l,Isc} \times T \rangle$ - set of control actions ($C_{l,Isc} \subseteq C_l \subseteq C$) on the interface at different points in time T from the FC control subsystem (IDS) at the level l. With its help, there is a rational distribution of l level resources $c_{l,Isc} : R_l \rightarrow \{R_{l,Isc}\}$, as well as control of the parameters of the interface functioning $O_{l,Isc}$.

When determining the efficiency of the interface, it is necessary to adhere to the following rule: all quality indicators must be higher or equal to the required values:

$$\left\{ q_{l,Isc} \geq q_{l,Isc}^{required} \right\} \in Q_{l,Isc}. \tag{1}$$

The compliance of all QoS indicators from the set $Q_{l,Isc}$ with the efficiency criterion (1) determines the quality of the interface.

The required level of the criterion can be provided by a variety of control actions $C_{l,Isc}$ that depend on the allocation of interface Isc_l resources $R_{l,Isc}$, or by changing the parameters $Pa_{l,Isc}$ of the algorithms $A_{l,Isc}$ on the basis of which the interface functions.

The software quality control subsystem, implemented by IDS through the interface Isc_l observation channel $Ch1_{l,Isc}^{(mon)}$, collects data on the interface state pos_{Isc} by fully or partially determining the output parameters $P_{l,Isc}^{(out)}$ and quality indicators $Q_{l,Isc}$ for the interface Isc_l.

At the same time, the purpose of monitoring is to determine the state pos_{Isc} of the interface Isc_l and control the impact accordingly $C_{l,Isc}$. Channel observation $Ch1_{l,Isc}^{(mon)}$ can be represented by a vector that includes many observed parameters:

$$Ch1_{l,Isc}^{(mon)} = \left\langle P_{l,Isc,Ch1}^{(out)} \times Q_{l,Isc,Ch1} \right\rangle, \tag{2}$$

where $P_{l,Isc,Ch1}^{(out)} \subseteq P_{l,Isc}^{(out)}$ and $Q_{l,Isc,Ch1} \subseteq Q_{l,Isc}$ – are the components of the sets $P_{l,Isc}^{(out)}$ and $Q_{l,Isc}$ are determined by the software quality control subsystem.

Current state of the interface pos_{Isc} under parametric narrowing conditions

$$pos_{Isc} \rightarrow \left\langle \{q_{l,Isc}\}, \{p_{l,Isc}^{(out)}\} \right\rangle \rightarrow \left\langle \{q_{l,Isc,ChI}\}, \{p_{l,Isc,ChI}^{(out)}\} \right\rangle \rightarrow ChI_{l,Isc}^{(mon)} \quad (3)$$

and the narrowing of the reaction due to the limitation of the observation time $0 \leq T \leq t^{observation}$ in expression (2) can be determined on the basis of the theory of observation of dynamic systems [12-14] and control systems [15].

Based on the foregoing, we will consider the features of diagnosing the operability of the FC software of the IS following dynamic model.

4. Development of a complex of dynamic models

Based on the results of the model for diagnosing the operability of the IS software, a dynamic model is constructed, the purpose of which is to describe the cause-and-effect relationships between the IS FC, taking into account the space-time dependencies.

To create a DM, we will use the apparatus of Petri-Markov nets [16, 17]:

$$M_d = \{\Psi, \Omega\}. \quad (4)$$

For the most complete description of the structure of the diagnostic process, it is advisable to use a Petri net

$$\Psi = \{I, O, A, Z\}, \quad (5)$$

where we introduce the following notation of finite sets: I – system inputs; O – system outputs; A – positions; Z – transitions.

The process that determines the probabilistic characteristics of software diagnostics is denoted as

$$\Omega = \{P, f(p), \Lambda\}, \quad (6)$$

where P is the probability matrix; $f(p)$ is the matrix of the probability distribution function; Λ is a matrix of logical conditions.

In fig. 2 shows a diagram of making diagnostic decisions. It displays the operation of the software diagnostics algorithm and risk assessment of the system operability [18, 19].

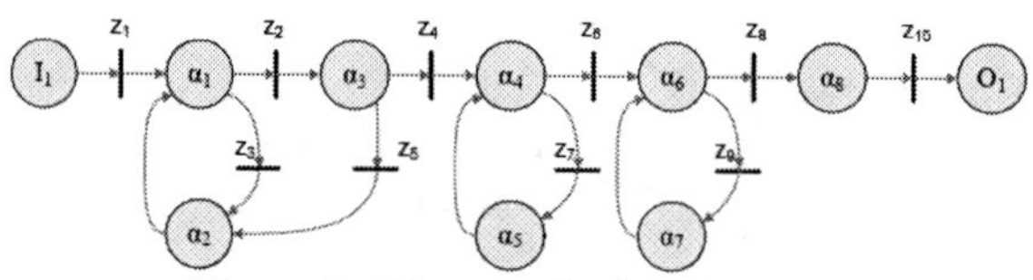

Figure 2. Diagnostic decision tree

The stages of the diagnostic process are the vertices of the graph. The sequence of actions is as follows: obtaining diagnostic data α_1; additional tests α_2; determination of functional diagnostic parameters α_3; α_4 data patterning; application of an alternative method of forming the α_5 data pattern; state classification of IC α_6; application of an alternative method for classifying the state of the IC α_7; interpretation of α_8 classification results.

The matrix of transition probabilities of the dynamic model can be represented in the following form:

$$P = \begin{pmatrix} p_{11} & p_{12} & p_{13} & 0 & 0 & 0 & 0 & 0 & 0 & 0 \\ 0 & 0 & 0 & 0 & 0 & 0 & 0 & 0 & 0 & 0 \\ 0 & 0 & 0 & p_{34} & p_{35} & 0 & 0 & 0 & 0 & 0 \\ 0 & 0 & 0 & 0 & 0 & p_{46} & p_{47} & 0 & 0 & 0 \\ 0 & 0 & 0 & 0 & 0 & 0 & 0 & 0 & 0 & 0 \\ 0 & 0 & 0 & 0 & 0 & 0 & 0 & p_{68} & p_{69} & 0 \\ 0 & 0 & 0 & 0 & 0 & 0 & 0 & 0 & 0 & 0 \\ 0 & 0 & 0 & 0 & 0 & 0 & 0 & 0 & 0 & p_{810} \end{pmatrix}. \quad (7)$$

In this case, the vertices of the graph ($\alpha_1,\ldots,\alpha_8$) form the rows of the matrix, and the transitions ($z_1,\ldots,z_{10}$) – the columns. Transition z_1 will be triggered with probability $p_{i1}=1$ in case of receiving input information I_1. Since the transitions z_2 and z_3 are asynchronous, then

$$p_{13}=1-p_{12}. \tag{8}$$

By analogy for asynchronous transitions z_4, z_5; z_6, z_7; and z_8, z_9 can be written:

$$p_{35}=1-p_{34};$$
$$p_{47}=1-p_{46}; \tag{9}$$
$$p_{69}=1-p_{68}.$$

Reachability condition for chain vertices

$$\sum_{ij} p_{ij}>0 \tag{10}$$

can be performed by introducing feedbacks for each step of the transition conditions. This leads to the fact that the sum of the probabilities of transitions along the columns of the matrix is always nonzero.

The resulting dynamic model can be used as the basis for a generalized classification algorithm.

In fig. 3 shows the decision-making logic for determining the degree of software performance in the form of a Markov chain.

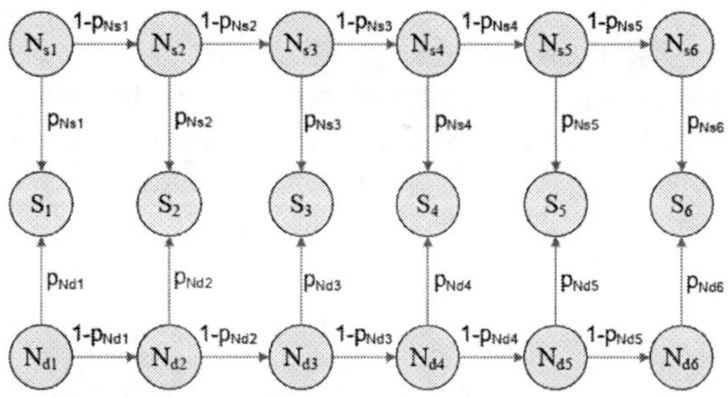

Figure 3. Graph for determining the degree of software performance

Here, the health of the software is determined by the health of each component. Accordingly, the condition for the transition of N to the next degree S is sequentially checked. When constructing the graph, the following designations are adopted: N_{s1} – the component failure level is minimal, that is, the time of the component i function f execution $t^i_f \le \min(t^i_f)$; N_{s2} – component failure level is higher than the minimum; N_{s3} – average component failure level; N_{s4} – component failure rate above average; N_{s5} – component failure level is high; N_{s6} – maximum level, when the time of the component i function f execution $t^i_f \ge \max(t^i_f)$; N_{d1}, N_{d2}, N_{d3}, N_{d4}, N_{d5} – respectively, automatic elimination of a component malfunction in a short period of time, for a sufficient period of time, for a non-critical period of time, for a significant period of time, for a long period of time, requiring an additional function; N_{d6} – the malfunction is not eliminated, a complete software replacement is required for a long period of time.

Thus, the probabilities of diagnosing software operability are respectively equal:

$$p_{S1}=p_{Ns1}+p_{Nd1};$$
$$\ldots, \tag{11}$$
$$p_{S6}=p_{Ns6}+p_{Nd6}.$$

where p_{S1}, p_{S2}, p_{S3} – the probability of determining, respectively, the optimal, normal and high levels of software performance; p_{S4}, p_{S5}, p_{S6} – the probability of determining, respectively, degrees 1, 2, 3 of software failure. In Fig. 4. the graph for determining the diagnostic results of the software operability is presented, corresponding to the decision logic.

Depending on the degree of damage to target components in accordance with the methodology [20], the following classification can be carried out: M_1 – the functions of the software components are not impaired or slightly impaired; M_2 – functions of software components are moderately impaired; M_3 – the functions of software components are significantly impaired.

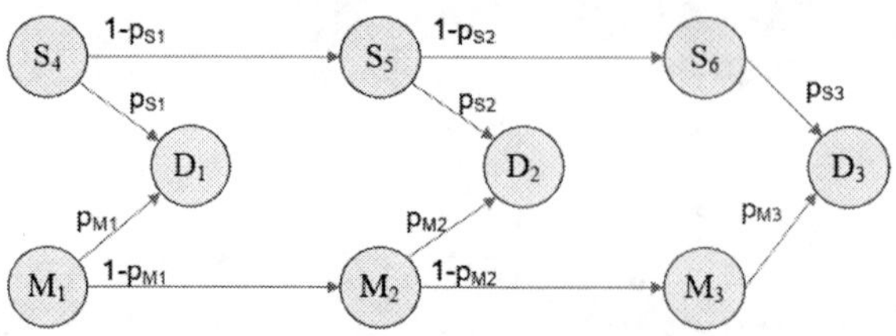

Figure 4. Graph for determining the diagnostic results of software performance

Accordingly, the probabilities of diagnosing the degree of software operability are respectively equal to:

$$p_{D1} = p_{S4} + p_{M1};$$
$$p_{D2} = p_{S5} + p_{M2};, \tag{12}$$
$$p_{D3} = p_{S6} + p_{M3}.$$

where p_{D1} – probability of software malfunction (stage I); p_{D2} – probability of software malfunction (stage II); p_{D3} – probability of software malfunction (stage III).

The decision-making logic in the form of a directed graph for determining the risk of software performance is shown in Fig. 5. At the same time, the following factors influence the level of risk in accordance with the methodology [21]: the degree of software performance in the presence of risk factors F; defeat of target components M; list of faults A of the IS associated with failure or partial failure of the corresponding components.

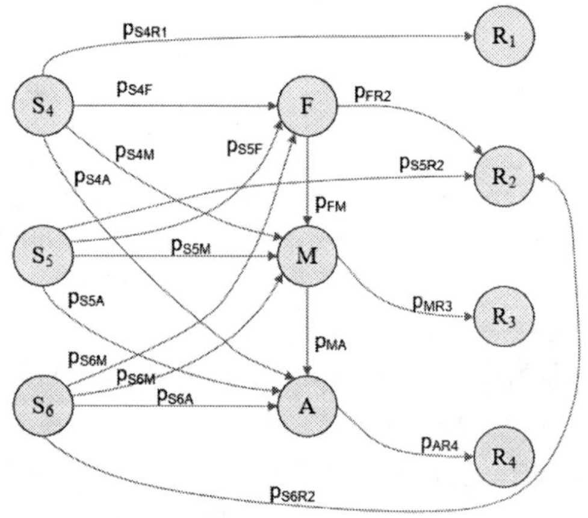

Figure 5. Oriented graph for determining the risk of software performance

The corresponding probabilities of determining the level of risk of software performance can be represented as follows:

$$p_{R1} = p_{S4R1};$$
$$p_{R2} = p_{S5R2} = p_{S6R2} = p_{S4F} + p_{FR2} = p_{S5F} + p_{FR2} = p_{S6F} + p_{FR2};$$
$$p_{R3} = p_{FM} + p_{MR3} = p_{S4M} + p_{MR3} = p_{S5M} + p_{MR3} = p_{S6M} + p_{MR3}; \tag{13}$$
$$p_{R4} = p_{MA} + p_{AR4} = p_{S4A} + p_{AR4} = p_{S5A} + p_{AR4} = p_{S6A} + p_{AR4}.$$

where p_{R1}, p_{R2}, p_{R3}, p_{R4} – respectively, the probabilities of determining risks 1, 2, 3, 4.

Thus, the logic of the operation of the intrusion detection system subsystems that control and diagnose is described by a set of proprietary dynamic models and makes it possible to make a decision on the operability of the software. Probabilistic estimates obtained in the course of calculations can be used to determine the prior probabilities of diagnostic estimates. Thus, the paper describes a set of models that allow making diagnostic decisions about the state of software performance.

5. Conclusion

This paper describes the algorithm and models for diagnosing the IS operability in the context of cyber attacks. The implementation of these proposals in the form of software products makes it possible to start designing new intrusion detection systems for ICs that allow continuous diagnostics of the software performance. A complex of mathematical models, with the help of which it is possible to describe the processes of making managerial decisions when diagnosing IS components, make it possible to determine the level of software performance; assess risks taking into account risk factors, the health of software components and the list of IC malfunctions associated with the failure or partial failure of the corresponding components. Thus, the paper describes a set of models that make it possible to make diagnostic decisions about the state of software performance in the context of cyber attacks.

References

[1] National institute of standards and technology & national security agency (USA), Communication security establishment (Canada), Communications-electronics security group (United Kingdom), Bundesamt fur sichereit in der informationstechnik (Germany), Service central de la securite des systemes d'information (France), National communications security agency (Netherlands) 1999 *Common criteria for information technology security evaluation* Version 2 1 – URL: http://www.commoncriteriaportal.org/files/ccfiles/ccpart1v21.pdf

[2] Sowa J F 2020 *Semantic Networks* – URL: http://www.jfsowa.com/pubs/semnet.htm

[3] Kostogryzov A, and Korolev V 2020 *Probability, combinatorics and control* (UK London: IntechOpen) p 322 – DOI: 10.5772/intechopen.79802

[4] Kostogryzov A 2018 *Probabilistic modeling in system engineering* (UK London: IntechOpen) p 278 – DOI: 10.5772/intechopen.71396

[5] Novoseltsev V I 2017 Tool and algorithmic diagnostic devices of operability of actuation mechanisms of automated control systems *Proceedings: 2nd International Ural conference on measurements* (Chelyabinsk) pp 193-198 – DOI: 10.1109/URALCON.2017.8120709

[6] Kasatkina T I, Novoseltsev V I and Ivanov S A 2018 An improved method for predicting the evolution of the characteristic parameters of an information system *Journal of Physics: Conference Series* **973** 012031 – DOI: 10.1088/1742-6596/973/1/012031

[7] Pavlov V A, Kasatkina T I and Shatovkin R R 2018 Algorithm for predicting the evolution of series of dynamics of complex systems in solving information problems *Journal of Physics: Conference Series* **973** 012035 – DOI: 10.1088/1742-6596/973/1/012035

[8] Noev A N and Gubin I A 2018 Method of optimum channel switching in equipment of infocommunication network in conditions of cyber attacks to their telecommunication infrastructure *Journal of Physics: Conference Series* **1015** 032101 – DOI: 10.1088/1742-6596/1015/3/032101

[9] Sumin V I and Smolentseva T E 2018 Mathematical Models to determine stable behavior of complex systems *Journal of Physics: Conference Series* **1015** 032136 – DOI: 10.1088/1742-6596/1015/3/032136

[10] Sumin V I, Grechishnikov E V and Ivanov S V 2019 Determining the reliability of network information systems *Journal of Physics: Conference Series* **1203** 012083 – DOI: 10.1088/1742-6596/1203/1/012083

[11] Goncharov I V, Goncharov N I and Parinov P A 2019 Opportunities to study the characteristics of information systems and manage them using modern technologies *International conference*

Actual problems of system and software engineering (Moscow) pp 95-105 – URL: http://ceur-ws.org/Vol-2514/paper55.pdf
[12] Kobzistyy S Yu and Markin P V 2019 Method of analysis the functional stability of the software of the infocommunication system in cyber-attack conditions *International conference actual problems of system and software engineering* (Moscow) pp 289-300 – URL: http://ceur-ws.org/Vol-2514/paper115.pdf
[13] Goncharov N I and Goncharov I V 2020 Mathematical modeling of the security management process of an information system in conditions of unauthor-ized external influences *International conference on control systems, mathematical modelling, automation and energy efficiency* (Lipetsk) pp 77-82 – DOI: 10.1109/SUMMA48161.2019.8947513
[14] Kobzistyy S Y, Novoseltsev V I and Smolyakova S D 2020 Algorithm and method for recognizing critical situations using semantic networks on critical information infrastructure facilities as a result of cyber attacks *IEEE Conference of russian young researchers in electrical and electronic engineering* (St. Petersburg and Moscow) pp 2066-2071 – DOI: 10.1109/EIConRus49466.2020.9039255
[15] Noev A N and Sumin V I 2020 Mathematical model for managing the dynamics of the development of information conflict in information systems *International conference on control systems, mathematical modelling, automation and energy efficiency* (Lipetsk) pp 88-93 – DOI: 10.1109/SUMMA48161.2019.8947546
[16] Dobryshin M M, Kochedykov S S, Novoselcev V I and Grechishnikov E V 2019 Algorithmic model of functioning of the system to detect and counter cyber attacks on virtual private network *Journal of Physics: Conference Series* **1203** 012064 – DOI:10.1088/1742-6596/1203/1/012064
[17] Goncharov N, Goncharov I and Dushkin A 2019 Assessment of the conflict stability of information systems using mathematical modeling *International Multi-Conference on Industrial Engineering and Modern Technologies* (Vladivostok) pp 1-5 – DOI: 10.1109/FarEastCon.2019.8934011
[18] Parinov P, Goncharov N, Goncharov I and Dushkin A 2019 Simulation and evaluation of conflict interactions in information systems *Journal of Physics: Conference Series* **1203** 012063 – DOI: 10.1088/1742-6596/1203/1/012063
[19] Parinov P, Goncharov N, Goncharov I, Dushkin A and Maximova M 2019 Modeling of information processes for modern information system security assessment *IEEE Conference of Russian Young Researchers in Electrical and Electronic Engineering* (Moscow) pp 1758-1763 – DOI: 10.1109/EIConRus.2019.8656828
[20] Kochedykov S, Orlova D, Grechishnikov E and Dushkin A 2019 The mathematical model of cyber attacks on the critical information system *Journal of Physics: Conference Series* **1202** – DOI: 10.1088/1742-6596/1202/1/012013
[21] Markin P, Kochedykov S and Dushkin A 2019 Express Assessment Method for the Risk of Impaired Functional Stability Information and Communication System in Conditions Cyber Attacks *IEEE Conference of Russian Young Researchers in Electrical and Electronic Engineering* (Moscow) pp 1754-1757 – DOI: 10.1109/EIConRus.2019.8657323

On the using Datamessage as evidence of Cybercrime

A Zharova [1,2] **V Elin** [1,3]

[1] National Research University Higher School of Economics. Moscow. Russia

[2] Department of Legal Security of the Fuel and Energy Complex, Russian State University of Oil and Gas named after I.M. Gubkin. Moscow. Russia

[3] Department of Integrated Safety of Critical Facilities, Russian State University of Oil and Gas named after I.M. Gubkin. Moscow. Russia

E-mail: elin_ vm@mail. ru

Abstract. The article reveals a number of features of using computer data in cybercrime cases. The contradictions between the European and Russian approaches to the question of the equivalence of an electronic document and a paper document are considered. The authors reflect the problem of using an electronic document as evidence in criminal cases in compliance with the principle of admissibility of evidence in the fight against cybercrime. The paper concludes that there is a need for comprehensive legal and technical regulation of the field of IT relations.

Keywords: IoT; Cybercrimes; Russian procedural law; Electronic document

1. Introduction

More and more public relations are formed using information technology (IT). Not any single developed country can imagine its existence without IT. It is expected that by 2030 "the number of active IoT devices will more than triple and amount to 24.1 billion against 7.6 billion registered at the end of 2019"[1]. This will create a communication system for devices connected to the Internet and transfer large amounts of data. In this regard, the question of the applicability of the technical models to solving the problem of proving cybercrimes is natural. At the level of technical regulation, the problems associated with the development of technologies that meet the requirements of ensuring the integrity and availability of data, reliability, and fault tolerance of information technologies have been solved for a long time. However, there is the problem of the reliability of computer data and it used as evidence in criminal proceedings remains.

With the proliferation of artificial intelligence systems, the number of such issues will increase, since the use of the Internet of Things (IoT), big data technologies used to ensure digital government, and predictive modeling is based on the processing of the information received. In this regard, there is a need to solve the problem of data reliability and information technology reliability.

2. The status of computer information in Russian procedural law

The UNCITRAL Model Law on Electronic Commerce uses the term "Data message", which means information generated, sent, received, or stored by electronic, optical, or similar means including, but not limited to, electronic data interchange (EDI), electronic mail, telegram, telex or telecopy. Electronic data interchange (EDI)" means the electronic transfer from computer to computer of information using an agreed standard to structure the information [2]. This point of view is also supported by Directive 2000/31/EC of the European Parliament and of the Council of June 8, 2000, on certain legal aspects of information society services, and in particular on electronic commerce in the internal market (Electronic Commerce Directive) [3].

However, Russian law determines that information in electronic form is recognized as an electronic document equivalent to a paper document signed with a handwritten signature only if it is signed with a qualified electronic signature [4][4]. At the same time, initially, an equal sign is not put between an electronic document and a paper document. The conditions determined under which an electronic document can be equated to a written document.

The Convention on Cybercrime [5] contains recommendations for improving procedural legislation and for including computer data in the list of evidence. However, the Russian Federation is ambivalent about the use of the Convention's recommendations.

Firstly, in 2005, Russia explicitly refused to participate in the work of the Convention on cybercrime. When Russia joined Convention, it stated that "the provisions of paragraph" b "of article 32 of the Convention is possible to interpret and apply a way that is inconsistent with the purposes and principles of the Convention, and this may harm the sovereignty and national security of States parties and also rights and legitimate interests of their citizens and legal entities.

In this regard, the Russian Federation "can participate in the Convention, subject to a possible revision of the provisions of article 32, paragraph (b) ". The reason for refusing to participate in the joint work was the wording of article 32 "b" that "Party may, without the authorization of another Party access by a computer system in its territory to stored computer data located in another Party, if the Party obtains the legal and voluntary consent of the person who has the legal authority to disclose the data to the Party through that computer system" [6].

Secondly, the norms of the Russian Criminal Code have correlated with the corresponding provisions of the Convention on Cybercrime. For example, the disposition of Art. 159.6 of the Criminal Code of the Russian Federation "Fraud in the field of computer information" actually coincides with the convention's disposition. It should also be borne in mind that the Criminal Code of the Russian Federation includes Chapter 28 "Crimes in the field of computer information." The chapter contains articles providing, including, criminal liability for:

- illegal access to computer information;
- creation, use, and distribution of malicious computer programs
- violation of the rules for operation means of storage, processing, or transmission of computer information (telecommunication networks data).

Thirdly, the Russian Federation does not actually implement the proposals of the Convention on Cybercrime in the field of procedural law. According to the authors, the most significant is the convention's concept of computer data what "means any representation of facts, information or concepts in a form suitable for processing in a computer system, including a program suitable to cause a computer system to perform a function". Concerning this type of evidence, various procedural actions can be performed.

The search and seizure of computer data are aimed at promptly ensuring the safety and partial disclosure of data on information flows. The process includes:

- seizure of a computer system, its part or media used for storing computer data, or in another similar way, ensuring their safety;
- making a copy of the relevant computer data;
- ensuring the integrity of the relevant stored computer data;
- inaccessibility or seizure of computer data located in the computer system.

Collecting data on information flows in real-time provides the ability to:

- collect or record data on information flows using technical means;
- oblige service providers, within the limits of their technical capabilities:
- collect or record using technical means;
- cooperate with competent authorities and help them collect or record in real-time data on information flows associated with specific messages transmitted through a computer system.

Interception of content data in real-time allows:

- collect or record using technical means,
- oblige the service provider, within the limits of his available technical capabilities:
- collect or record using technical means;
- cooperate with the competent authorities and assist them in collecting or recording in real-time data on the content of these messages on its territory, transmitted using computer systems.

Russian Criminal Code understands computer information as information (messages, data) presented in the form of electrical signals, regardless of the means of their storage, processing, and transmission [11] Art. 272]. The Criminal Procedure Code of the Russian Federation does not directly include computer information in the exhaustive list of admissible evidence [10]. The use of information and computer technologies as evidence is possible if they are categorized as: conclusions and testimonies of experts (specialists), material evidence, or other documents.

According to Federal Law "On Information" an electronic document is documented information presented in electronic form, i.e. in a form suitable:

- for human perception using electronic computers;
- for transmission over information and telecommunication networks;
- for processing in information systems [7], Art.2, p.11-1].

In Russian, documented information is understood as information recorded on a tangible medium by documenting information with requisites that make it possible to determine such information or, in cases established by the legislation of the Russian Federation, its tangible medium. The result of documenting information is a number of legal consequences:

- the expressed subject form of documented information, which allows exercising ownership and other property rights to tangible media containing documented information [8];
- the ability to perform orderly actions in relation to documented information,
- on the perception, use of information, transmission and provision of information during the interaction of various subsystems during storage, processing, and extraction of information ;
- the presence of certain rules for organizing document flow, creation, provision, storage, and accounting of information carriers;
- the highest efficiency of legal, organizational, and technical protection measures aimed at ensuring the protection of information [9], [10];
- the ability to formulate requirements for the composition of the details of documents and forms of documents; - the existence of requirements for the characteristics of document management systems, including reliability, integrity, complexity and consistency.

3. Electronic document as evidence. Properties and characteristics

The criminal procedure of the Russian Federation determines that proof consists in the collection, verification, and assessment of the evidence in order to establish the event of a crime, the guilt of a person in the commission of a crime, and other significant circumstances [10, Art. 85,73]. The evidence requirements are correlated with the requirements for data security.

In accordance with GOST R ISO / IEC 27000-2012, information security is associated with ensuring the properties of confidentiality, integrity, availability of information, authenticity, accountability, and non-repudiation for the information system [13]. Thus, from a technological point of view, information security is ensured by maintaining confidentiality, integrity, information availability, authenticity, accountability, and non-repudiation. However, the problem of data reliability is not determined, as solving such a problem by technological processes means is a rather difficult task [14].

Reliability is defined as "collecting data from the primary source of information or receiving them from authorities that collect data directly from the primary source of information or organizations that introduce registers with the necessary data if there are data collection standards and their aggregation, as well as the lack of interest of the information source in its content. There is a confirmation that the data has not been changed when performing any operation on them, be it transfer, storage, or publication. "

Accuracy is determined through "quality control of data collection and generation, as well as lists of assumptions on the use of the collected data. The description of the data provides for their main characteristics: coverage, collection period, update frequency ".

The completeness of the data is determined through the coverage of data associated with all "counterparties from which information is to be obtained" [15].

It can be concluded that concepts from technical regulation penetrate into the area of legal regulation IT relations.

It should be borne in mind that in 2019 the requirements for ensuring the reliability, integrity, and availability of data in the development of technology for distributed ledger systems were determined. The Ministry of Digital Development, Communications and Mass Media of the Russian Federation signed the "Roadmap for the development of end-to-end digital technology of the Distributed Register System to ensure the implementation of the strategic directions of state development – "Digital Economy of the Russian Federation" and "Digital Technologies" [16]. There are components of the National Program "Digital Economy of the Russian Federation", which defines the legal requirements for the processes implemented by distributed registry systems [17].

Legal regulation of the reliability of the processed information is associated, among other things, with solving the problem of forming a reliable digital profile of a person [18], and the accuracy of decisions made based on the analysis of the data obtained. In this regard, the methods of legal support for the reliability of data must be investigated in conjunction with the methods of technical assurance of the integrity, security, confidentiality of data, since they determine the parameters of the models for the implementation of technological processes of data processing and the use of information systems, which in turn is associated with ensuring information security [19], [20].

4. Conclusion

In the educational and scientific legal literature, it is argued that the norms of technical regulation and norms of legal regulation are independent areas of regulation. We believe that for the legal regulation of relations arising from the creation and use of IT, this position should be adjusted since the formation of public relations is actually determined by IT developers. And the lack of a coherent system of legal and technical regulation of this sphere of public relations leads to problems of improvability of cybercrimes and, in General, to the problem of ensuring information security.

We can note that the integrity and reliability of the information, the uninterrupted operation of information technology, and its fault tolerance become legal concepts with which the legal regulation of information security is associated. The need to ensure the reliability of data is recognized by the state, but this process requires the involvement of various categories of subjects and the establishment of their rights, duties, and responsibilities, as well as the definition of requirements for the information technologies used for these purposes. To solve this problem, a system of legal regulation in the field of ensuring the reliability of information technologies that process data must be determined. It is also necessary to resolve the issue of amending the criminal procedure legislation of the Russian Federation in connection with the need to include electronic evidence in the list of evidence in criminal cases.

5. Acknowledgments

The reported study was funded by project group of HSE and by RFBR (research project 18029-16013 "Research of conceptual approaches to the formation of a system of legal regulation of information security in the face of great challenges in the global information society).

References

[1] The Number of IOT devices will triple by 2030, Retrieve from https://safe.cnews.ru/news/top/2020-06-30_chislo_ustrojstv_interneta

[2] Model Law on Electronic Commerce of the United Nations Commission on International Trade Law. 1996, Adopted in New York 28.05.1996 - 14.06.1996 at the 29th session of UNCITRAL.

[3] Directive No. 2000/31/EC of the European Parliament and of The Council "On Certain Legal Aspects Of Information Society Services, In Particular Electronic Commerce, In The Internal Market", Retrieve from http://eur-lex.europa.eu/

[4] Federal Law of 06.04.2011 No 63-FZ "On Electronic Signature" *The legal reference system" ConsultantPlus".*

[5] Convention On Cybercrime (Budapest, 23.XI.2001) Retrieve from https://www.europarl.europa.eu/meetdocs/2014_2019/documents/libe/dv/7_conv_budapest_/7_conv_budapest_en.pdf

[6] Order of the President of the Russian Federation of November 15, 2005 N 557-rp "On Signing the Convention on Cybercrime" *The legal reference system" ConsultantPlus".*

[7] Federal Law No. 149-FZ of 27.07.2006 "On Information, Information Technologies and Information Protection" *Collection of legislation of the Russian Federation, No. 31 (1 part), Art. 3448.*

[8] Zharova A and Elin V M The Internet of Things in the Russian Federation: Integrated Security *Deep Learning Strategies for Security Enhancement in Wireless Sensor Networks.* (IGI Global) p 274-290. DOI: 10.4018/978-1-7998-5068-7.ch014

[9] Resolution of the RF Government 28.07.2005 No 452 "About Typical regulations of the internal organisation of Federal enforcement authorities", *Collection of legislation the Russian Federation, 01.08.2005, No. 31, Art. 3233*
[10] Decree of the Government of the Russian Federation of 19.01.2005 No. 30 "On Standard regulations for interaction of Federal Executive authorities", *Collection of legislation of the Russian Federation, 24.01.2005, No. 4, Art. 305.*
[11] Criminal Code Russian Federation, *The legal reference system" ConsultantPlus"*
[12] Criminal Procedure Code of the Russian Federation, *The legal reference system" ConsultantPlus"*
[13] GOST R ISO / IEC 27000-2012 Information technology (IT). Methods and means of ensuring security. Information security management systems. Overview and terminology *The legal reference system" ConsultantPlus"*
[14] Andreev A G, Kazakov G V and Koryanov V V 2019 Method for evaluating the reliability of output data prepared by means of an automated system for preparing flight data for aircraft, *Engineering journal: science and innovation* No 4 (88).
[15] Methodological recommendations for assessing the reliability, completeness and accuracy of data used in calculating performance indicators and the effectiveness of control and Supervisory activities" (approved by the minutes of the project Committee meeting dated 09.11.2017 No. 73 (13)), *The legal reference system" ConsultantPlus"*
[16] Passport of the National Project "National Program" Digital Economy of the Russian Federation "(approved by the Presidium of the Presidential Council For Strategic Development And National Projects, protocol of 04.06.2019 No 7), Retrieve from https://digital.gov.ru
[17] Zharova A 2020 The protect mobile user data in Russia, International *Journal of Electrical and Computer Engineering*. vol. 10. No 3. Pp. 3184-3192.
[18] Zharova A 2019 Ensuring the information security of information communication technology users in Russia *International Journal of Cyber Criminology*. vol. 13. No 2. Pp. 255-269
[19] Chebotareva A A, Chebotarev V E and Rozanov A S 2018 Communication society and security: Current threats and legal maintenance *Digital Communication Management. InTech Open Science* (Croatia, Spain) pp. 135–182. DOI: https://dx.doi.org/10.15405/epsbs.2018.09.02.72.
[20] Chebotareva A A, Danilina E I and Chebotarev V E 2020 Electronic Passports of Citizens as a Personal Essential Attribute During the Pandemic. Proceedings of the Research Technologies of Pandemic Coronavirus Impact (RTCOV 2020). *Advances in Social Science, Education and Humanities Research,* vol 486. DOI https://doi.org/10.2991/assehr.k.201105.083.

IOP Conf. Series: Materials Science and Engineering 1069 (2021) 012038 doi:10.1088/1757-899X/1069/1/012038

Hardware, Biometric and Passwordless Authentication: Vulnerability and Cybercrime Issues

Anna A. Chebotareva[1], Vladimir E. Chebotarev [1]

[1]Russian University of Transport, Moscow, 9b9 Obrazcova str., 127994, Russia,

E-mail: anna_galitskaya@mail.ru

Abstract: The problem of identification and authentication is relevant. In conditions of crisis manifestations, including in connection with the pandemic, the number of fraudulent actions on the network is increasing, aimed at taking over information - personal data, bank card data, other important information. This raises the question of identification and authentication. At present, in Russia and in the world, some approaches to methods, protocols, and standards of authentication have been developed. Hardware, biometric and passwordless authentication are some of the most promising areas. However, no holistic approaches have been developed to the essence and main directions of development of authentication aimed at reducing vulnerability, which does not contribute to overall security on the Internet. In general, this issue assumes an interdisciplinary approach that takes into account both technical, social, political, economic factors, and, of course, an effective legal mechanism. In search of a solution to the problem, approaches to hardware, biometric and passwordless authentication in Russia and abroad are being investigated.

Keywords: cybercrime, hardware, biometric and passwordless authentication, security.

1. Introduction

This study focuses on emerging models (approaches) for hardware, biometric and passwordless authentication against the backdrop of vulnerability and cybercrime issues. In the direction of this problem, some positive research experience has been accumulated to date, which became the basis for finding answers to the following questions: 1) How relevant is the issue of developing a systematic approach to methods, protocols, and standards of authentication? 2) What are the features and priorities of hardware, biometric, and passwordless authentication at the present time? 3) How justified is the introduction of a stable system of hardware, biometric and password-free authentication in the context of the development of health passports (immune passports), biometric passports, etc.)?

Over the past 10-15 years, the problem of authentication has been one of the demanded topics in the research of scientists around the world. A number of studies are devoted to the problems of global data transmission, data security, and privacy, a significant number of works are devoted to the study of the features of biometric data, with an emphasis on their security and the risks of losing confidentiality [1; 2; 3; 4; 5]. Researchers are also developing issues of multifactor authentication as a multi-level security system and, as a result, reducing the risk of unauthorized access to wireless networks in general [6; 7].

In the publications of 2020, in turn, against the background of the coronavirus epidemic, there is a tendency of increased attention to the transition in relation to various spheres of public life, for example, in the field of education, transport, from traditional approaches to passwords to more promising ones, including biometric authentication. [8; 9]

Our Contribution. This article presents some approaches to exploring hardware, biometric, and passwordless authentication. The study of this issue will allow solving the important task of ensuring information security as a whole as one of the institutions of the digital state as a whole, the development of which we have discussed a lot in a number of previous studies [10; 11]

2. Informal statement of the problem

The attention of all states to all components of digitalization is unprecedented. In 2021-2023, a Federal Information System for Biometric Accounting will be created in Russia, including the software necessary for the formation, maintenance, and use of a data bank. This will allow the procedure for identifying a person, as well as unidentified bodies by facial images, fingerprints, and genomic information (including when interacting with the combined search federal system for genetic identification). Testing of ATMs with face recognition function has begun in Russia since 2020. These and many other processes indicate an increase in attention to the problems of authentication and identification in general.

3. Formalization and analysis of hardware, biometric and password-free authentication

In general, the attention to the topic of passwords, which is one of the priority targets for hackers who professionally use social engineering, phishing, stolen credentials, is increasing.

In the context of the transition to digital technologies and the Internet of Things, the volume of corporate and personal data that is stored in cloud systems is inevitably increasing, which leads to a significant risk of increasing data leaks, especially from cloud servers. Among the easy targets for cyberattacks are also smart home devices that collect and store personal data that are mostly barely protected. And peripherals are equipped with extremely simple software to protect confidential data and run on basic operating systems.

This is why there has been a major rise in hardware, biometric, and passwordless authentication. Today, the password as the main factor of identification, in general, will gradually give up positions in favor of biometric (using, for example, Secure Enclave and Trusted Platform Module (TPM) technology for Touch ID or Windows Hello applications) and password-free authentication based on WebAuthn standards (Web Authentication) and CTAP (Client-to-Authenticator Protocol). Thus, the WebAuthn standard is based on two basic methods, both requiring a secure connection. They are used by WebAuthn to register credentials with the server and then use them to "authenticate" the user. Navigator.credentials.create creates credentials either when registering an account or to associate a new asymmetric key pair with an existing account. Navigator.credentials.get uses the already known credentials to authenticate to the service.

Hardware and biometric authentication methods are becoming more accessible and accurate, which, together with the advent of password-free standards, contributes to the growing popularity of this method. So, in 2018-2019, the demand for voice recognition systems as the most popular form of biometric authentication grew rapidly. As you know, in the spring of 2020, China introduced updated standards for applications that collect biometric data, including for face recognition systems. At the same time, paradoxically, it is China that cannot boast of a good system for protecting such data. For example, case of November 2019, when the first lawsuit concerning facial recognition systems was filed in China when law professor Guo Bing sued a wildlife park for introducing this technology without the notice and consent of visitors.

Serious steps forward have been made in Russia towards the formation of a digital profile infrastructure as a transitional step towards the implementation of the idea of electronic passports; Federal Law No. 747513-7 "On Amendments to Certain Legislative Acts (in terms of clarifying identification and authentication procedures)" has been developed. There should be established rules and basic approaches to regulating the issues of remote identification and authentication of persons. It is important to create a clear mechanism for establishing identifiers and (or) information about a person, comparing this information with an identifier or verifying this information, as well as verifying a person for belonging to a person of identifier (identifiers) by comparing the identifier (identifiers) with available information about the person and establishing the validity of a person's use of the identifier (s).

Highlighting the priority of information security issues, the draft law emphasizes one of the most important aspects - the issue of strict compliance with the requirements for organizational and technical measures to protect information.

The period of the pandemic serves is a kind of "trigger" for legislative activity, in the next 2-3 years; it will contribute to the final formation of the infrastructure of the digital profile, and bring the issue of developing the digital Code of Russia to the agenda.

The federal law No. 168-FZ "On the Unified Federal Information Register Containing Information on the Population of the Russian Federation" was adopted, that establishes the organizational and legal framework for the formation and maintenance of a unified federal information register containing information on the population of the Russian Federation. The scope of the law specifies its target, aimed at creating a mechanism for the formation and maintenance of a federal information register, including the organizational and legal component of ensuring information security in the formation of a population data register.

In the wording of the principles of the law, it is important to emphasize the importance of measures to ensure the safety of information about the population of personal data, as well as aspects of the completeness, relevance, and reliability of such information.

The analyzed law establishes that in order to ensure the protection of information contained in the federal register of information about the population, including personal data, during the creation, operation, and development of the state information system, the operator of the state information system carries out a set of measures: in particular, the use of certified in accordance with the legislation of the Russian Federation the procedure for information protection means, including means of cryptographic information protection, ensuring the security of interdepartmental information interaction, developing and implementing a system for protecting information contained in the federal register of information about the population, attesting the state information system for compliance with information protection requirements, ensuring the invariability of information contained in the federal register of information about the population, during their storage and transmission, etc.

We discussed these problems in the previous study "Electronic Passports of Citizens as a Personal Essential Attribute during the Pandemic" [12].

All these questions are relevant, especially against the background of the development of concepts for electronic, biometric passports. As we remember, for the first time the concept of biometric passports began to develop after the events of September 11, 2001 and is based on the possibility of storing special biometric information in a microchip. 118 countries that signed the New Orleans Agreement a year later recognized the prospects for the development of biometric documents, primarily in order to ensure security.

Today there is information about the beginning of testing digital health passports in various countries. The concept of such a passport in Estonia is based on software that can help increase trust within the teams of companies during the exit from quarantine. Its developers explain that an authorized person - a doctor - will be able to enter data on the results of a person's coronavirus test into the program. After that, the application will generate a unique QR code with the result data. The user, on his own initiative, can share it with the employer, school administration, or his family [13]. As a result, the protection of such data, primarily due to the system of passwords - hardware, biometric, password-free - will be increasingly updated. It is no coincidence that the researchers Sheena I. Sapuay, Bobby D. Gerardo, Alexander A. Hernandez emphasize the increase in the importance of authentication in information systems containing data about people [14].

Bernard Chan in the pages of the South China Morning Post [15] proposing the creation of a well-structured system of an international digital health passport, discusses the so-called single digital "health declaration" in order to simplify passenger checks at airports and border control points so that travelers do not have to bring sheaves of documents to the airport check-in counter, and the airline personnel did not waste time checking them. At the same time, the construction of such a unified digital "health declaration" can also be very seriously vulnerable, and must be accompanied by both a legal mechanism for data protection and well-developed identification mechanisms.

4. Conclusion

In general, it is possible to formulate a number of preliminary conclusions that will form the basis for further study of the problems of hardware, biometric and passwordless authentication, vulnerability issues, and countering cybercrime.

The pandemic is contributing to the acceleration of digitalization processes in the field of digital transformation in all spheres of public life, and the direction of authentication is gaining in importance. By now, serious developments in these issues have already been created, at the same time, agreed approaches are also needed.

As for Russian research in relation to solving problems of identification and authentication, while the process of developing an appropriate regulatory framework is ongoing [16], [17], in particular, a corresponding federal law is planned to be adopted in the near future.

5. Acknowledgment.

The research is carried out within the framework of the functioning of the Laboratory of Information Technologies in the Field of Law and Economics of the Law Institute of the Russian University of Transport, as well as with financial support from the Russian Foundation for Basic Research (research project 18029-16013 "Research of conceptual approaches to the formation of a system of legal regulation of information security in the face of great challenges in the global information society).

References

[1] Taban Habibu, Anael E Sam 2018 *Assessment of vulnerabilities of the biometric template protection mechanism* International Journal of Advanced Technology and Engineering Exploration vol 5 (45) DOI: http://dx.doi.org/10.19101/IJATEE

[2] Mahesh Josh, Bodhisatwa Mazumdar Somnath Dey 2018 *Security Vulnerabilities Against Fingerprint Biometric System* DOI: https://arxiv.org/pdf/1805.07116.pdf

[3] Ratha N K, Connell J H, Bolle R M 2001 *Enhancing security and privacy in biometrics-based authentication systems* Ibm Systems Journal vol 40 (3) pp 614-634 DOI: 10.1147/sj.403.0614

[4] Scheirer W J, Boult T E 2008 *Bio-cryptographic protocols with bipartite biotokens* Preprint of a paper to appear at the 2008 IEEE Biometric Symposium at the Biometric Consortium Conference. Biometrics Symposium. INSPEC DOI: 10.1109/BSYM.2008.4655516

[5] Awalkar Kirti V, Kanade Sanjay G, Jadhav Dattatray V, Ajmera Pawan K 2015 *A multi-modal and multi-algorithmic biometric system combining iris and face* International Conference on Information Processing (ICIP). DOI: 10.1109/INFOP.2015.7489434

[6] Muchamad Rusdan, Daniel TH Manurung 2020 *Designing of User Authentication Based on Multi-factor Authentication on Wireless Networks* Journal of Advanced Research in Dynamical and Control Systems vol 12 (1) pp 201-209 DOI: 10.5373/JARDCS/V12I1/20201030

[7] Washek Matyáš, Zdeněk Říha 2010 *Security of Biometric Authentication Systems* International Journal of Computer Information Systems and Industrial Management Applications. vol 3 pp 174-184 DOI: 10.1109/CISIM.2010.5643698

[8] Ayesha S Shaikh, Vibha D Patel 2020 *Significance of the Transition to Biometric Template Protection: Explore the Future* International Journal of Image and Graphics. DOI: 10.1142/S021946782150025X

[9] Shradha D Jamdar, Yogesh Golhar 2017 *Implementation of unimodal to multimodal biometrie feature level fusion of combining face iris and ear in multi-modal biometric system* Trends in Electronics and Informatics (ICEI) International Conference pp 625-629 DOI: 10.1109/ICOEI.2017.8300778

[10] Al Wahshat Z M, Chebotareva A A, Chebotarev V E, Naser Abdel Rakheem A A 2020 *The right to self-determination of the people in the conditions of current information environment as a threat to the security of the state* Journal of Critical Reviews vol 7 (14) pp 3821-3824
[11] Chebotareva A.A., Chebotarev V.E., Rozanov A.S. 2018 *Communication society and security: Current threats and legal maintenance* (InTech Open Science Croatia) Digital Communication Management pp 135-182 DOI: https://dx.doi.org/10.15405/epsbs.2018.09.02.72
[12] Chebotareva A A, Danilina E I, Chebotarev V E 2020 *Electronic Passports of Citizens as a Personal Essential Attribute During the Pandemic* Proceedings of the Research Technologies of Pandemic Coronavirus Impact (RTCOV 2020). Advances in Social Science, Education and Humanities Research vol 486 DOI https://doi.org/10.2991/assehr.k.201105.083
[13] *ImmunityPass survey* (accessed 20.11.2020) URL: https://docs.google.com/forms/d/e/1FAIpQLSdL2qv-zTQ_Bzp9kbYYsKcpqAa9hDNI7kncPtnTKMPmcK5Tfw/viewform
[14] Sheena I Sapuay, Bobby D Gerardo, Alexander A Hernandez 2019 *Dynamic Third-Factor for Enhanced Authentication in Human Resource Information System* (IEEE 7th Conference on Systems, Process and Control) DOI: 10.1109/ICSPC47137.2019.9068077
[15] Bernard Chan (accessed 20.11.2020) *Living with coronavirus: global travel cannot restart without common rules to certify those free of Covid-19* (South China Morning Post) URL: https://www.scmp.com/comment/opinion/article/3098968/living-coronavirus-global-travel-cannot-restart-without-common
[16] Zharova A 2020 *The protect mobile user data in Russia* International Journal of Electrical and Computer Engineering vol 10 (3) pp 3184-3192
[17] Zharova A 2019 *Ensuring the information security of information communication technology users in Russia* International Journal of Cyber Criminology vol 13 (2) pp 255-269

Software model for the synthesis of increased volumes of systems of discrete orthogonal code sequences

A V Studenikin[1], A P Zhuk[1]

[1]North-Caucasus Federal University, Stavropol, Pushkin str., 1, 355017, Russia

E-mail azhuk@ncfu.ru

Abstract. To ensure high structural secrecy of information transmission systems with MA and CDM, it is necessary to use many ensembles of stochastic discrete orthogonal code sequences differing from each other. The purpose of the article is to develop a program for studying the correlation and statistical characteristics of ensembles of stochastic discrete orthogonal sequences described by the eigenvectors of symmetric matrices, with random filling of the diagonal coefficients. Assessment of the statistical and correlation indicators of sequence ensembles obtained on the basis of the developed program showed their random nature.

Keywords: stochastic discrete orthogonal sequences; structural secrecy; code division multiplexing systems.

1. Introduction

In accordance with GOST 50922-2006 in the field of information security, which regulates information security issues, information security is understood as the state of information security, in which its confidentiality, availability and integrity are ensured.

In the modern world, radio channels are present everywhere, the prospects of their use are undeniable. The advantages and disadvantages of radio communication are also known. However, due to the specificity of the propagation of radio waves in space, one of the problems in transmitting information over radio channels is to ensure its confidentiality.

It is known that the confidentiality of message transmission over radio channels can be achieved by providing [1,2,3,10]:

- energy secrecy of signals - carriers of information.
- structural secrecy of these signals.
- information secrecy of the message itself.

A special place among wireless systems for information transmission is occupied by systems with multiple access (MA) and code division multiplexing (CDM). This is because they provide efficient use of the allocated frequency range, simultaneously providing information exchange in the interests of a large number of users [11-16]. These systems are subject to threats typical of wireless information transmission systems, therefore, they require additional information protection mechanisms, taking into account the group nature of the use of these systems. A comparative analysis of possible means of protecting information in networks with multiple access and code division of channels showed that it is advisable to use not all of the above criteria for the security of systems with MA and CDM, but only those that relate to the assessment of structural secrecy [7,9,10, 18].

It is possible to provide high structural secrecy of information transmission systems with MA and CDM on the basis of using a large number of different ensembles of stochastic discrete orthogonal code sequences, which are randomly replaced when used [8,18].

In this case, an important condition of the systems under consideration is that in addition to the statistical characteristics of the system of discrete orthogonal code sequences, they must also have correlation and spectral characteristics, taking into account the parameters of the communication channel in which they will be used [1-6].

In papers [8,17,18], approaches to the synthesis of discrete orthogonal code sequences are proposed, taking into account the requirements for their number and dimension, however, the issue of automating the procedure for assessing their statistical and correlation characteristics in them is not fully resolved.

The purpose of the article is to develop a program for studying the correlation and statistical characteristics of ensembles of stochastic discrete orthogonal sequences (ESDOS) described by the eigenvectors of symmetric matrices, with random filling of diagonal coefficients.

2. The solution of the problem

The criteria for assessing the structural secrecy of the ITS with CDM are:

1. Difficulty of solving the structure of sequence ensembles.
2. Sufficiency of the number of structures of ensembles of sequences for their automatic change and the presence of a single algorithmic basis for their formation.
3. The presence of "good" correlation characteristics of signal ensembles.

The analysis showed that the most rational option for increasing the structural secrecy of the ITS with MA and CDM is to use the vector synthesis method to obtain signal ensembles [10, 18]. Taking this into account, the developed software model is designed to study the correlation and statistical characteristics of the ESDOS described by the eigenvectors of symmetric matrices, with random filling of the diagonal coefficients.

The software model for the synthesis of increased volumes of systems of discrete orthogonal code sequences is based on the stochastic modeling algorithm ESDOS.

Let us consider the algorithm of stochastic modeling of ESDOS using the example of a bidiagonal symmetric matrix A of the 4th order described by expression (1). It should be noted that the diagonal coefficients of the symmetric matrix A of the form (1) have the following property $a_{ij} = a_{ji}$.

$$A = \begin{bmatrix} 0 & a_{12} & 0 & 0 \\ a_{21} & 0 & a_{23} & 0 \\ 0 & a_{32} & 0 & a_{34} \\ 0 & 0 & a_{43} & 0 \end{bmatrix}. \quad (1)$$

At the first stage of the algorithm, a procedure for generating random numbers is implemented for the subsequent assignment of random values to the coefficients of the second diagonal of the bidiagonal symmetric matrix A, which is illustrated as follows. The random number generator at time t_i performs the procedure for generating the i-th set of random numbers

$$rnd_i \rightarrow \{..., rnd_{i1}, rnd_{i2}, rnd_{i3}, ...\}. \quad (2)$$

At the second stage of the algorithm, the generated i–th set of random numbers is assigned to the coefficients of the second diagonal of the matrix A.

$$rnd_{i1} \rightarrow a_{12}, rnd_{i2} \rightarrow a_{23}, rnd_{i3} \rightarrow a_{34}. \quad (3)$$

At the third stage of the algorithm i–th set of eigenvectors of the matrix A is calculated.

$$\vec{x}_i = \begin{bmatrix} x_{i,11} & x_{i,12} & x_{i,13} & x_{i,14} \\ x_{i,21} & x_{i,22} & x_{i,23} & x_{i,24} \\ x_{i,31} & x_{i,32} & x_{i,33} & x_{i,34} \\ x_{i,41} & x_{i,42} & x_{i,43} & x_{i,44} \end{bmatrix}. \quad (4)$$

The set of eigenvectors of the form (4) of the matrix A has the property of orthogonality and is the ESDOS model.

At the fourth stage of the algorithm, the procedure for calculating the correlation and statistical characteristics of the obtained i-th ESDOS model, such as the autocorrelation function $R_i(\tau)$, correlation dimension, correlation entropy and Hurst H exponent, is implemented.

At the fifth stage of the algorithm, the ESDOS is selected, the correlation and statistical characteristics of which meet the specified requirements and criteria.

At the sixth stage of the algorithm, the display of the selected ESDOS and its correlation and statistical characteristics is realized.

At the seventh stage of the algorithm, the selected ESDOS and its correlation and statistical characteristics are recorded into the computer database.

Depending on the required limit value of the number of ESDOS K structures required to use ensembles of stochastic discrete orthogonal code sequences differing from each other during a communication session, further repetition of the main stages of the algorithm described above is performed. The algorithm will be implemented until K ensembles of stochastic discrete orthogonal sequences with given correlation and statistical characteristics are generated in the computer database. After that, from the database, at the user's request K SDOS ensembles are issued for use in the information transmission system with MA and CDM.

In a generalized form, the main stages of the proposed algorithm can be presented as shown in Figure 1.

Analysis of Fig. 1 shows that the ESDOS stochastic modeling algorithm includes two main blocks:

Block 1. Development of a software complex for modeling ESDOS.

Block 2. Assessment of ESDOS properties using the Fractan program.

The first block includes the following stages:

1. Input of diagonal coefficients - loading from a file, or manual input of diagonal coefficients of a symmetric matrix.
2. Generation of a pseudo-random sequence - loading from a file or using one of the five proposed algorithms for generating pseudo-random numbers to obtain the sign structure of the coefficients of the symmetric matrix.
3. Calculation of an ensemble of stochastic discrete orthogonal sequences.
4. Conversion of calculations of ensembles of discrete orthogonal sequences for subsequent analysis in the Fractan program.

The second block includes:

1. Calculation of the correlation dimension, which is used to check for the presence of a chaotic component in a time series.
2. Calculation of the correlation entropy, which shows the degree of divergence of close phase trajectories and allows you to estimate the amount of information required to predict the behavior of the time series in the future.
3. Calculation of the Hurst exponent used in the analysis of time series. This amount decreases when the delay between two identical pairs of values in the time series increases.
4. Building a graph of ensembles of stochastic discrete orthogonal sequences and displaying a graphical implementation.

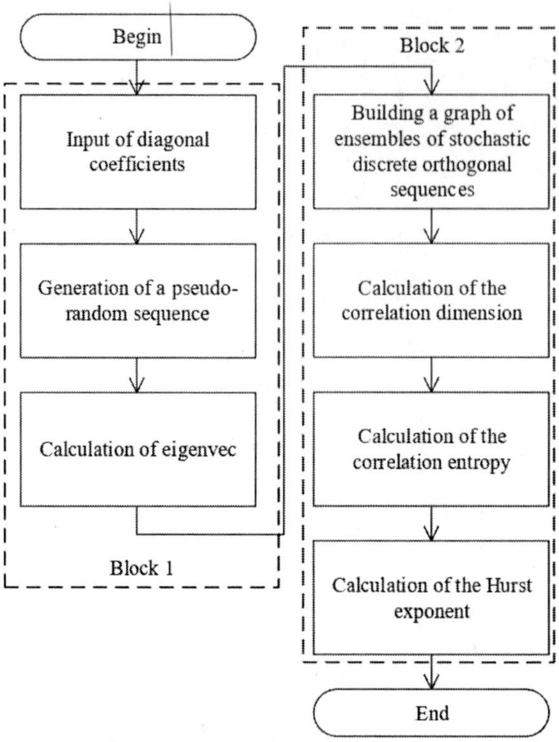

Figure 1. The main stages of the stochastic modeling algorithm ESDOS

Let us describe the program interface, the procedure for entering initial data and presenting the calculation results.

The interface of the main menu of the program is shown in Figure 2.

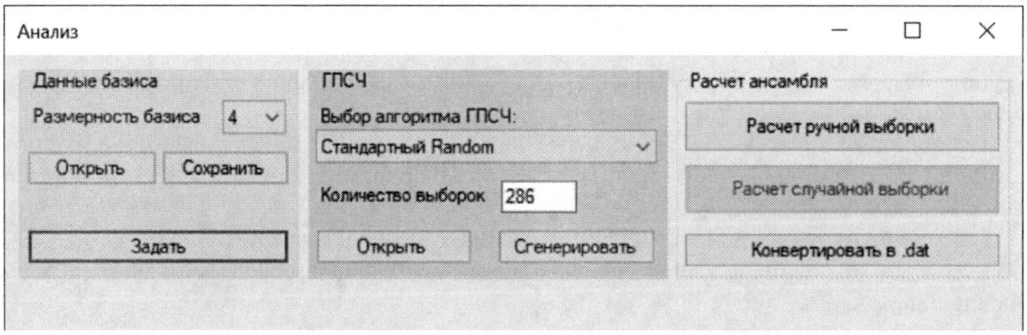

Figure 2. The main menu of the software model for the synthesis of increased volumes of ESDOS

With the help of this menu, in the first window, the dimension of the matrix A of the form (1) and, accordingly, the dimension of the ESDOS basis are set. In the second window, a pseudo-random number generator (PRNG) is selected and its parameter is set - the number of samples. The third window is intended for starting calculations by manually or randomly setting the coefficients of the matrix A and converting the data.

When entering the initial data in one of the selected ways, the entered data is displayed in a special program window as shown in Figure 3.

IOP Conf. Series: Materials Science and Engineering 1069 (2021) 012039 doi:10.1088/1757-899X/1069/1/012039

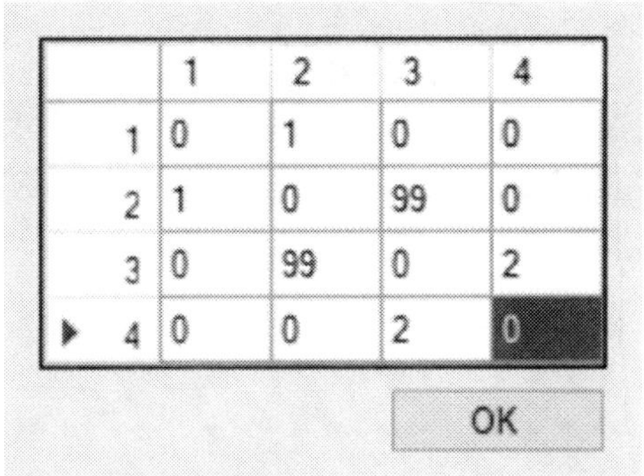

	1	2	3	4
1	0	1	0	0
2	1	0	99	0
3	0	99	0	2
4	0	0	2	0

Figure 3. Illustration of the procedure for setting the initial data

After specifying the initial data, the software model for the synthesis of systems of discrete orthogonal code sequences of increased volumes carries out the statistical and correlation indicators of the model of the ensemble of sequences and displays the results of the calculations.

Figure 4 illustrates the results of calculating the correlation dimension of the obtained sequence model.

Figure 5 illustrates the results of calculating the correlation entropy of the obtained sequence model.

Figure 6 illustrates the results of calculating the Hurst exponent of the resulting sequence model.

For the initial random sequence of numbers, which served to generate symmetric matrices, using the developed program, the following indicators were calculated:

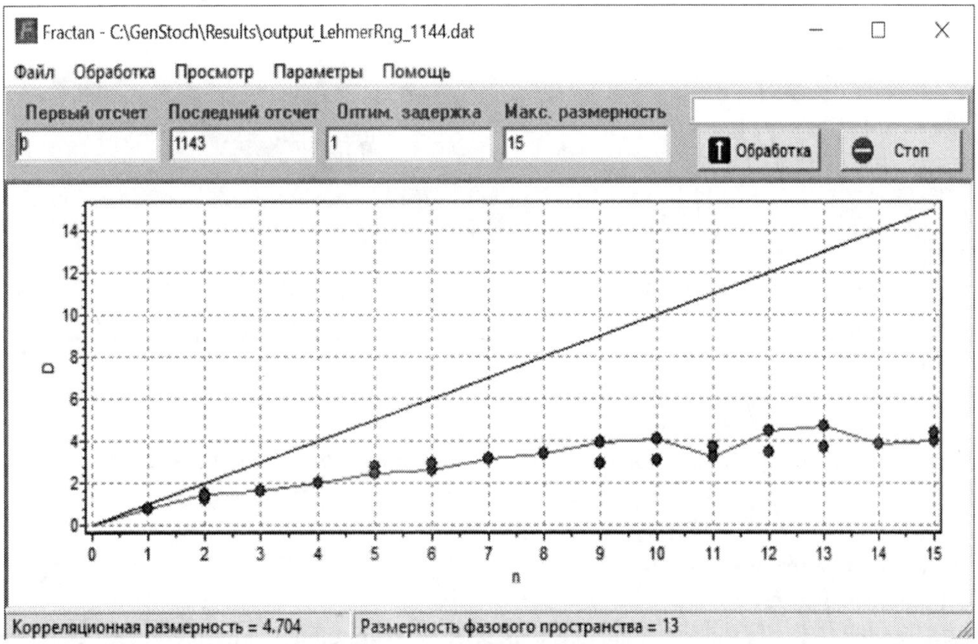

Figure 4. Illustration of the results of calculating the correlation dimension

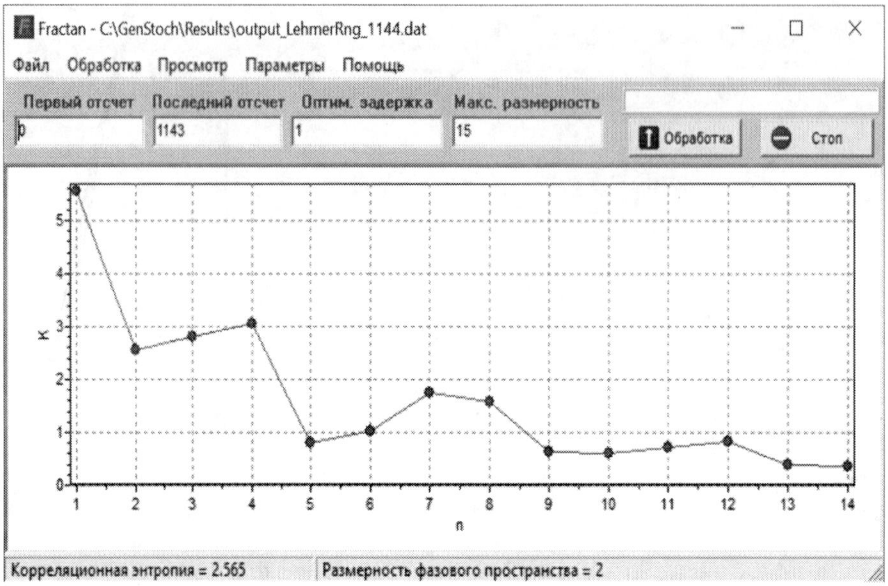

Figure 5. Illustration of the results of calculating the correlation entropy

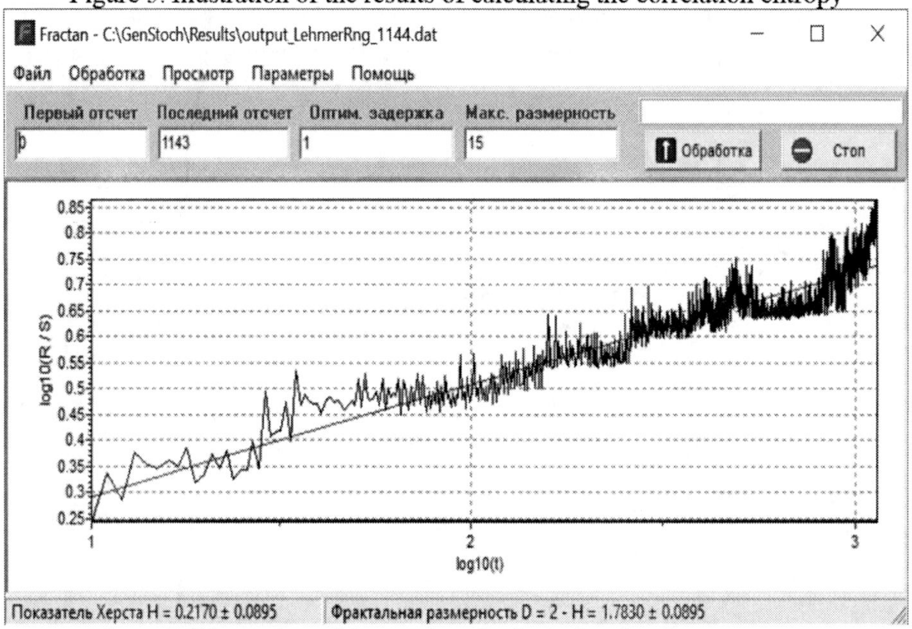

Figure 6. Illustration of the results of calculating the Hurst exponent

- Correlation dimension D = 8,02.
- Correlation entropy K = 0,69.
- Hurst exponent H =0,53.

Based on the initial random sequence, ESDOS was simulated and the main indicators were calculated:

- Correlation dimension D = 4,70.
- Correlation entropy K = 2,56.
- Hurst exponent H = 0,21.

The data obtained indicate the random nature of the representation of the eigenvectors of symmetric matrices obtained on the basis of the initial random sequence of numbers.

This conclusion is encouraging, however, for a detailed assessment of the statistical and correlation indicators of sequence ensembles obtained on the basis of the developed program, additional research is needed.

Conclusions:

1. To solve the problem of developing a software model for the synthesis of increased volumes of systems of discrete orthogonal code sequences, the known numerical methods for calculating the eigenvectors of symmetric matrices were investigated and the most acceptable of them was chosen.

2. Created an algorithm for modeling ensembles of stochastic discrete orthogonal signals of increased volumes.

3. A software model for the synthesis of increased volumes of systems of discrete orthogonal code sequences has been developed, which allows calculating the statistical and correlation indicators of sequence ensembles and displaying the calculation results.

4. The results of a preliminary assessment of the statistical and correlation indicators of sequence ensembles obtained on the basis of the developed program showed the random nature of the sequences obtained, and further research is needed for a more accurate assessment of these indicators.

3. Conclusion

t is possible to provide high structural secrecy of information transmission systems with MA and CDM based on the use of a large number of different ensembles of stochastic discrete orthogonal code sequences, which are randomly replaced when used.

In this article, a program has been developed for studying the correlation and statistical characteristics of the ESDOS described by the eigenvectors of symmetric matrices, with random filling of the diagonal coefficients.

A preliminary assessment of the statistical and correlation indicators of sequence ensembles obtained on the basis of the developed program showed their random nature.

4. Acknowledgment

The reported study was funded by Russian Ministry of Science (information security), project number 29/2020.

References

[1] Viterbi A J 1995 *CDMA: Principles of spread spectrum communication* (Reading, MA: Addison-Wesley)

[2] Proakis J 2001 *Digital Communications* (N. Y.: McGraw-Hill)

[3] Sklar B. 1988 *Digital Communications. Fundamentals and Applications* (Upper Saddle River NJ, Prentice-Hall)

[4] Varakin L E 1978 *Complex signals theory* (Moscow, Soviet radio)

[5] Petrovich N T, Razmahin M K 1969 *Telecommunication systems with noise-like signals* (Moscow, Soviet radio)

[6] Golomb S W, Gong G 2005 *Signal design for good correlation for wireless communication, cryptography, and radar* (Cambridge: University press)

[7] Dyatlov A P, Dyatlov P A, Kulbikayan B H 2004 *Electronic warfare with satellite radio navigation systems* (Moscow, Radio and communications)

[8] Orel D V 2013 *Modeling of stochastic systems of binary quasi-orthogonal code sequences based on the method of functional transformations* (Ph.D. thesis on 05.13.18 Mathematical modeling, numerical methods and software packages. Stavropol, North Caucasus Federal University)

[9] Pashintsev V P, Kalmykov I A, Zhuk A P, Orel D V, Zhuk E P 2018 Formation Algorithms and Properties of Binary Quasi-Orthogonal Code Sequence of Modern Satellite Systems *International Journal of Engineering & Technology* vol 7 (4.38) pp 1205-1209 DOI: 10.1117/12.2324249

[10] Zhuk A P, Ivanov A S 2011 Increasing the structural secrecy of the information transmission system with code separation of channels *High-tech technologies in Earth space research* No1 pp 26-28

[11] Golomb S 1964 *Digital communications with space applications* (Upper Saddle River NJ, Prentice-Hall)

[12] Golomb S 1967 *Shift Register Sequences* (San Francisco: Holden-Day)

[13] Ziemer R. E., Peterson R. L., Borth D. E. Introduction to spread spectrum communications. – Englewood Cliffs, NJ: Prentice Hall, 1995. – 695 p.

[14] Nunn C J, Coxson G E 2009 *Polyphase pulse compression codes with optimal peak and integrated sidelobes* IEEE Transactions on Aerospace and Electronic Systems Vol. AES-45, No 2. pp 775-781

[15] Rushanan J 2006 *Weil sequences: a family of binary sequences with good correlation properties* (Seattle convention center. Seattle)

Assessment of methods online forensics in alleviation of the consequences of the attacks of extortioners

F B Tebuyeva, M G Ogur, M M Gasparyan, A I Bliznetsova

[1]North-Caucasus Federal University, Stavropol, Pushkin str., 1, 355017, Russia

E-mail mogur@ncfu.ru

Abstract This article is devoted to disclosure of essence of a concept online forensics. On the basis of the analysis of practical activities and works of domestic and foreign experts in article the author's concept characterizing estimated judgment of methods in an issue of alleviation of the consequences of the attacks of extortioners reveals.

Keywords: forensic; criminalistics; forensic methods.

1. Introduction

There is no secret, that with development of new technologies also the illegal actions posing in themselves potential threat not only narrow-minded but also world scale actively progress. The amount of information grows in a geometrical progression that inevitably conducts to a compromise of a classical triad of information security: confidentiality, integrity and availability.

In a type of complication of the process of development and deployment of any given IT product the criminalistic examination accepts ambiguous character: not only the equipment and ways of investigation, but also a possibility of malefactors are improved. From there is an urgent need in assessment of the standard methods of a forensic for the purpose of ease of consequences of the attacks of extortioners.

2. Problem research:

The historical semantic origin of a forensic originates in Ancient Rome when one of favorite processes the people had judicial debate and performances which are characterized by the word "foren" that means "the speech to a forum".

However, emergence of this term in Russian is caused by loan from English: the concept "forensic science", or the "judicial science" understood in our country as criminalistics began to be used in a short form thanks to what got narrower circle of application and from the category of the general criminalistics passed into the category computer [1].

Taking into account a modern variety of the executed functions, there is no uniform comprehensive concept of a forensic, the number of interpretations is high, but the essence is invariable, we will consider a little from them.

Owing to the fact that the forensic as applied activity received the popularity in our country relatively recently, we will dilute a Russian-speaking segment of definitions English-speaking.

Michael A. Kaloyannides understands set of the methods and tools used for search of the data in the computer which can be used to the detriment of its user [1] as forensics. Addressing Henry B. Vulfi's works, it is possible to claim that the forensic is a methodical number of receptions and procedures of collecting proofs, since computer facilities and various memory devices and finishing with digital data carriers which can be presented in court in a coherent and substantial format [2].

It is necessary to mention also one of the fullest and popular Russian-speaking sources "Forensic - computer criminalistics" N.N. Fedotov.

Relying on this author, one may say, that the forensic (computer criminalistics) is applied science about disclosure and investigation of the crimes connected with computer information about methods of receiving and a research of the proofs having the form of computer information (so-called digital proofs) about the technical means used for this purpose [3].

In spite of the fact that many things described in this work are outdated, Fedotov's interpretation is used in works of various level to this day.

Generalizing the above, it is possible to draw a conclusion that the forensic, in a generalized view, is the fundamental and most volume section of digital criminalistics comprising set of methods and means on identification, investigation and prevention of computer crimes.

3. Methods of computer criminalistics

At the moment the mankind only begins to plunge into an era of high technologies, but already now it is possible to notice how digitalization influences tenor of life habitual to us: personal computers, laptops, smartphones, etc. became an integral part of our life, separately in itself they are less subject to any given threats, but at connection to network and the Internet everything cardinally changes and this situation does us is significantly more vulnerable at any given attacks on devices [4].

Overwhelming part of information is on electronic media, from there is a certain risk of hit of all information base to the third parties.

As it was already told, the forensic comprises methods on identification and investigation already happened or planned cyber-attacks, then it is worth allocating two main methods of examination: static analysis (static analysis) and dynamic analysis (live analysis). Let's consider them in more detail.

3.1. Static analysis

This method is traditionally directed to creation of dumps (molds) and a research of copies of a disk that then to take contents of cells of memory. Remote files, Web browser history, files and fragments of open access, network connections and so on can be an example of that.

They are taken in order that it was possible to create the accurate schedule and statistics with the periods of time giving to the criminalist an idea of use of the device to restore the course of early events, that is the essence of a method consists in creation of the copy partial or a summary of any activity executed in the device of the victim before its reset or blackout.

The relevance of this method comes to the forefront only when there is an opportunity in due time to react to identification of an incident.

For direct unloading of data in the static analysis different types of instruments of program and technical support are used, presence or absence at malicious software of protection against removal of molds of the contents can serve as an important factor.

Data for a forensic are collected on various types of external stores, such as USB, external hard drive or CD/DVD carriers and then data are sent to specialized crime lab for carrying out by the investigator or expert of all necessary procedures for the analysis of evidentiary data.

As advantages of this method, it is possible to allocate that chances of modification of data are insignificant also an opportunity for extraction of dangerous materials small. Among shortcomings there is a work directly with cryptographic ways, dangerous data of the software, gigabytes of data for the analysis, the lack of the standardized procedures, practical and lawful restriction, proofs can easily pass into a condition of inaccessible.

3.2. Dynamic analysis

The sense of this analysis is revealed already in the name: if literally to translate "Live analysis", then we will receive "the live analysis". That is work, unlike the static analysis, happens directly on the device of the victim without further power off [5].

It turns out that collecting, the analysis, generation in reports of necessary information happens at that time, the system so far compromised remains functional.

The tools which are used at this method of the analysis can provide a clear picture of dumps of memory, the started processes, open network connections and unencrypted versions of the encrypted files. As when using the static analysis, the contents of memory cannot be received properly fully, the dynamic analysis provides coherence and integrity of criminalistic data. Information collected by such method can be used in various ways for obtaining proofs necessary for investigation or idea of suspicious activity and operations performed directly by the user or existence of remote access to the compromised system [6].

Advantages of this method are search of harmful information and that fact that collected information is limited to relevant. Also, it should be noted shortcomings: modification of information in real time, indistinct images, limited volume of collected information, each computer installation is unique and bears in itself difficulties in work with it, it becomes more difficult to prove authenticity and reliability, the tool kit does not correspond to criminalistic.

4. Assessment of methods in alleviation of the consequences of the attacks of extortioners

The ideal system capable to fully protect any given device from various threats, does not exist therefore at stages of prevention and detection of the attacks most often attacks of criminals are crowned with success that forces users to liquidate their consequences [7].

One of the most topical issues of this subject are programs extortioners - a kind of malicious software which at hit on the device blocks to the owner partial or full access to its data, demanding to pay a certain sum for receiving the decoder.

Often such programs get on the device in several ways: harmful spam and advertising with use of technologies of social engineering, external foreign stores, phishing letters and so on.

Today this phenomenon gets new ways and tricks: if earlier extortioners asked repayment for the encrypted data, then now they replaced a business model with more flexible - threats of violation not only availability, but also integrity and confidentiality of the infected data are added to standard requirements. The specified scheme of succession of events nonpluses the victim and it does not have other choice how to pay, swindlers have access to the stolen information and can vary dexterously the sums and terms of transaction.

If after all program of this sort appeared on the device, then first of all it is worth isolating it from other devices if those are available, in order to avoid loss of full access to all database. Understanding and acceptance of that fact that it is impossible to go on similar transactions at all has to be the following step and to try to redeem access or data, it, on the contrary, will only aggravate a situation. When the swindler understands that you already on a hook, he will make sure that you and other users can be manipulated quietly, and it will lead to even big losses not only with yours, but also the foreign country.

The attempt of restoring access to some encrypted files by means of free programs decoders can become the following step, but because of use by criminals of difficult algorithms of enciphering not for all families of programs extortioners it is possible to find the decoder. But unintentionally not to aggravate the already deplorable state with the incorrect decoder and once again to encrypt the data, it is the best of all to address experts in the sphere of cyber security or information technologies [8].

Here we also approached consideration of the above-stated methods of a forensic.

Let's begin with traditional - static. As it was already told, this method will be the most effective at rapid response to intervention of the stranger ON. As the dump (mold) is created and the device is switched-off, the probability that not all files will be subject to infection increases. On the isolated platform there is a possibility of a research of the program extortioner of this type and selection of the corresponding decoder, and existence of backup copies of cells of memory guarantees safety of the encrypted file systems. Further at selection of the decoder it will be possible to carry out this operation several times and not to be afraid for additional encoding of files. Act as essential minuses of this method

need in cryptographic methods of interpretation and volumes of data of a dump, quite large for the analysis. But, despite this, the probability that you will go to the transaction with criminals, and this unconditional advantage over the essence of programs extortioners is reduced [9].
Speaking about a dynamic method of a forensic it is worth specifying first of all that the most part of files is saved and gives wider picture of the event as work happens directly in the infected functioning system. Availability of the relevant information but which is not written down final will be indisputable advantage that is an opportunity to trace in real time the center of distribution and according to its consequence is given. On the other hand, this method is characterized by a number of powerful shortcomings: in real time the risk of modification of all data on the device is high that leads to expansion of a zone of defeat by the program extortioner and strengthening of influence of swindlers on the victim; the device in itself is unique and there is not always an opportunity to competently process a flow of harmful information [10].

5. Conclusion

Accumulating information stated above it is possible to draw the following conclusions:

1. The efficiency of any given method, in the solution of a question of programs extortioners, cannot be hundred percentage: each of them has a number of advantages and shortcomings, but it does not cancel existence both a human factor on both sides of crime, and refinement of swindlers;

2. Decoding of important data independently or with the expert cannot bring fruitful results, and the stolen information can become object of blackmail (violations of confidentiality and dissemination of data to the fourth persons) there are two options of an outcome of events: either to pay, or to find the criminal;

3. To prevent hit of similar programs on the device it is worth following some simple actions:

- Installation of the reliable antivirus software,
- Timely SOFTWARE updating and other components of a system,
- Creation reserve or backup copies of data,
- the Attentive relation to advertising mailings, letters from unknown senders,
- Use of safe VPN connection at connection through public WI-FI.

References

[1] Caloyannides M.A. 2004 *Privacy Protection and Computer Forensics (Second Edition).* (Artech House Publishers)
[2] Irfan Shakeel 2015 *Introduction to Computer Forensics and Digital Investigation* (INFOSEC INSTITUTE)
[3] N.N. Fedotov 2007 *Forensic - computer criminalistics* (M.: Legal world,. - 360 pages of ISBN 5-91159-013-1)
[4] Henry B. Wolfe 2001 *An Introduction to Computer Forensics: Gathering Evidence in a Computing Environmen* (Informing Science, Special Series: Expanding the Focus vol 4 No 2) pp 47-52
[5] Mamoona Rafique, M.N.A.Khan 2013 *Exploring Static and Live Digital Forensics: Methods, Practices and Tools* (International Journal of Scientific & Engineering Research vol 4, Issue 10) pp 1048-1056
[6] D.N. Shevchenko 2018 *Methods of digital criminalistics and a computer forensic for*

investigation of incidents of information security
[7] Kalber, S.; Dewald, A.; Freiling, F.C. 2013 *Forensic application-fingerprinting based on file system metadata* (Proceedings of the Seventh International Conference on IT Security Incident Management and IT Forensics, Nuremberg, Germany) pp. 98-112
[8] Bang, J.; Yoo, B.; Kim, J.; Lee, S. 2009 *Analysis of time information for digital investigation* (Proceedings of the Fifth International Joint Conference on INC, IMS and IDC, Seoul, Korea, 25-27 August 2009) pp. 1858-1864.
[9] Chabot, Y.; Bertaux, A.; Nicolle, C.; Kechadi, M.T. 2014 *A complete formalized knowledge representation model for advanced digital forensics timeline analysis* (Digit. Investig. 2014, 11, 95-105)
[10] Carrier, B. 2019 *Open Source Digital Forensics Tools*

Judicial technical expertise methods for investigation of cybercrimes

V I Petrenko, A S Bereznitsky, M G Ogur, E A Nekrasova

North-Caucasus Federal University, Stavropol, Pushkin str., 1, 355017, Russia

E-mail mogur@ncfu.ru

Abstract Computer technical expertise is rather new sort of the expertise which are carried out within criminal, civil and arbitration proceedings. And though separate expertise of computer means was carried out and earlier, "computer crimes" laid the foundation for all-round development of judicial computer technical expertise as independent sort of examination. Its emergence is caused by growth of use of computer and information systems in our country in recent years.

Keywords: forensic; criminalistics; computer technical expertise; investigation of incidents.

1. Introduction

The purpose of conducting computer technical expertise (further - CTE) is protection of the rights and freedoms of citizens and the interests of the state by means of evidence-based examinations and expert researches. The special knowledge used in process carrying out CTE is knowledge in the following scientific areas: electronics, electrical equipment, information systems and processes, radio engineering and communication, computer facilities (including programming) and automation. A basis a procedural form of use of special knowledge in the field of computer technologies is judicial computer technical expertise [1-7].

2. Problem research:

Computer technical expertise as a separate sort of judicial examinations it began to be formed spontaneously in the mid-nineties, way as it usually and happens, productions of separate unique examinations. During this period the subject and problems of JCTE, its terms framework, specific division were not still formulated. There was no uniformity even in the name of examination.

One of the first publications in Russia on the matter was S.A. Katkov, I.V. Sobetsky and A.L. Fedorov's article who suggested to call this examination program and technical, however such name was too narrow and did not answer all circle of the tasks solved by this examination. There were also other names of judicial examination of this sort: information and technological examination, computer and technological examination, engineering and computer, judicial and cybernetic examination, etc. In 1996 it was offered to E.R. Rossinskaya to call this sort of judicial examinations computer technical expertise and to include it in a class of technical examinations as the computer facilities are obliged by the birth to technical sciences. In the presents time this term became standard and the discussion about the name of this sort of judicial examinations can be considered complete. Confirmation of the end of a discussion on the matter can be considered addition of "computer technical expertise" in the list of new childbirth of the expertise which are carried out by the judicial and expert institutions (JEI) of the Ministry of Justice of the Russian Federation according to order N 237 of December 27, 2012. "About the approval of the list of childbirth (types) of the examinations which are carried out in the public judicial and expert institutions of the Ministry of Justice of the Russian Federation, and the list of expert specialties on which the right of independent production of judicial examinations in the public judicial and expert institutions of the Ministry of Justice of the Russian Federation is granted". According to this order the computer technical expertise is officially regulated for carrying out in the SEU system on the instructions of vessels, prosecutor's office, bodies of the Ministry of Internal Affairs and so forth.

Thus, judicial computer and technical examination - the independent sort of judicial examinations belonging to the class technical. JCTE is carried out for: determination of the status of an object as computer means, identification and studying its role in case in point and also gaining access to information on carriers of data with the subsequent its comprehensive investigation. A subject of scientific bases of JCTE are regularities of formation and a research of computer systems and the movement of computer information.

Patrimonial subject JCTE are the facts and circumstances established on the basis of a research of regularities of development and operation of the computer means providing realization of information processes which are recorded in materials of a criminal, civil, arbitration case and case of administrative offense.

Specific classification of JCTE is organized on the basis of providing a component of any computer means (hardware (technical), program and information support). Proceeding from it in judicial computer technical expertise are allocated [1-7]:

- Hardware and computer examination:

The essence of this examination consists in carrying out a research of technical (hardware) means of a computer system. A subject of JCTE of this type are the facts and circumstances established on the basis of a research of regularities of operation of hardware of a computer system - material data carriers about the fact or an event of a criminal or civil case.

- Program and computer examination:

Subject are regularities of development (creation) and application (use) of the software of the computer system presented on a research for establishment of the truth on business. The purpose of program and computer examination is studying functional purpose, characteristics and the realized requirements, an algorithm and structural features, current state of the software of a computer system presented on a research.

- Information and computer examination (data):

The purpose of information and computer examination is search, detection, the analysis and assessment of the information prepared by the user or generated (created) programs for the organization of information processes in a computer system.

- Computer and network examination:

Subject of computer and network examination a research of the facts and circumstances connected with use of network and telecommunication technologies.

Such division into types of JCTE most fully covers technological features and operational properties of the objects of examination shown for a research. This classification allows already at early stages of formation of examination differentially to approach developments of methods and techniques of an expert research. At the same time such approach to JCTE can be effectively used during the assigning of complex examinations and the decision of the big list of tasks.

Practice shows that the main types of judicial computer technical expertise described above by production of the majority of expert researches are applied in a complex and, most often, consistently. Therefore, now, the patrimonial name of examination is specified in resolutions and definitions on production of judicial examination, i.e., "to appoint judicial computer technical expertise".

It is possible to distinguish from the main objectives resolved by judicial examinations:

- Identification tasks, i.e., identification of an object for its displays.

- Diagnostic - consist in identification of the mechanism of an event, time, a way and the sequence of actions, events, the phenomena, causal relationships between them, the nature, quality and quantity characteristics of the objects, their properties and signs which are not giving in to direct perception, etc.

3. The tasks solved by computer technical expertize

When determining diagnostic tasks, not only properties and a condition of an object of judicial computer technical expertise, but also the mechanism, processes and actions by results of application (use) of computer means at crime commission are investigated. And identification problems of CTE aim to establish the fact of individual and concrete identity or the general group accessory of the presented objects of examination. Let's consider tasks solvable each of the types of computer technical expertise described above.

1. Standard research problems of computer hardware objects:

- definition of a look (type, brand), properties of the hardware and also its technical and functional characteristics;
- determination of actual state and serviceability of the hardware, existence of physical defects;
- determination of structure, the mechanism, a circumstance of an event by its results due to use of hardware;
- definition of conditions of application of hardware;
- establishment of a causal relationship between use of concrete opportunities of hardware and results of their application;
- definition of conditions of application of hardware, recovery of the chronological sequence of their use, scene of action and functioning.

2. Standard research problems of program objects:

- definition of the main characteristics of the operating system, identification and research of functional properties and software configurations, time of its installation;
- determination of actual state of a program object;
- identification of signs of the relation ON to system or applied;
- establishment of an initial condition of the program;
- diagnosing of an algorithm of the software product, types of the tools used at its development and also types of the supported hardware-software platforms;
- definition of the purposes and conditions of change of properties and a state ON.

3. Standard research problems of information objects:

- establishment of properties and a type of the provided information in a computer system at its direct use;
- determination of actual state of information, clarification of existence or absence of deviations in it from a standard object of JCTE;
- establishment of an initial condition of information on the carrier of data;
- definition of time, chronological sequence of impact on information;
- definition of conditions of change of properties of the studied information.

4. Standard research problems of the computer network and its component:

- determination of properties and characteristics hardware and software;
- identification of properties and characteristics of the computer network, establishment of its architecture, configuration;
- definition of the reasons of change of properties of the computer network;
- the researches connected with Internet-technologies.

It should be noted that all tasks can be detailed as well for each concrete stage of an expert research.

Expert problems of judicial computer technical expertise are concretized by production of each examination. They are not identical to the questions formulated in the resolution or definition.

Sometimes several questions, raised before the expert, conduct to the solution of one expert task. To the contrary, one question can require the solution of two and more independent tasks.

The essence of the judicial computer technical expertise belonging to the class of engineering technical expertise consists in studying properties and conditions of objects of examination, a research of mechanisms, processes and actions by results of use of computer means in any given crime (including computer) and also to materials of a civil dispute, administrative offense.

The objects investigated by the expert of JCTE differ not only on the natural and technical characteristics, functional purpose, but also in the status. Besides, for considerable number of objects of JCTE (the computer - the hardware block - the program - data) this situation is defined insufficiently accurately and differently in various legal acts. Any concrete object is individual and unique, it defines specifics of a certain expert research. Applicable JCTE can formulate the following definitions:

Patrimonial (specific) object of JCTE - a certain category of the objects having the general signs and relating to computer means. One of important features of an object of JCTE is its compound character.

Concrete object of JCTE - the certain computer means investigated in the course of this examination, having signs of identity and originality as defines specifics of a concrete research [8].

Now the system of objects of judicial computer technical expertise is divided according to the specific division and looks as follows:

Class of hardware objects:

- personal computers;
- peripheral devices;
- network hardware (servers, workstations, active equipment);
- the integrated systems (organizers, pagers, mobile phones);
- any components all specified a component.

Class of program objects:

- system software;
- auxiliary programs;
- development tools and debugging of programs;
- application software.

Class information objects:

- the text and graphic documents made by means of computer means;
- given in multimedia format;
- information in formats of databases and other applications.

Class of objects of the computer network and its component:

Subject to application of special knowledge can be quite different

- from computers of the users connected to Internet to various resources of the supplier of network services (provider) and the information services provided to them (e-mail, service of electronic announcements, teleconferences, WWW-service and so forth)
- servers; - workstations; - active equipment.

The special interest is of conducting judicial computer technical expertise within arbitration process where CTE is shown in new contents - as normative and technical examination. However, recently, the trend of conducting similar examinations and within criminal proceedings was outlined. A subject of such examinations, in most cases, this establishment of the facts of compliance (or discrepancies) technical documentation of a system - documentation regulating its development according to state standards on information technologies. A basis of this examination is the contract signed on production or processing of a computer system or on performance of other work with transfer of its results to the customer. If the customer after acceptance of a computer system, found in it any given shortcomings and between the customer and the performer there was a dispute, then upon the demand of any of the parties JCTE can be appointed [9].

3.1. The issues resolved within judicial computer technical expertize

By production of judicial computer technical expertise, the following issues can be resolved:

The approximate questions resolved by judicial hardware and computer examination is the following:

Whether the presented device belongs to hardware computer means?

To what type (brand, model) does the hardware belong? What its technical characteristics and parameters?

What functional purpose of the hardware?

What role and functionality of this hardware in a concrete computer system?

Whether this hardware belongs to the presented computer system?

Whether this hardware for the solution of a specific functional objective is used?

What actual state it (is serviceable/is faulty, efficient/is disabled) the presented hardware? Whether there are in it defects?

Whether malfunction of this means is a consequence of violation of certain service regulations?

Whether the presented hardware is the data carrier?

What type (type, model, mark) of the presented data carrier?

The approximate questions resolved by judicial program and computer examination is the following:

What general characteristic of the presented software of what component (software) it consists?

Whether traces of overcoming protection have software?

What name, type, version, type of representation (obvious, hidden, remote) software?

Whether this software for the solution of a certain functional task is used?

What actual state of the software, its working capacity on realization of separate (concrete) functions?

Whether the software has protective opportunities (program, hardware-software) from unauthorized access and copying?

Judicial information and computer examination (data) is made for permission of the following questions:

What properties, characteristics and parameters (volumes, dates of creation change, attributes, etc.) have data on the data carrier?

What look (obvious, hidden, remote, archive) information is available on the carrier?

To what type do the revealed (certain) data belong (text, graphic, the database, the spreadsheet, multimedia, etc.)?

What data on the owner (user) of a computer system (including names, passwords, access rights and so forth) are available on data carriers?

What initial condition of data on the carrier (in what look what contents and with what characteristics, attributes there were certain data before their removal or modification)?

3.2. Requirements, to the questions which are taken out on permission of judicial computer technical expertize

1. At statement of question, it is necessary to avoid slangy and semi-professional terms, they have to contain the settled conceptual framework.
2. The formulation of a question has to be accurate and unambiguous.
3. The formulation should not concern stages of carrying out a research (for example, it should not be specified that to copy the found information on a disk as it is an obligatory stage of information research).
4. Questions should not have legal character and go beyond competence of the expert.
5. Questions should not have help character.
6. Questions have to correspond to the technical base which is available in expert establishment and modern level of methodical providing.
7. Questions have to correspond to the physical evidence produced on a research.
8. Questions have to be directed to establishment of specific circumstances of the investigated event.
9. Statement of questions is allowed in that look that at their decision the costs of their decision (financial, technical, temporary and so forth) have to be minimum.

4. Perspective of investigation of information crimes

Crime in the field of computer information represents special type of crime connected with use of the automated systems, software for illegal receiving, transfer or destruction of information resource. The complexity of their investigation consists in lack of tangibility and visibility of specific responsible for crimes. Besides the staff of investigating authorities more often should face the types of crime new, not known earlier having not clear origin and possibilities of causing damage.

To define a circle of the problems connected with crimes in the sphere of computer information once their category investigates which in detail are considered in Chapter 28 of the Criminal Code of the Russian Federation. This head carries to the considered group of crimes:

- Illegal use and access to information resource;
- Development, use, distribution and introduction of virus electronic programs;
- Violations in the field of storage, transfers and processing of information flows;
- Illegal actions for the purpose of infliction of harm to a critical information system and structure of the Russian Federation.

The first category of criminal actions represents corpus delicti which is covered by Article 272 of the Criminal Code of the Russian Federation. At the same time this article defined more concrete list of the crimes connected with computer information.

In their structure, in particular, it is allocated:

- destruction of the information massif, change, blocking. The fact of recovery of data has no legal value subsequently.

- blocking of the information massif. The lawful user has no opportunity in an operating mode to carry out work.

- modification - illegal impact on parameters of information resource.

- copying - preservation and transfer of the classified information on the isolated object of storage (copying, photo fixing).

During creation of the automatic copy a system, in the absence of intention, the person is not subject to prosecution. The person who committed a crime by negligence can be brought to responsibility at commission of this group of crimes or it is deliberate.

The law enforcement official at the same time, establishes connection between deliberate act and criminal result. At its absence the sentence cannot be imposed to the person. It is worth explaining that the considered sphere is regulated also by other laws.

So since 01.01.2018 FZ-187 "About safety of critical infrastructure of the Russian Federation", also in the Criminal Code of the Russian Federation took effect, Article 274.1 of the Criminal Code of the Russian Federation which violation attracts responsibility for the guilty person was enacted. This article brought explanation that object of crime in the sphere of computer information - the public relations which provide safe functioning of electronic systems of the Russian Federation.

The guilt is expressed both by active actions of the user, and his inaction. The special subject made responsible on the qualifying sign - the lawful user allowed to an electronic system owing to official position.

The complexity in investigation of this group of crimes consists first of all in their qualification as investigating authorities have to possess special knowledge of the field of electronic technologies.

Specialists of investigating authorities besides have to have knowledge connected with the forbidden introduction in Internet network, copying, violation of ways of storage of information, breaking of systems, development, introduction of special programs.

The harmful product influences information resource of infrastructure of institutions and organizations of the Russian Federation with any form of ownership for the purpose of infliction of harm or generation of profit.

When carrying out investigative actions on this category of crimes it is necessary to consider the following nuances:

- in advance it will be prepared for withdrawal of the data carrier on which this information as it, as a rule, contains penetration traces which study will help to establish authentically a way and time was stored, and also the place from which penetration is made is possible.

- to arrive to the place as soon as possible to limit access for all persons to this device. It is connected with the fact that, as a rule, the staff of the company is either criminals, or accomplices of crime, and can try to hide traces of crime.
- to disconnect the device from network to protect it from repeated intervention through remote connection.
- to try not to conduct independently before withdrawal any manipulations with the equipment if it is impossible to be sure of result of any given action. Withdrawal of the carrier has to be made according to requirements of the Criminal Procedure Code of the Russian Federation. After the equipment which underwent unauthorized access was seized it is necessary to pay attention to personnel of the company, especially to persons who are engaged in service of the COMPUTER, maintenance of network and information security.

As the main suspects at the first stages of preliminary investigation those employees who meet the following criteria act:

- knew or could know that on this device information which is of a certain value is stored;
- have the high skill and special knowledge which are allowing or presumably giving the chance of commission of this crime.

Special attention experts who are engaged in information security from unauthorized access as they, as a rule, independently "write" the programs directed to information security deserve and know everything their weak and strengths. Besides the employees working in the company, it is necessary to pay also attention to those who worked in it earlier and presumably corresponded to those criteria which were listed above, both under the employment contract, and on civil. As a rule, there will be enough these measures to solve this crime. Nevertheless, cases when the criminal has extensive experience and the latest technologies sometimes meet, never worked in this company, and worked, for example, by the order or aiming. Such hackers in Russia of unit, but, nevertheless, they are.

For disclosure of the crimes committed by them hardly law enforcement agencies will have enough experience and opportunities. The consequences of violation of the rules of storage, use, information transfer similar to the previous articles - harm to the victim in the form of loss of information, violation of privacy of data, blocking and change of volume.

By the first part of Article 274 of the Criminal Code of the Russian Federation the punishment term up to 2 years, on condition of causing large damage is provided. At qualification of act the investigating authorities define what norms of criminal law on information security were violated by the guilty person and what consequences his actions entailed, establish a fault form.

Errors of the investigation and court at prosecution and qualification of actions lead to the termination of criminal cases and release from responsibility of guilty persons.

So, for example, the guilty convict according to Article 272 of the Criminal Code of the Russian Federation was exempted for access to official information and causing damage of the organization from liability by superior court, having specified that the special subject - the person who has lawful access to the COMPUTER and an information system is brought to responsibility. Negative consequences were caused by failure to follow security guidelines and neglect of the duties. Actions need to be qualified according to Article 274 of the Criminal Code of the Russian Federation.

In this situation the system administrator of establishment during the work did not use a security system. Copying of information with personal data of employees and personal contacts of commercial partners was as a result carried out. Criminal case was mistakenly brought according to Article 272 of the Criminal Code of the Russian Federation and brought to trial. The decided adjudication was repealed by supervising instance, criminal case is dismissed, on circumstances of the expiration of terms of criminal prosecution. Investigating and judicial authorities did not consider that a crime was committed by a special subject - the lawful user. Actions needed to be qualified according to Article 274 of the Criminal Code of the Russian Federation.

The given examples prove complexity of disclosure of computer crimes for law-enforcement bodies. There is a need of production of information and technological examinations.

In expert way also existence of corpus delicti in actions of persons as neither the court, nor for participants of process often has special knowledge of the field of modern electronic technologies is possible to establish the legal facts.

It is obvious that the edition of special instructions which will allow to understand effectively features of this type of crime is necessary and will give the criminalistic analysis of these acts as it was made at the time for other types of crime.

For today such crimes are rather a nonsense, than routine work of law enforcement agencies. Nevertheless, in criminalistic science exist already today, though very poor, based on experience of the advanced countries data allowing to sketch a technique of investigation of these crimes. On the basis of knowledge gained during clarification of the above-mentioned questions the investigator has to have a clear picture of crime.

The circle of subjects which are capable to get unauthorized access to the closed servers or networks of the company is extremely limited because the criminal in this case has to have huge knowledge base in the field of computer science and information security [8- 10].

5. Conclusion

Investigative practice shows that the more difficult technically the way of penetration into a computer system or network, is easier to allocate those the suspect as the circle of the experts having the corresponding abilities is usually limited even more. Solvability of computer crimes in the Russian Federation now quite low. It is connected with the fact that law enforcement agencies meet this type of crime extremely seldom and therefore real practical experience and a technique of investigation of these types of crime by them it is not developed yet.

At the same time, in most the companies there are no expensive modern programs and other means of information protection that does them vulnerable for the attacks of hackers. Most likely, the situation with these crimes will shortly change.

At the present stage in Russia everywhere electronic document flow, the electronic auction, online of the request for rendering services actively takes root. All this is attracted, on the one hand, by acceleration of processes of filing of applications, and with another - transfers them to less protected reality - virtual that enhances risks of information leakage from the computer in bases of malefactors.

References

[1] Constitution of the Russian Federation (of 12.12.1993).

[2] Federal law of the Russian Federation "About Information, Information Technologies and on Information Security" (No. 149-F3 of 27.07.2006).

[3] Doctrine of information security of the Russian Federation (No. Pr-1895 of 06.09.2000).

[4] Federal law of the Russian Federation "About the State Secret" (No. 5485-1 of 06.10.1997).

[5] Decree of the Russian President "List of Data of Confidential Character" (No. 188 06.03.1997).

[6] Federal law of the Russian Federation "About Personal Data" (No. 152 of 27.07.2006).

[7] Federal law of the Russian Federation "About the Digital Signature" (No. 1-FZ of 26.12.2001).

[8] Carrier, B. 2019 *Open Source Digital Forensics Tools*

[9] Olajide F, Savage N, Ndzi D, Al-Sinani H 2019 *Forensic Live Response and Event Reconstruction Methods in Linux Systems*

[10] Thomas, D.S.; Forcht, K.A. 2004 *Legal methods of using computer forensics techniques for computer crime analysis and investigation* (Issues Inf. Syst) pp 692-698.

[11] Lillis, D.; Becker, B.A.; Scanlon, M. 2016 *Current Challenges and Future Research Areas for Digital Forensic Investigation* (Proceedings of the 11th ADFSL Conference on Digital Forensics, Security and Law, Daytona Beach, FL, USA, 13 April 2016) pp. 9-20.

[12] Prasad, M.A.R.; Satish, Y.N. 2013 *Reconstruction of Events in Digital Forensics* (Int. J. Eng. Trends Technol. 2013, 4) pp 3460-3467. [CrossRef]

[13] Carvey, H. 2009 *Windows Forensic Analysis DVD Toolkit 2E* (Syngress: Burlington, MA, USA)

[14] Harrell, C. 2011 *What's a Timeline.* Available online:

http://journeyintoir.blogspot.com/2011/09/whats-timeline.html (accessed on 28 May 2019).
[15] Esposito, S.; Peterso, G. 2013 *Creating Super Timelines in Windows Investigations* (Proceedings of the 9th International Conference on Digital Forensics, Orlando, FL, USA, 28-30 January 2013) pp 135-144.

Using of live response technologies during the blocking of the attacks and investigation of incidents in real time

V I Petrenko, M G Ogur, M V Zubkov, O E Solgalova

North-Caucasus Federal University, Stavropol, Pushkin str., 1, 355017, Russia

E-mail mogur@ncfu.ru

Abstract. Acquisition of a clear idea of the events and artifacts occurring eventually is a difficult task which needs to be solved in digital criminalistics. Recovery of chronology of events and artifacts, allows experts to understand chronology of digital crimes and to interpret a conclusion in the form of digital proofs with use of live-response technologies.

Keywords: forensic; criminalistics; investigation of incidents, live-response technologies

1. Introduction

The digital criminalistics became new area of researches in the last two decades in connection with the explosive growth of the Internet, use of electronic devices, fast innovations in technology and the growing sizes of memory devices and also high growth of number of digital or computer crimes. The purpose of digital criminalistics is search and interpretation of origin of the event or an artifact found in a computer system. Besides, the digital criminalistics is engaged in restoration and investigation of digital proofs; in case of need it is often connected with (cybercrime). Attribution, identification of leak in the organization and assessment of the possible damage caused during violation are the most common causes of conducting digital criminalistic examination. The field of digital criminalistics is subdivided into several divisions depending on type of the digital device which is object of investigation. The computer criminalistics, network criminalistics, criminalistic examination of data and criminalistics of mobile devices belong to them. Digital investigators study data and devices to learn as much as possible about the violation or crime connected with use of digital devices. in the form of digital proofs. The carrier classifies digital criminalistics by three main stages, namely: receiving, analysis and representation [1-3].

The purpose of a stage of data collection is preservation of a state the Digital system allowing to take and analyze subsequently proofs in controlled conditions. Following a phase, i.e., the analysis, consists in studying the obtained data for identification of signs of harmful activity. The stage of the presentation is completely based on precepts of law and rules which change depending on country jurisdiction where there are digital proofs, and, thus, are beyond the present document. As a rule, at an acquisition phase the first step initiated by the investigator is obtaining images of devices of storage of the confiscated computer. The following step after an acquisition phase includes definition of proofs which can be connected with studying the data concerning investigation. In digital criminalistic investigation, as well as at usual restoration on the crime scene, it is important to have an opportunity to make the schedule of work of the file system in a computer system.

The chronology is used for crime reconstruction, and it is favorable in terms of providing to the user access to a target system, implementation of certain applications and also identification of system files and data to which access which were modified or removed during the certain periods of time which are of interest to investigation was got. To provide collected proofs from the confiscated computer in court, process by means of which the evidence was obtained has to be accurately shown with observance of the adopted procedures and the practician. In legal proceedings this practice is known as "a chain of deliveries". To prove authenticity of proofs, chains of justification have to be brought into court about the scientific methods accepted in extraction of proofs. Experts face various problems, such as high speed and volumes of data, development of standards, maintaining confidentiality of investigations and growth of anti-criminalistic methods when digital or electronic devices take from them proofs [4,5].

Abstract approach to assistance to experts during the investigation by recovery of chronology of events and artifacts which are clear for investigators is presented in this article and reduce time necessary for collecting digital proofs by means of digital devices [6].

2. Problem research.

Prasad & Satish defined digital criminalistics as the procedure of identification, collecting, preservation, the analysis and submission of digital proofs thus which is legally recognized by court. For effective, effective and during the limited period of time of collecting digital proofs from digital devices the reconstruction of terms is required. In "Review" the time frames are defined as "means of identification or a linking of the sequence of events so that the persons who are responsible for response to incidents could visualize and understand easily". Harrell defined that "the analysis of time frames is an excellent equipment for determination of activity which occurred in a system during a certain period of time in a system". Thus, "creation of a timeline of various events which took place during the incident is one of the key tasks which are carried out by the expert in digital criminalistics". The analysis of time frames has to be a key component of any investigation as time frames of events are almost always relevant. The main source of chronological information are metadata of the file system. File systems trace various temporary tags and have nuances which have to be considered when carrying out the criminalistic analysis [7,8].

The Inglot & Liu company specified that in principle there are two approaches to the analysis of a timeline. First, there are applications which were specially created for the analysis of a timeline, and they to focus on visual representation of a timeline, for example, Cyber Forensics Time Lab (CFTL), Zeitline, Encase, tool kit for forensic medicine and many other things. Secondly, there is a combination of tools of a command line and applications for work with spreadsheets which are labor-consuming, for example, Log2timeline and excel together. Guðjónsson developed the tool for extraction of temporary tags from various files found on the ordinary computer. and to aggregate them. The Log2timeline tool is a part of the plazo with the back engine on the basis of Python. Purpose of these systems is to receive temporary tags in one place for computer criminalistic examination. Such timeline sometimes is called "Super-timeline" [5].

Sitompul, Handoko, and Rahmat claim that to have a clear idea of events which took place during the certain period of time it is difficult to reach in digital investigation. The restoration of events allowing the expert to understand chronology of commission of crime is one of the most important stages of digital investigation. This difficult task demands a research of a large number of events that is connected with frequent technological innovations, heterogeneity, huge number of data and also manual process of reconstruction of events which is inefficient and expensive [9-11].

Bang, etc. discussed how creation time, time of the last record and time of the last access to the file or the folder are important factors which can indicate the events which affected a computer system. They analyzed changes in time of information on files and folders for various operations of the FAT and NTFS file systems and tried to restore actions of the user. Besides, they show use of information on time for the digital analysis of proofs on the example of a specific case. Shabo, etc. was defined by two main problems (heterogeneity and volume of data) with reconstruction of events. In them the approach supported with theoretical concepts which can help investigators throughout all process, including construction and interpretation of the events describing business is stated [12,13].

Hargreaves and Patterson considered the possibility of use of the Log2timeline tool for reconstruction of a timeline., but was decided to construct an own prototype on Python. The basis allowing to write the analysis which can lead to the event of high level based on presence of one or several events of low level became other contribution of research. A lack of this approach was the fact that analyzers which were focused on certain events have to be written in advance [14].

Brady, Overill and Keppens suggested to use digital semantic ontology of proofs (DESO) which allows the expert to find out quickly what artifacts can be available on the device. DESO is constructed on the ideas of Gene ontology (GO). The general principle which is the cornerstone of DESO is bidirectional [15,16]:

1. Experts use a certain form of classification or the system of tags which allow experts to estimate easily what artifacts are available;
2. after artifacts are taken from various sources, DESO provides means for their comparison.

The main idea is in having an opportunity to compare the artifacts taken from different sources. Brady & Overill continued development of ontology of DESO. A main objective - performance of functions of a repository and the qualifier of artifacts of the digital evidence for a possibility of comparison of the data obtained from heterogeneous sources. Research objectives are established and carried out by statement of simple questions, based on the one who that when where why and how? ("5WH"). Only a subclass "That" it was detailed. Classes "why" and "as" were not discussed at all. Realization of DESO in general was not opened.

Debinski, Breytinger and Mokhan claim that reconstruction of events is a fundamental step for experts to understand the main point, and the important tool for generation of time frames is Log2timeline. In spite of the fact that these schedules give big proofs and help to understand the event, they are difficult and demand additional tools and also educational scenarios. Moreover, some of the main restrictions of Log2timeline are also specified in them, for example, the fact that there is no simple tool in use which beginners/investigators could use for the analysis of the generated time frames, and is not present a free training material which would also allow the practicing experts to study and improve the acquaintance to Log2timeline and also to instruments of visualization.

Authors developed Timeline2GUI for support of researchers - the independent tool written in the Python language which supports the analysis of a timeline of CSV (result of Log2timeline). The tool is similar to widely used sheet Excel that provides easy transition for the practicing experts. Besides, three educational cases which are in free access were developed and can be used for improvement of skills of the analysis of time frames of the investigator or by means of the graphic Timeline2GUI interface, the sheet Excel or any other tool. Performance of Timeline2GUI can be increased due to process acceleration, for example, due to removal of irrelevant fields or association of fields in future work. Soltani, Seno & Yazdi offered the scheme of reconstruction of events which defines whether the application was started on the compromised system. The offered structure constructed a signature or TPFSM-A (temporary template of modification of the file system of the application) and TPFSM-D (temporary template of modification of the file system of the hard drive). Besides, the framework presented a distance metrics which is used for calculation of distance between a signature of the application and TPFSM-D of the hard drive. The instrument of decision-making of a framework used the calculated distance for making decision on whether the application was started on the compromised system. Zhao and Kopliyen said that success of applications of symmetry in various scientific disciplines induced them to study the concept of symmetry in the software. Besides, authors explained that in object-oriented designs of language the concept of symmetry is used, and studied symmetry in object-oriented language and also gave some examples of symmetry outside object-oriented programming to show that reasons of symmetry are beyond object-oriented programming and extend to other fields of development of the software [17].

The review of literature shows that the set of tools and techniques was developed for support of the criminalists who are engaged in the analysis of time frames, but any of them is not able to solve a problem of difficult creation of chronology of events and artifacts in real time. The following belongs to number of the main causes of inefficient adjustment of terms:

1. huge number of data,
2. heterogeneity of data,
3. fast innovations of technologies,
4. rapid growth of the Internet and integrity of the schedule.

The purpose of work is development and introduction of methodology of recovery of chronology of events and artifacts which is capable completely and it is clear to restore the timeline based on abstraction level with use of technologies of real time and the so-called "live-response technologies" [18].

3. Technique of creation of a timeline in real time

In the section of the review of literature various problems connected with the huge volume of data when the timeline is taken from an image of a disk, in particular, when approaching "Super-chronometrical" are covered. This research is based on the data provided to Log2timeline in the form of the file which is sorted by means of the Psort tool, as shown in Figure 1.

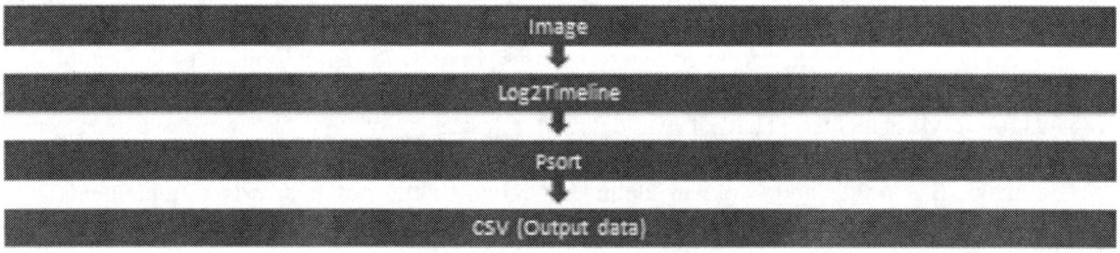

Figure 1. Formation of a timeline.

Besides, Psort allows to transform the file to usual formats, such as CSV. The timeline contains 17 fixed fields as it is specified in table 1. Then data from the CSV file are imported to the sheet Excel. One of the most often found problems on the timeline taken by means of Log2timeline is repeated repetition of the same unit of time. For example, unit of time "11:49:53" in the field "time". Repeated repetitions correspond to various data sources, such as "WEBHIST" and "LNK" in the field "source" showing that emergence of the same event caused reflection in various magazines and various artifacts, as shown in Table 2. Such scenarios formulate chronology big, difficult and difficult analyzed for experts. So, the corresponding scenarios were taken from the schedule and replaced with the corresponding methods (for example, on one event per unit of time for duplication elimination) to reconstruct the simple, compact, recognizable and structured timeline for experts.

Table 1. Fields in the CSV file

Field	Description
Date	Date of approach of an event
Time	Time when there was an event
Time zone	The time zone which was used for a tool call
MACB	Modification, access, creation and origin
Source	Short name of a source, for example, records of the register - REG
Source type	Description of a source
Type	Time tag type, for example, last access or last record
User	What user name is associated with an event if that is available
Host	what name of a host is associated with record
Short description	contains the field of the short description in which the text is stored
Description	here the most part of the sorted information is stored

Version	specifies the version number of a tag of time
Code number	sets code number of the processed file
Notes	extra space for storage of information for some modules of input
Format	the module of input which was used for analysis
In addition	The sorted information which is integrated and is stored here. These parts information make a super timeline which is created by Log2Timeline.

The methodology by means of which it is expedient to represent chronology of events and artifacts is presented using the concept of abstraction for expansion of a timeline of events and artifacts on four various levels of abstraction, i.e., events: High level; events: Low level; location of an artifact: High level; and location of an artifact: Low level, as shown in figure 2, for creation of a timeline, readable and clear for criminalists. The ideas which are the cornerstone of breakdown of the terms presented to Log2timeline. at four levels of abstraction give different idea to information. For each level the structure and also separate levels of specification of information has to be specified to reduce complexity of the schedule, to pass undesirable details, to provide correctness of the schedule, and submission only of that information which will help to distinguish and understand specific actions. it is carried out by users by the analysis of various sources and fields.

Table 2. A timeline with repetition (duplication) of the same unit of time

Date	Time	MACB	Source	Sourcetype	Type
19 November 2017	11:49:53	.A..	WEBHIST	Firefox History	Last Visited Time
19 November 2017	11:49:53	.A.B	LNK	Windows Shortcut	Creation Time; Last Access Time
19 November 2017	11:49:53	MA.B	LNK	Windows Shortcut	Content Modification Time; Creation Time; Last Access Time
19 November 2017	11:49:53	.A.B	LNK	Windows Shortcut	Creation Time; Last Access Time
19 November 2017	11:49:53	.A..	WEBHIST	Firefox History	Last Visited Time
19 November 2017	11:49:53	MA.B	LNK	Windows Shortcut	Content Modification Time; Creation Time; Last Access Time
19 November 2017	11:49:53	MA.B	LNK	Windows Shortcut	Content Modification Time; Creation Time; Last Access Time

In a forensic the event can be characterized as the activity which is carried out by users with use of digital devices. On the other hand, "artifact" has no formal definition in limits area a cyber / digital criminalistics now that results in lack of the standardized reporting and linguistics for mutual understanding between professionals.

As a rule, the artifact can be defined as a part of data which can have or not have relations to investigation/reaction.

For example: register keys, files, temporary tags and logs.

In other words, artifacts can be defined as something observable in scientific research or an experiment which is made as a result of the preparatory or investigative procedure.

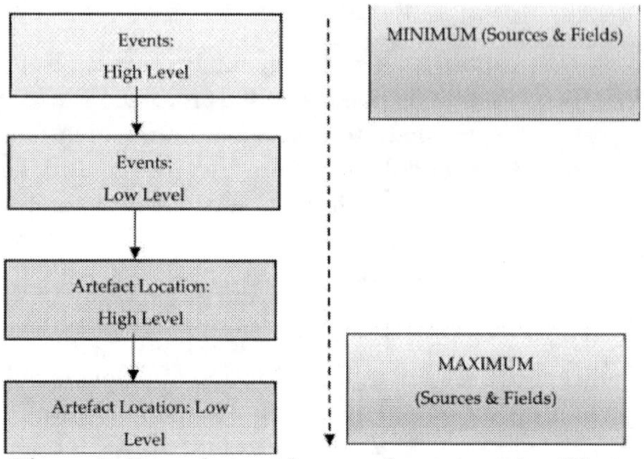

Figure 2. Four abstract layers of events and artifacts

The chosen methodology of a research assumes development of four various modules, each of which corresponds to one of levels of abstraction of chronology of events and artifacts. At each level of methodology various types of sources, parameters and fields for drawing up advanced structure and representation of a timeline are considered. On actions: information of high level connected with the activity which is carried out by users such as creation of files, access to web pages and loading of information in the form of files of various formats is provided (jpeg, pdf and docx). Such information with higher level of abstraction is taken by the analysis of six various sources - "LNK", "LOG", "META", "OLECF", "PE" and "WEBHIST" - and five certain areas, namely: "date", "time", "source", "visit" and "exile" that allows to obtain necessary information for recognition of concrete types of activity. Besides, the quantity of the sources and areas considered for the analysis depends on the type of information and the level of its specification demanded at the concrete level of a time scale.

For obtaining detailed information the different types of operations connected with files are carried out namely: modification, access, change and origin and also actions on web surfing which concern concrete web pages to which the user got access, and loading of files by the user [12].

The second level of abstract events: Low level, is reconstructed with additional details in comparison with the previous events: High level, for the analysis is required additional quantity of sources and fields. At this level nine various sources and seven area with expansion of three new sources, including "FILE", "REG" and "RECBIN" and also a new field, "additional", respectively, are studied for reconstruction of a time scale. The first two levels of abstraction of methodology can provide the smallest volume of information, related to what types of operations are performed by the user. To get a clear idea of a timeline, additional information is required. Here the list of all .exe-files started by the user, applications, and files which often address authors of files, the list of files enters. Namely, automaticDestinations-ms provides information on the used application which is attached to the taskbar of the user, temporary tags and also ways to elements (documents, web pages, images, etc.) which the program addressed recently while .msp provides updating of information relating to the Windows operating system and other Microsoft .msi programs containing information on installation for the specific program, such as files expecting installation. Besides, the detailed information concerning various events held on the Internet is provided. For example, to what application program (web browser) and as (the LINK user clicked on the link, GENERATED - chose record from the list, RELOAD user rebooted the page and gathered TYPED user. The URL address in a line of the URL address) each concrete web page is available. This detailed information is provided on an artifact: High level: "sourcetype" and "desc" (description) with lower value abstraction level is reconstructed by studying nine various sources and ten fields with addition of three new fields of "MACB". At various levels of abstraction, the relevant information joins and undesirable details for reconstruction of the corresponding timeline taking into account various number of sources and the field are removed. At the end the timeline with the smallest level of abstraction from all levels is reconstructed: Low level - by the analysis of the available all nine sources

and seventeen fields by means of a method and also addition of 9 fields, namely: "time zone", "type", "user", "host", "version", "file name", "code", "notes". and "format" for providing the full detailed information concerning all operations performed by the user.
The offered methodology consists of four various levels of abstraction of events and artifacts. Each level provides various information connected with various actions, carried out by the user, with different levels of abstraction of information, since minimum and finishing with the maximum details to reconstruct a relevant and recognizable timeline for the best understanding by the practicing digital expert.

4. Conclusion

In digital criminalistics of one of the main tasks the reconstruction and the analysis of terms, events and artifacts for interpretation of digital proofs in connection with a huge number and a variety of data is. Researches of scientific and technical literature show that various approaches were developed to show it, but most of them are not able to resolve an issue finally. In this work the flexible approach based on four various levels of abstraction, chronology of events and artifacts, such as events was considered: High level (new records and web surfing), events: Low level (web surfing, actions for modification), arrangement of artifacts: High level (including all .exe-files), and arrangement of artifacts: Low level. At each level of the schedule the unique structure which gives information relating to the concrete operation performed by the user is observed. For a timeline various parameters, fields and criteria (for example, on one event in unit of time), and various relevant structures of data excluding irrelevant for reconstruction of a compact and effective time scale with the accurate level of abstraction from details are considered to show the structured timeline which helps the investigator to understand and quickly to analyze information on incident. For realization of the offered approach and check of results various experiments are made. Results show that abstract approach is capable to recreate the chronological sequence of events and artifacts effectively, irrespective of quantity and heterogeneity of data, in comparison with the existing methods.

References

[1] Carrier, B. 2019 *Open Source Digital Forensics Tools*
[2] Olajide, F.; Savage, N.; Ndzi, D.; Al-Sinani, H. 2018 *Forensic Live Response and Event Reconstruction Methods in Linux Systems*
[3] Thomas, D.S.; Forcht, K.A. 2004 *Legal methods of using computer forensics techniques for computer crime analysis and investigation* (Issues Inf. Syst. 2004, 5) pp 692-698.
[4] Harrell, C. 2011 *What's a Timeline*
[5] Esposito, S.; Peterso, G. 2013 *Creating Super Timelines in Windows Investigations* (Proceedings of the 9th International Conference on Digital Forensics, Orlando, FL, USA, 28-30 January 2013) pp. 135-144.
[6] James, J.I.; Gladyshev, P. 2015 *Automated inference of past action instances in digital investigations*)Int. J. Inf. Secur. 2015, 14) pp 249-261. [CrossRef]
[7] Inglot, B.; Liu, L. 2014 *Enhanced Timeline Analysis for Digital Forensic Investigations*. (Inf. Secur. J. Glob. Perspect. 2014, 23) pp 32-44. [CrossRef]
[8] Guðjónsson, K. 2010 *Mastering the Super Timeline Who Am I?*
[9] Sitompul, O.S.; Handoko, A.; Rahmat, R.F. 2018 *File Reconstruction in Digital Forensic* (TELKOMNIKA Indones. J. Electr. Eng. 2018, 16) pp 776-794. [CrossRef]
[10] Cho, G.S. 2013 *A computer forensic method for detecting timestamp forgery in NTFS* (Comput. Secur. 2013, 34) pp 36-46. [CrossRef]
[11] Kalber, S.; Dewald, A.; Freiling, F.C. 2013 *Forensic application-fingerprinting based on file system metadata* (Proceedings of the Seventh International Conference on IT Security Incident Management and IT Forensics, Nuremberg, Germany, 12-14 March 2013) pp 98-112.
[12] Bang, J.; Yoo, B.; Kim, J.; Lee, S. 2009 *Analysis of time information for digital investigation* (Proceedings of the Fifth International Joint Conference on INC, IMS and IDC, Seoul, Korea, 25-

27 August 2009) pp 1858-1864.
[13] Chabot, Y.; Bertaux, A.; Nicolle, C.; Kechadi, M.T. 2014 *A complete formalized knowledge representation model for advanced digital forensics timeline analysis* (Digit. Investig. 2014, 11) pp 95-105. [CrossRef]
[14] Hargreaves, C.; Patterson, J. 2012 *An automated timeline reconstruction approach for digital forensic investigations* (Digit. Investig. 2012, 9) pp 69-79. [CrossRef]
[15] Brady, O.; Overill, R. 2015 *DESO: Addressing volume and variety in large scale criminal cases* (Digit. Investig. 2015, 88) pp 72-825. [CrossRef]
[16] Brady, O.; Overill, R.; Keppens, J. 2014 *Addressing the increasing volume and variety of digital evidence using an ontology* (Proceedings of the 2014 IEEE Joint Intelligence and Security Informatics Conference, Hague, The Netherlands, 24-26 September 2014) pp 176-183.
[17] Debinski, M.; Breitinger, F.; Mohan, P. 2019 *Timeline2GUI: A Log2timeline CSV parser and training scenarios* (Digit. Investig. 2019, 28) pp 34-43. [CrossRef]
[18] Soltani, S.; Seno, S.A.H.; Yazdi, H.S. 2019 *Event reconstruction using temporal pattern of file system modification* (IET Inf. Secur. 2019, 13) pp 201-212. [CrossRef]

A method for controlling groups of unmanned aerial vehicles in a virtual environment using haptic gloves

V O Antonov[1], D A Arustamov[1], U V Zavolokina[1], A A Apurin[1]

[1]Institute of Mathematics and Information Technologies (named after Prof. Nikolay Chervyakov), North Caucasus Federal University, Stavropol, 2 (Edifice 9) Pr. Kulakova, 355000, Russia.

E-mail llllamorolll@gmail.com

Abstract. The development of control of unmanned aerial vehicles (UAVs) is caused by the lack of developments that allow you to perform a fully autonomous flight to the target position with automatic bypass of static and dynamic obstacles. To ensure the correct execution of the flight task, many methods and algorithms have been developed, however, almost all of them are designed only for performing the UAV flight task in manual or semi-automatic mode. However, there are no widely available developments that allow you to perform a fully autonomous flight. Also, existing UAV control systems have low interactivity, due to the fact that the operator uses a two-dimensional map to control the UAV and also most often one joystick to control one UAV. Based on this, the purpose of this work is to increase interactivity and visibility in the management of UAV groups by developing a system for forming a flight task and control in a virtual environment using haptic gloves. The creation of a system for forming a flight task and managing groups of UAVs will allow the use of UAVs in various areas of human activity, including in conditions that are dangerous to human health and life. The description is presented, both mathematical and software. The advantages, disadvantages and limitations of the proposed method are analyzed.

Keywords: haptic glove, virtual reality glasses, a group of unmanned aerial vehicles.

1. Introduction

Scientific and technological progress in recent years has greatly changed the way of life of mankind. The needs of society have also changed, which has led to the need to create technologies that meet new needs and requirements. The development of transport has led to the fact that it has become a multifunctional tool [1]. The advent of flying technology and the development of wireless communications led to the idea of an unmanned aerial vehicle. UAVs have a number of advantages. They can be used in various fields: geodetic companies using UAVs for topographic surveying and support of construction work, agricultural companies for inspecting their lands, security companies for monitoring in hard-to-reach places, rescue services for determining emergency zones and coordinating rescue and search operations, city and dispatch services to monitor the situation in the city, energy companies and regional dispatch services using UAVs to monitor power lines, pipelines and for technical inspection of solar panels, construction companies using UAVs to inspect remote, hard-to-reach and dangerous objects, transport companies for transportation cargo to hard-to-reach places .. They allow you to save human lives, since in an emergency situation only the UAV itself is lost. Also, the economic costs can be lower, since the training of highly qualified pilots and other crew members is not required [2].

To ensure effective remote control in various tasks, an autonomous UAV operation is required based on data received from sensors in real time, since remote control depends on the quality of communication, which does not always allow effective control. The main task in this area is to obtain new methods of effective control when the UAV moves over rough terrain, including taking into account atmospheric conditions, stationary and dynamic obstacles [3].

2. Informal statement of the problem

The aim of this work is to increase interactivity and visibility when controlling UAV groups by developing a system for generating a flight task and control in a virtual environment using haptic gloves. The method proposed in this article is aimed at solving the following problems:
1) increasing interactivity;
2) raising the level of the methodological apparatus for managing UAV groups;
3) increasing the level of automation of the control process of UAV groups.

3. A method for controlling groups of unmanned aerial vehicles in a virtual environment using haptic gloves

To control the UAV groups, it is proposed to use the operator's virtual glasses and a haptic glove connected via Wi-Fi to a computer. An operator in virtual reality on a three-dimensional map can select an area to select several UAVs, or one UAV, then he indicates the point at which the UAV needs to arrive, a flight task is built, input data such as flight altitude are set, as well as the data that is needed collect, this data is transferred to the cloud storage. From the cloud storage, the UAVs receive flight missions and begin to carry them out. After its execution, the data collected by them are transferred to the cloud storage so that the operator can analyze them and make a decision based on them, after which the UAVs return to the home station in automatic mode, where they are recharged or the battery is replaced [4].

The proposed ergatic system generally consists of: an operator, a cloud storage that stores information, a flight controller represented by a haptic glove, one or several UAVs with a flight controller that decrypts and executes commands coming from the cloud storage. The operator processes the measurement results and, on their basis, adjusts the target coordinates of the UAV by generating control commands in the format transmitted via the transmission channel from the haptic glove to the computer and received by the computer. Based on the data received, the PC regulates the rotation speed of the UAV engines. Telemetry and Measurement System (STI) performs measurements and telemetry reception.

A functional diagram of the method for controlling UAV groups in a virtual environment using haptic gloves is shown in figure 1.

The haptic glove, in turn, consists of the simplest modules: a bend sensor, Arduino nano, a lithium-polymer battery, a nine-axis sensor needed to determine the hand in space, a Wi-Fi module and a level converter [5]. The haptic glove can be connected both via Bluetooth and Wi-Fi, but this requires replacing the Wi-Fi module.

The basic functional diagram of a haptic glove with a Wi-Fi module is shown in figure 2.

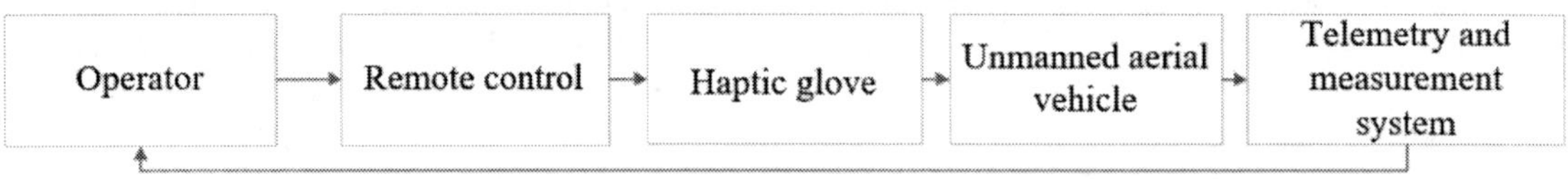

Figure 1. Functional diagram of the method for controlling UAV groups in a virtual environment using haptic gloves

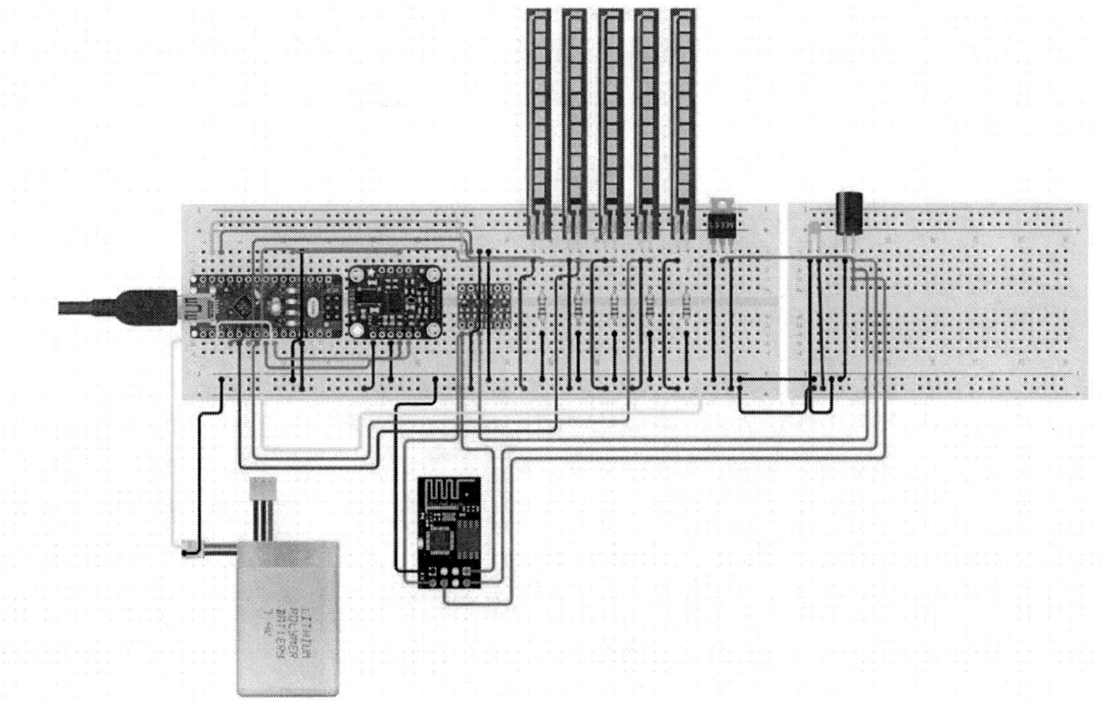

Figure 2. Basic functional diagram of a haptic glove with a Wi-Fi module

DJI Googles are supposed to be used as virtual reality glasses.

When constructing a flight task, we will consider the planning problem as a task of finding a path on a graph of a special type - the graph of regular decomposition of the grid. Formally, the grid is a triple:

$$Gr = \langle A, ADJ, d \rangle, \quad (1)$$

where A is a set of cells, which is a matrix $A_{m\times n} = \{a_{ij}\}: a_{ij} = 0 \vee a_{ij} = 1, \forall i, j: 0 < i < m, 0 \leq j < n,\ m, n \in N; ADJ \subseteq A \times A$ – a relation defining adjacency on a set of cells (a set of adjacent cells); d is a metric on the set A (a function that specifies a method for determining the distance between cells). Hereinafter, we will periodically use the notation $a(i; j)$ or simply $(i;\ j)$ to denote the cell a_{ij}. In those cases when the index coordinates of the cells are not important, we will denote them with letters of the Latin alphabet: a, b, c, etc. Two different grid cells $(i_1; j_1)$ and $(i_2;\ j_2)$ will be considered adjacent if $|i_1 - i_2| \leq 1 \vee |j_1 - j_2| \leq 1$.

In fact, adjacent cells correspond to adjacent elements of the matrix $A_{m\times n}$ [6].

The path as a sequence of cells is shown in figure 3.

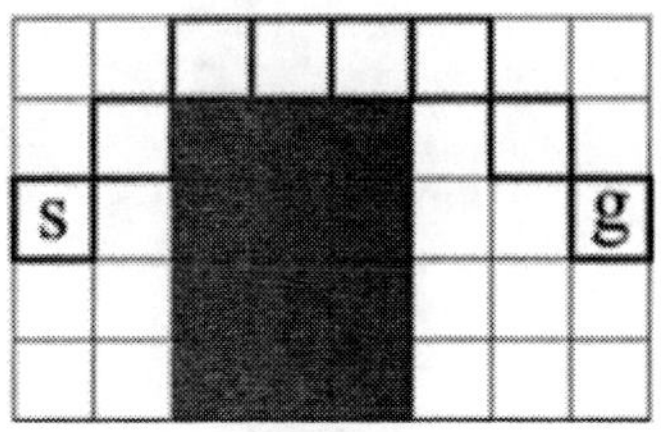

Figure 3. Path as a sequence of grid cells

We will assume that the Euclidean metric is used as a metric on the grid.

A grid cell will be called passable if $a_{ij} = 0$, impassable if $a_{ij} = 1$. The set of cells adjacent to a_{ij} will be denoted as $adj(a_{ij})$ or as $adj(a(i;\ j))$ [7].

The set of pairwise adjacent impassable cells will be called an obstacle: $Obs = \{a(i_0;\ j_0), a(i_1;\ j_1), \dots, a(i_s;\ j_s) \mid a(i_k;\ j_k) = 1, a(i_k;\ j_k) \in adj(a(i_{k-1};\ j_{k-1})), k = 0,1,\dots,s, s \in N\}$.

Constructing a grid of constructing a graph for a given on plane of the region U: $x_{min} \le x \le x_{max}, y_{min} \le y \le y_{max}$, $(U = U_{free} + U_{obs})$ is performed according to the following algorithm:

Step 1. Put $i = 0, j = 0, p_x = x_{min} + l, p_y = y_{max} - l$.

Step 2. If the area $[p_x - l, p_x + l] \times [p_y - l, p_y + l]$ contains at least one impassable point from U, then add a cell $a_{ij} = 1$ to A, otherwise - a cell $a_{ij} = 0$.

Step 3. Put $j = j + 1;\ p_x = p_x + 2l$.

Step 4. If point (p_x, p_y) lies in the working area, then go to Step 2.

Step 5. If $p_x > x_{max}$ and $p_y < y_{min}$, then terminate the algorithm.

Step 6. Put $j = 0;\ i = i + 1;\ p_y = p_y - 2l;\ p_x = x_{min} + l$ and go to Step 2.

In fact, the algorithm enumerates the points of the working area that are spaced from each other at a distance of $2l$, "from left to right, from bottom to top." For each of the points, it is checked whether the square with side $2l$ and centered at the point (p_x, p_y) contains at least one impassable point. If yes, then an impassable cell is added to the grid, otherwise - a passable one [8].

The operation of the algorithm is shown in figure 4.

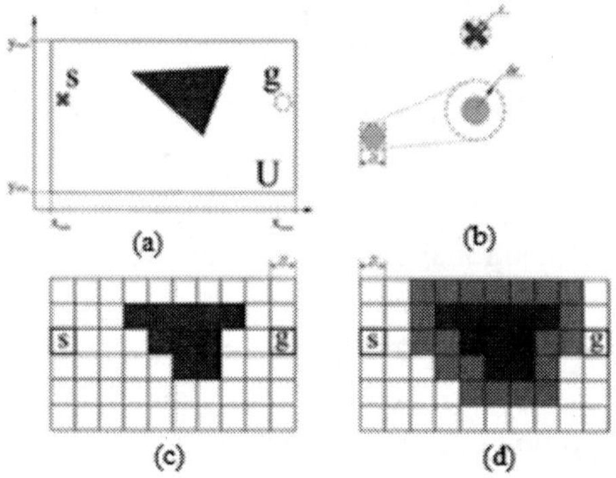

Figure 4. Work area (a), control object (b) and the corresponding hydraulic engine (c-d)

To visualize the proposed solution, the proposed interface of the software package is presented in figure 5.

IOP Conf. Series: Materials Science and Engineering 1069 (2021) 012043 doi:10.1088/1757-899X/1069/1/012043

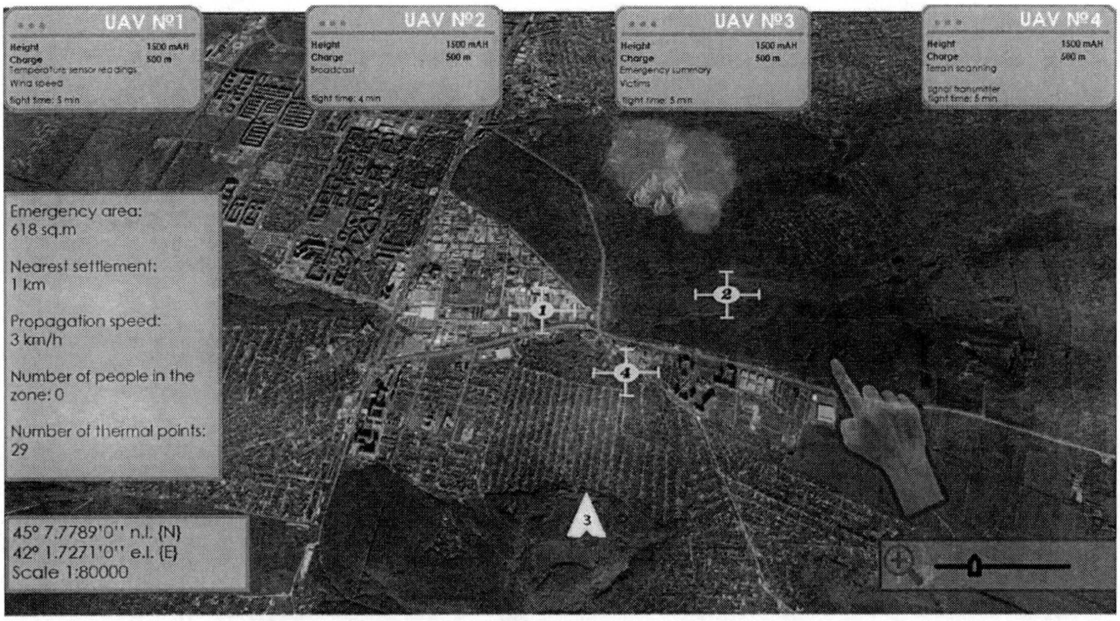

Figure 5. Software package interface

Thus, the proposed solution will lead to increased interactivity, methodological apparatus and automation in the control of UAV groups, which will reduce the number of operators required to control the UAV, and will allow for control in automatic and semi-automatic modes [9, 10].

4. Conclusion

A clear advantage of the proposed solution is the high level of interactivity provided by the use of virtual reality glasses and a haptic glove. In addition, the scope of this method of controlling UAV groups can be widely used in various fields of human activity. The proposed method can be applied in the following areas: geodesy, agriculture, security companies, the Ministry of Emergency Situations, city and dispatch services, energy companies, construction companies, transport. However, not every task posed can have a solution given the constraints. In addition, in some special cases, solving a problem may require significant computational resources.

Thus, this article proposes a method to increase interactivity and visibility when controlling UAV groups by developing a method for forming a flight task and control in a virtual environment using haptic gloves. The description of both the design solution and the mathematical component of the method is given.

5. Acknowledgment

The organizing committee thanks RFBR for assistance in holding the conference.

References

[1] Cherif, Nesrine & Jaafar, Wael & Yongacoglu, Abbas 2020 On the Optimal 3D Placement of a UAV Base Station for Maximal Coverage of UAV Users.

[2] Mandal, Murari & Kumar, Lav Kush & Vipparthi, Santosh 2020 MOR-UAV: A Benchmark Dataset and Baselines for Moving Object Recognition in UAV Videos. 2626-2635. 10.1145/3394171.3413934.

[3] Regmi, Yogesh & Manandhar, Gautam 2020 APPLICATIONS OF uavs AND DIGITAL PHOTOGRAMMETRY.

[4] Pathak Nidhi, Misra Sudip, Mukherjee Anandarup, Roy Arijit, Zomaya Albert 2020 UAV Virtualization for Enabling Heterogeneous and Persistent UAV-as-a-Service. IEEE Transactions on Vehicular Technology. PP. 1-1. 10.1109/TVT.2020.2985913.

[5] Petric Jasna, Saska Martin 2020 Mission planning for cooperative construction by UAV teams.

[6] Barbedo Jayme, Koenigkan Luciano, Santos Patricia 2020 Cattle Detection Using Oblique UAV Images. Drones. 4. 10.3390/drones4040075.

[7] Patil Ishwaragowda V 2020 Design And Fabrication Of An Uav.

[8] Pandey Ashish 2020 Introduction to UAV and its purposes.. 10.13140/RG.2.2.24296.70402.

[9] Intellectual algorithm for managing a group of unmanned aircraft // nauchkor.ru URL: https://nauchkor.ru/pubs/intellektualnyy-algoritm-upravleniya-gruppoy-bespilotnyh-appravleniya-gruppoy-bespilotnyh-apps

[10] Development and research of a complex of models and methods for flying sensor networks // URL:https://www.psuti.ru/sites/default/files/field/attachments/2017/12/kirichek_dissertaciya.pdf (date of access: 17.11.2020).

IOP Conf. Series: Materials Science and Engineering 1069 (2021) 012044 doi:10.1088/1757-899X/1069/1/012044

Consensus achievement method for a robotic swarm about the most frequently feature of an environment based on blockchain technology

V I Petrenko, F B Tebueva, S S Ryabtsev, I V Struchkov

North-Caucasus Federal University, Stavropol, Pushkin str., 1, 355017, Russia

E-mail nalfartorn@yandex.ru

Abstract. The technological development and popularity of swarm robotics actualizes the issues of increasing the efficiency and safety of consensus achievement between the swarm elements. Currently, most studies have little or no regard for information security issues when consensus achievement among swarm elements in the presence of robots with arbitrarily malfunctioning or malicious behavior. However, in many practical applications, when exploring the external environment, a swarm of one or more malicious robots may be sufficient to allow the current consensus mechanisms to fail. One of the promising ways to ensure information security in collective decision-making is the use of blockchain technology as a distributed system designed to work in conditions of lack of trust between the parties. The use of distributed ledger technologies, due to their complexity, leads to reducing efficiency in consensus achievement. The aim of the work is to improve the safety of the process of consensus achievement in swarms of robots through the use of blockchain technology in collective design making while maintaining the effectiveness of consensus achievement between robots. Increasing the safety of decision-making by a swarm will increase the stability and the possibilities of practical application of swarm robotic systems for solving problems in aggressive environments.

Keywords: Swarm robotics; Consensus achievement; Collective design making; Information security; Blockchain

1. Introduction

Swarm robotics is a promising direction in the development of intelligent systems and offers promising approaches to solving problems that require covering large areas and large volumes of space, including in unknown or hazardous environments. The spheres of application of swarms are: military objectives, search and monitoring, rescue operations, as well as many other applications.

The literature analysis on the use of swarm [1] showed that for the use of swarm in practice, it is necessary to solve a number of problems, one of which is to ensure information security of consensus achievement [2]. It should be noted that, with the declared high fault tolerance of swarm, most studies of swarm practically do not consider the security problems of collective design making (CDM) arising from the presence of robots with incorrect behavior (malicious robots), which can be a consequence of both a malfunction and a purposeful attacks. However, in many practical applications, one or more of these malicious robots (MRs) may be sufficient to disrupt the entire swarm by imposing bogus or inappropriate alternatives during consensus achievement in a collective design making process. These features actualize the development of secure algorithms for swarm consensus achievement.

The formulation and classification of information security issues in swarms and comparison of swarms with similar technologies are given in the research [3]. In the article [4], the analysis of algorithms for the functioning of the swarm is supplemented by their stability check from the point of view of information security. In the paper [5], the main threats relevant to the swarm with various principles of construction and operation are analyzed.

One of the promising approaches to solving the problems of protecting the collective design making process from the impact of malicious robot is the use of technologies based on blockchain [6–8]. The advantages of using blockchain are due to the similarity of the basic principles of the organization of swarms with the principles of building distributed storage and data processing systems, such as decentralization, peer-to-peer, scalability. The objectives of using consensus achievement algorithms in blockchain systems are also similar to the objectives of the collective design making in the swarm. Firstly, the blockchain is a distributed design-making system designed to work in conditions of a lack of trust between the parties, which corresponds to the conditions for the functioning of the swarm. Secondly, blockchain systems have built-in mechanisms for maintaining the relevance of information, and, thus, swarm built using such technologies do not need additional nodes to confirm records. Third, the loss of an individual robot should not pose a threat to the consensus- achievement process, as well as the loss of a separate node of a decentralized blockchain system.

The practical difficulties of using blockchain in collective design-making in swarm are discussed in papers [2,9], as a rule, they consist of a significant laboriousness of algorithms for consensus achievement and hardware costs in the blockchain and, as a consequence, a decrease in the speed of consensus achievement [10].

In addition, it is worth noting the limitations of the computing power of swarm and individual robots [11,12].

In this connection, the algorithms for consensus achievement in the swarm, despite a similar approach, require their modification for the tasks of the collective design making in the swarm. Thus, currently, there are a number of problems associated with the need to ensure an appropriate level of safety of swarm on the one hand, and with the impossibility of technical implementation of a number of promising solutions in this area, on the other hand. The aim of the paper is to increase the safety of the process of consensus achievement in the swarm through the use of blockchain technology in the collective design making while maintaining the effectiveness of consensus achievement between robots.

This work is devoted to the modification of the previously developed by the authors of the method of reaching a consensus on the attribute of the external environment [13] for swarm through the use of blockchain technology while maintaining the efficiency of the collective design making. The results obtained will make it possible to use the developed method in the future as a basis for the development of a wide range of solutions for ensuring the safety of swarm. For example, to extract a decision chain in order to investigate and analyze the collective design making process to identify harmful effects.

2. Informal statement of the problem

We will consider the issues of collective design making in swarm robotics on the example of the problem of reaching consensus using the classical formulation, as such an approach is used the study [14].

In this case, the purpose of the swarm will be to make a collective design and reach a consensus on the most common color of the mosaic (in all our experiments - white) of the black and white grid.

1. Robots are limited to a square room size $X \times X$ м^2 , which is bounded by four walls. The grid consists of |B| black and |W| white tiles. Each tile is X × X м2.

The difficulty of the problem can be varied by changing the ratio between the percentage of black tiles and B white tiles W and the size of one tile.

In a relatively simple task, the difference between the percentage of white and black tiles should be large (for example, $B = 34\%, W = 66\% \rightarrow \approx 0.52$), while in a relatively complex problem the difference is small (for example, $B = 48\%$, $W = 52\% \rightarrow \approx 0{,}92$).

2. At the end of a successful launch, all individual swarm robots will have the same opinion corresponding to the most common color.

Robots can communicate with each other only if there is no other robot interfering with communication and if the distance between robots is less than d_n = 50 sm. This is a constructive limitation in order to simulate the capabilities of the data transmission channel in the swarm, which will facilitate testing with real robots in future studies.

To simulate a swarm, we used the ARGoS software with the ARGoS-Epuck add-on installed [15]. An experimental scene was constructed containing 20 e-puck robots moving on a surface marked with black and white cells and able to perceive the color of the surface under them, Figure 1 (a) [16]. The E-puck robot has a diameter of 7 sm, a wheeled platform with a maximum speed of 16 sm/s, RGB LED lighting, 8 proximity sensors, a sensor for detecting surface color, and a module for local information exchange, consisting of 12 IR transceivers.

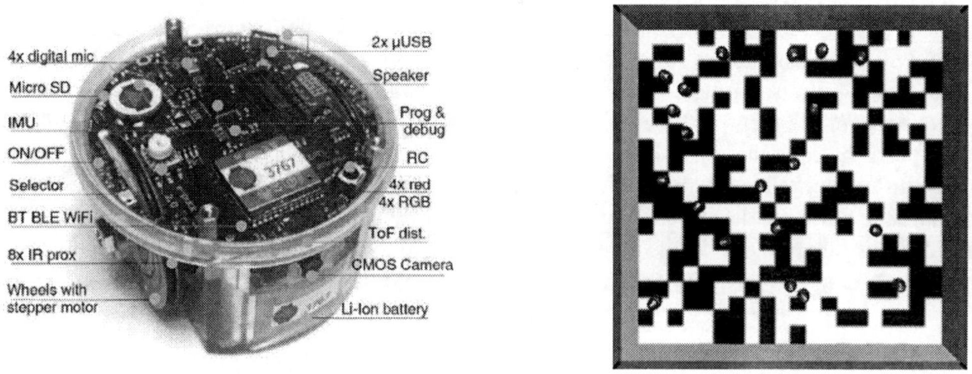

a) E-puck robot b) Experemetal scene

Figure 1. An example of the initial configuration of the experiment with $S = 2, N = 20, P_b = 40$

Before starting the experiment, robots are randomly placed inside the arena. An example of the initial configuration of an experiment for an arena measuring 2 by 2 meters ($S = 2$), the number of robots equal to 20 ($N = 20$) and the percentage of black cells $P_b = 40$ is shown in Figure 1 (b).

Each robot's trajectory is a broken line as the robot alternates between straight-ahead movement and rotation in place. Each of the two phases lasts for a random period of time, the value of which obeys an exponential distribution with a given average value. The direction of rotation is also randomly selected. Each robot is equipped with a rangefinder to determine the distance to other robots and obstacles. When an obstacle appears in the field of view, the robot turns around and continues to move in the opposite direction from the obstacle.

The task of robots is to achieve a consensus regarding the prevalence of the number of black or white cells. The key role in the implementation of the method is made by a smart contract, implemented in the "solidity" language, figure 2. The main purpose of the contract is decentralized aggregation and processing of data received from the elements of the swarm.

The contract stores data using a DataUnit structure, which records the number of black and white measurements, the robot's current opinion of the dominant feature, as well as the number of times the robot has updated information in the decision chain, that is, in the blockchain. Objects of this structure are stored in an associative array, the key of which is the robot's address in the blockchain.

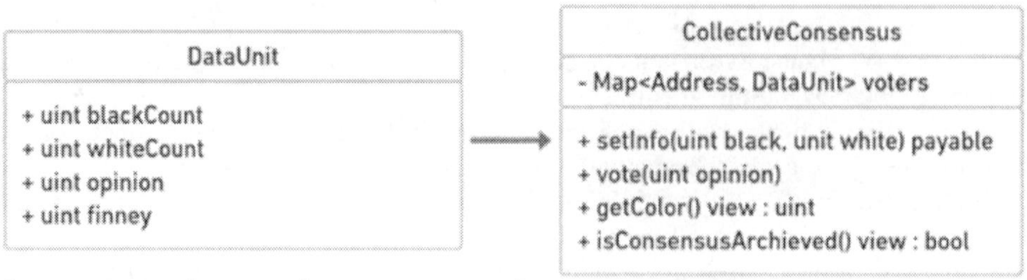

Figure 2. UML diagram of a smart contract for consensus achievement by swarm elements

To interact with the elements of the swarm, the contract provides 4 functions. The setInfo function adds or updates the data sent by the robot about the measurements of the external environment to the blockchain. In order to ensure the safety of the design-making process, the smart contract ignores "expired" data published more than 25 blocks ago, data from a robot that published more than 50 transactions in one block, as well as data from a robot with a different version of the blockchain.

To perform this function, the robot needs to send 900 finney, which corresponds to 0.9 ether. This restriction will not allow uncontrolled changes to the data stored in the blockchain and thereby protect them from unscrupulous or out-of-order elements of the swarm. The amount of 900 finney was assigned taking into account the fact that the reward for the created block during mining is 1 ether, which will be enough to perform this function, taking into account its price, as well as the commission provided for in the Ethereum network. It is worth noting that we are talking exclusively about the private Ethereum network and the amounts described above are not monetary equivalent.

The getColor function allows the elements of the swarm to find out the sign prevailing in the external environment, taking into account the data placed in the blockchain by all the elements of the swarm. The dominant feature in the external environment is formed as a weighted sum of all measurements stored in the blockchain:

$$Q(a) = \frac{\sum_{i=0}^{N} q_i(a)}{\sum_{i=0}^{N} c_i},$$

where c_i – total number of measurements made by the i-robot; q_i(a) – the number of measurements obtained for the feature a by the i-robot.

The final result of the function is the number of the attribute of the external environment with the highest score $Q(a)$.

The *vote* function updates the robot's opinion of the dominant feature in the blockchain. Unlike *setInfo*, when executing this function, the robot does not need to send additional funds, which allows the swarm elements to quickly maintain its current status.

When the *isConsensusArchieved* function of the smart contract is called, the robot receives current information about whether the entire swarm has reached a consensus. This decision is based on the entire set of information posted on the blockchain. As part of the experiment, the decision was made based on the opinion published by the rest of the robots about the most common feature of the environment.

Figure 3 shows a block diagram of the behavior algorithm of the swarm element. Before starting the experiment, using the *geth* program, the local version of the *Ethereum* blockchain is initialized and launched, in which the smart contract described above is published. During the experiment, the elements of the swarm are connected to this process to publish the data they have collected, as well as receive and process the data collected by other elements of the swarm. In addition, throughout the experiment, each robot is mining the cryptocurrency it needs to send data to the blockchain. This action is not shown in the flowchart as it is performed continuously in the background. The difficulty of opening a new block depends on the computing power of the swarm and is set so that an average of 2 new blocks appear per minute.

After initialization of the blockchain, robots perform a series of actions, shown in Figure 3, until a consensus is achieved between the elements of the swarm about the most common feature of the external

IOP Conf. Series: Materials Science and Engineering 1069 (2021) 012044 doi:10.1088/1757-899X/1069/1/012044

environment. So, by the first action, the robot determines the color of the surface underneath, and stores the result in its memory, and also increases the counter of perfect measurements:

$$q(id, a) = q(id, a) + 1,$$

where id – unique identifier of the robot; a – measurement result (white or black).

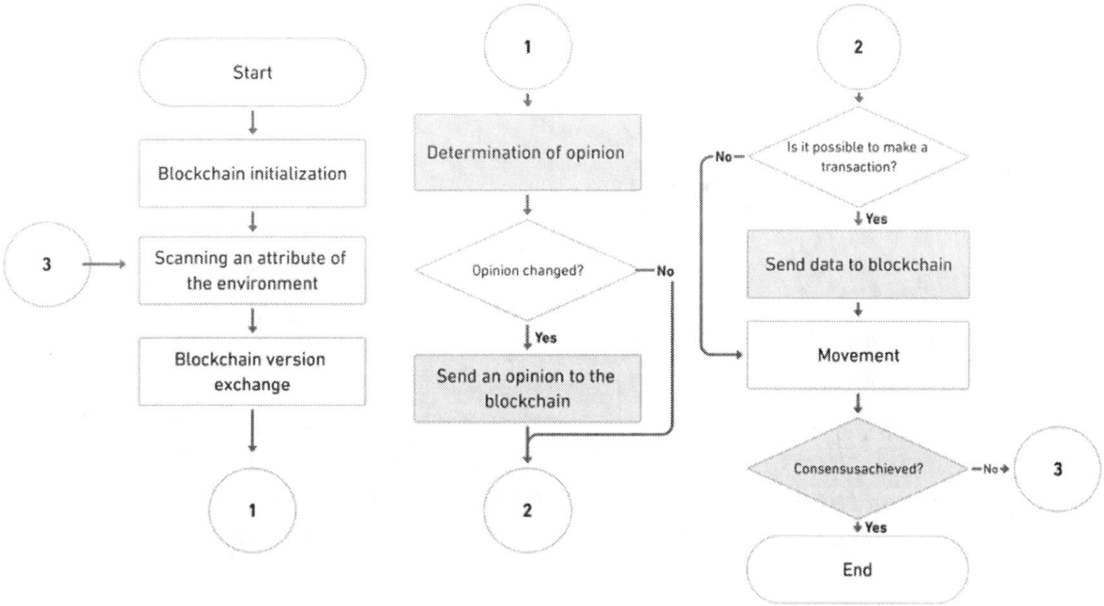

Figure 3. Flowchart of the swarm element behavior algorithm

The next step is to try to establish a connection and synchronize the blockchain with other elements of the swarm. As part of the experiment, the maximum range of wireless communication between robots was limited to 50 centimeters.

After completing this step, the robot calls the *getColor* function of the smart contract to check if its own opinion about the most common feature of the environment differs from the prevailing opinion of the swarm, according to its version of the blockchain. In case it differs, the robot initiates the transaction and sends its updated opinion using the vote function of the smart contract.

The next step for the robot is to check if it has enough funds to publish its measurements of the external environment on the blockchain. If so, it initiates a new transaction using the *setInfo* function of the smart contract. Otherwise, the robot skips this step and moves around the arena. As part of the experiment, the maximum speed of the robot in a straight line was limited to 10 cm / s. The movement algorithm consists of two phases, a phase of rectilinear movement and a phase of rotation in place. During the straight motion phase, the robot moves straight for an average duration of 40 seconds. After this period of time, or when it encounters an obstacle, the robot enters the phase of rotation in place, the average duration of which is 4.5 seconds, the direction of rotation, each time being chosen randomly. After completing the turn, the robot resumes straight motion.

At the last step, the robot calls the *isConsensusArchieved* function of the smart contract to find out if a consensus has been reached in the swarm. In the event that the function returns a negative result, the robot repeats the cycle again, if the function returns a positive result, the work of the algorithm is considered complete.

3. Results

To assess the effectiveness of the used as an analog and blockchain approaches, the following metrics were used:

– probability of completion (E_N), i.e. the probability of making the best decision, calculated as the proportion of runs
– time of results completion ($T_N^{correct}$): this is the number of seconds it takes for the robots in the swarm to have a "white" opinion. The metric is calculated over all experimental runs converging to "white"; chains converging to "black" are not counted.
The experimental results are shown in the figure 4a and 4b.

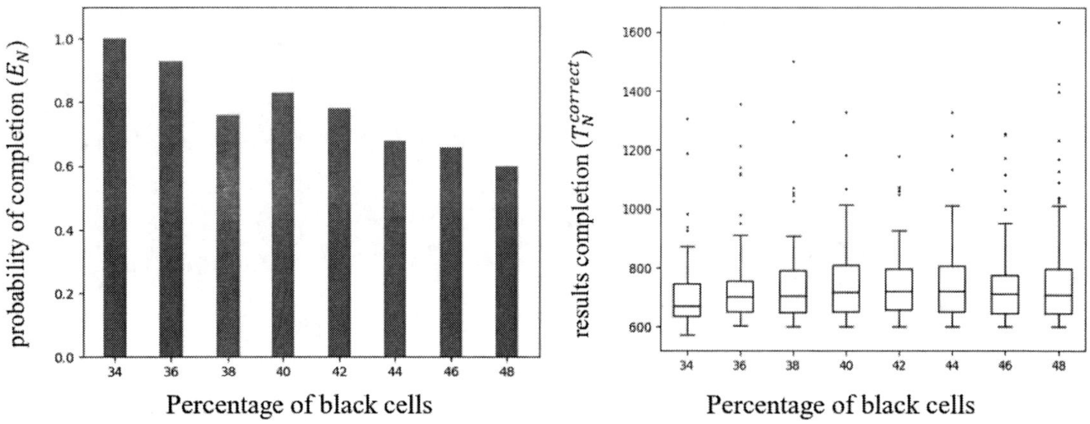

a) b)

Figure 4. The results of the algorithm with S=2, N=20

A group of experiments demonstrates the results of the algorithm when the difficulty of the task changes, i.e., when the ratio of white and black cells on the field changes. To carry out these experiments, the following parameters were used: a field measuring 2 by 2 meters, 20 e-puck randomly placed in the arena before the start of the experiment. The percentage of black cells ranged from 34% to 48% in 2% increments. A series of 50 experiments was performed for each configuration. In Figure 4 (a), the abscissa shows the percentage of black cells placed on the field, and the ordinate shows the probability of making a correct decision about the dominant feature in the environment by all robots at the end of the experiment. It can be seen from the graph that in most cases robots make the correct decision about the dominant feature, but the probability decreases markedly with increasing complexity of the task. So, with 48% black cells, the probability of consensus achievement with the correct result $E_N = 0.6$, which is slightly above the threshold level $E_N = 0.5$ – the probability of a random correct choice of one of the two signs of the environment.

In Figure 4 (b), the ordinate is the time to consensus achievement in seconds. The graph shows the tendency for the time to reach consensus to increase, as well as its spread with the increasing complexity of the problem.

It is worth noting that the probability of consensus achievement has decreased by 10-15% compared to a similar method, without using the blockchain, described in [16]. This fact may be associated with a decrease in the frequency of information updates in a distributed swarm information environment. In this experiment, this frequency is regulated by two parameters: the reward for the open block and the cost of sending data to the blockchain network.

4. Conclusion

This work makes it possible to develop the approach proposed earlier by the authors [13] of consensus achievement for the further provision of information security, collective design making, swarm based on blockchain in the following directions:

1. Analysis of malicious robot's behavior in order to identify the most effective strategies for the attacker.

2. Reducing the computational costs of the blockchain for more profitable use in the swarm and for efficient management of the system logging system in the blockchain, which will allow processing information security incidents in the swarm based on the analysis of the collective design making transaction chain.

3. Analysis of vulnerabilities of swarm from blockchain to attacks on swarm and development of approaches to protection against them.

The presented approach can be applied not only in swarm, but also in other similar technologies, the tasks in which may require a decentralized approach to management and consensus achievement.

5. Acknowledgment

The reported study was funded by Russian Ministry of Science (information security), project №10/2020.

References

[1] Bayindir L. A review of swarm robotics tasks // Neurocomputing. 2016. Vol. 172. P. 292–321.

[2] Beskopylny A. Blockchain in Robotic Distributed Multi-Level Systems // Adv. Robot. Mech. Eng. 2018. Vol. 1, № 4. P. 59–61.

[3] Higgins F., Tomlinson A., Martin K.M. Threats to the swarm: Security considerations for swarm robotics // Int. J. Adv. Secur. F., Tomlinson, A., Martin, K. M. (2009). Threat. to swarm Secur. considerations swarm Robot. Int. J. Adv. Secur. 2(2), 288–297. https//pdfs.semanticscholar.or. 2009. Vol. 2, № 2. P. 288–297.

[4] Komarov I.I. et al. Postanovka zadachi obespecheniya informacionnoj bezopasnosti roevy`x robototexnichskix sistem // Sci. Bus. Dev. ways. 2015. Vol. 3, № 45. P. 66–72.

[5] Basan A.S., Basan E.S., Makarevich O.B. Analiz problem obespecheniya informacionnoj bezopasnosti dlya sistem podvizhny`x ob``ektov s ispol`zovaniem besprovodny`x texnologij // Izvestiya Kabardino-Balkarskogo nauchnogo centra RAN. 2018. Vol. 6, № 86. P. 30–43.

[6] Ferrer E.C. The Blockchain: A New Framework for Robotic Swarm Systems. 2018. P. 1037–1058.

[7] Nguyen T.T., Hatua A., Sung A.H. Blockchain Approach to Solve Collective Decision Making Problems for Swarm Robotics // International Congress on Blockchain and Applications. 2020. P. 118–125.

[8] Strobel, V., Ferrer, E.C., Dorigo M. Managing byzantine robots via blockchain technology in a swarm robotics collective decision making scenario: Robotics track // Proc. Int. Jt. Conf. Auton. Agents Multiagent Syst. AAMAS. 2018. Vol. 1. P. 541–549.

[9] Odelu V. Blockchain Approach to Solve Collective Decision Making Problems for Swarm Robotics // International Congress on Blockchain and Applications. Springer International Publishing, 2020. Vol. 1010, № January. 1–10 p.

[10] Hamann H. Introduction to Swarm Robotics // Swarm Robotics: A Formal Approach. Cham: Springer International Publishing, 2018. P. 1–32.

[11] Swarm robotic platforms [Electronic resource]. 2019. P. 9–13.

[12] Navarro I., Matía F. An Introduction to Swarm Robotics // ISRN Robot. Hindawi, 2013. Vol. 2013. P. 1–10.

[13] Petrenko VI, Tebueva FB, Ryabtsev SS, Gurchinsky MM, Struchkov IV. Consensus achievement method for a robotic swarm about the most frequently feature of an environment // IOP Conference Series: Materials Science and Engineering. 2020. Vol. 919, № 4.

[14] Valentini G. et al. Collective perception of environmental features in a robot swarm // Lecture Notes in Computer Science (including subseries Lecture Notes in Artificial Intelligence and Lecture Notes in Bioinformatics). Springer Verlag, 2016. Vol. 9882 LNCS. P. 65–76.

[15] Pinciroli C. et al. ARGoS: A modular, parallel, multi-engine simulator for multi-robot systems // Swarm Intell. 2012. Vol. 6, № 4. P. 271–295.

[16] E-puck robots [Electronic resource]. Access mode: http://www.e-puck.org/ (date of access: 11.11.20)

Task decomposition and distribution method for heterogeneous multi-robotic systems

V I Petrenko[1], F B Tebueva[1], V O Antonov[1], N Y Svistunov[1]

[1]North-Caucasus Federal University, Stavropol, Pushkina str., 1, 355017, Russia

E-mail: svistunovn4@gmail.com

Abstract. This article presents a method for decomposition and distribution of tasks in a multi-robotic system (MRS). The proposed implementation of the method of elementary operations distribution is based on maximizing the expected efficiency of each agent and collective decision making (CDM). To compare the effectiveness of various methods, computational experiments were carried out. According to the results, the implementation of the CDM mechanisms allows reducing the time for performing a global task by MRS agents by more than 10%. For the initial conditions described in the work, the use of the proposed method for maximizing the individual efficiency of an agent in combination with the CDM algorithm allows reducing the average execution time of a global operation by 21% in comparison with the method of minimizing the individual execution time without CDM.

Keywords: Multi-robotic systems; Decentralized control; Task decomposition; Tasks distribution; Collective decision making.

1. Introduction

Currently, methods of constructing complex adaptive systems based on multi-agent methods are being actively developed. These technologies can be used to control multi-robotic systems (MRS) capable of performing complex composite tasks in uncertain environments. Robotic devices for various purposes, called agents or group members [1-3], are used as executive elements of the MRS.

The taxonomy of MRS is given by Iocchi, Nardi, Salerno in [4]. The proposed taxonomy was used and developed in [5]. MRS in this context is a system containing several robots that can be combined into heterogeneous groups consisting of homogeneous subgroups.

The main research areas of MRS are centered around modeling and management, planning and decision-making, technological and methodological issues [6].

The intensive growth in the number of publications and projects in these areas of research indicates the high popularity and relevance of this topic among researchers [7-11]. The main issues of the development of MRS, described in these works, are summarized in the topic of implementing a decentralized control system for a complex heterogeneous MRS.

According to the research [7-11], a number of problems can be identified in the construction of MRS control systems:

- a complex hierarchy of interrelationships among MRS agents;
- requirements for a high level of autonomy of MRS;
- limited communication range of agents;

– limited agents' sensory capabilities;
– insufficient optimization depth of subtasks distribution;
– high computational complexity of control algorithms;
– lack of a system for evaluating and optimizing time parameters for performing complex tasks.
– the complexity of organizing a centralized control system, which is associated with a number of the above factors and the need to solve complex optimization problems for the distribution of tasks between agents, etc.

The difficulties can be partially solved by modifying the design of agents, increasing their computing power, which will affect their cost, or by building a decentralized group management system and a system for collective decision making.

This article deals with the problem of constructing a scenario for performing tasks by agents of MRS with a decentralized control system.

Decentralized control can be roughly divided into two groups:

– control systems, where each element of the MRS is not controlled by other agents, but at the same time random interactions lead to global positive effects, which is a manifestation of the emergence effect;
– control systems, which assume the presence of a control action that “directs” the MRS agents and is the source of some of the input data. Moreover, control signals can come both from a control system external to the group and from a “local” leader (agent) within the group [12].

The main problems of decentralized management arise at the stage of building such a system in view of the particular complexity of the structure of interaction of all elements.

Diagrams of the processes of performing tasks of the centralized and decentralized control systems of the MRS are presented in figure 1 [8-11].

Data from the control center is transmitted to the MRS agents for further distribution. In turn, the agents collect information about the environment and send the data directly or through a connected element for processing to the control center.

In this case, the control center decides to perform a complex composite task, for which it is necessary to decompose it into a group of elementary subtasks.

Analysis of external data on the environment and available information on the state of the MRS allows you to divide a complex task into groups of simple subtasks.

The MRS agents receive a set of subtasks and perform the procedure for their distribution within the group in such a way as to minimize or maximize the values of one or several selected optimization criteria (time, energy consumption, path length, etc.).

Each MRS agent orders an individual chain of subtasks, which are then recursively decomposed to elementary operations.

In this system, there is a complex hierarchical structure of relationships when performing global complex tasks, taking into account the non-determinism of the environment and the instability of the elements of the group. Consider the classic and modern systems of decomposition and distribution of tasks between the MRS.

2. Materials and methods

According to the results of the analysis of some works devoted to the methods of decomposition and distribution of tasks in multi-agent systems, some drawbacks of the existing approaches were revealed.

The paper [15] describes a method for multi-agent assessment of the state of moving objects based on a cyclic search algorithm of stochastic approximation. The disadvantages of the described approach include insufficient optimization depth of the choice of subtasks during their distribution, as well as high computational complexity, since the regulation of the search cycle must perform a large number of passes to obtain an optimal solution.

The project [16] uses algorithms for simultaneous localization and mapping for a heterogeneous group of robots in a natural disaster. However, according to the specifics of the project, in the proposed system, the MRS group is the results of the formation of several swarms, underwater, ground and air.

Thus, task groups for each swarm are identical and cannot in any way be distributed among agents of another swarm.

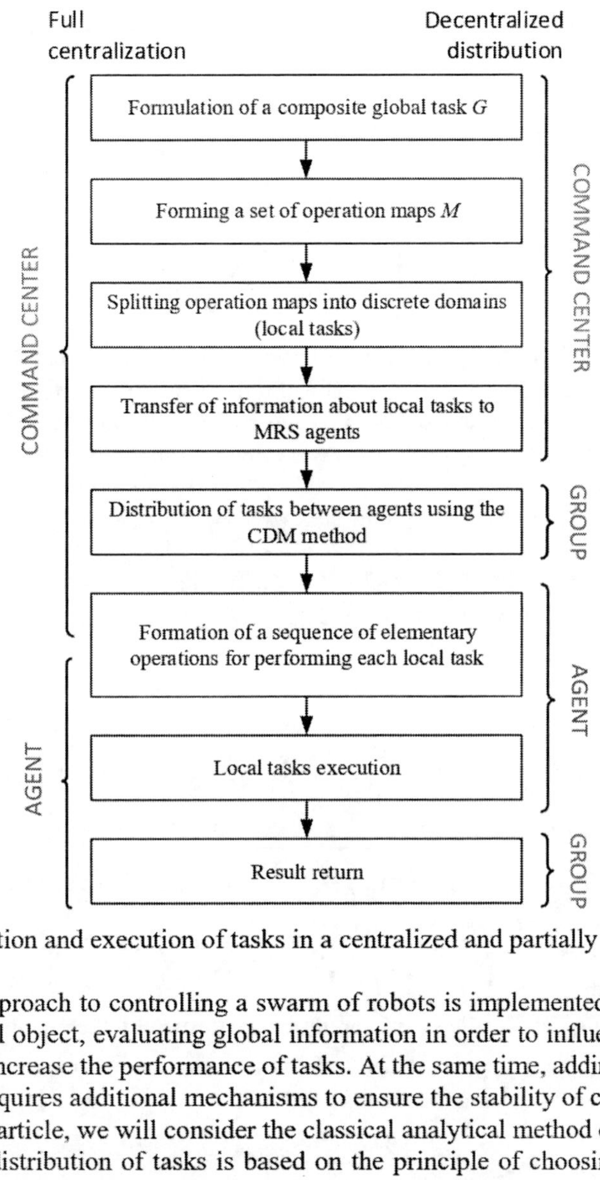

Figure 1. Distribution and execution of tasks in a centralized and partially decentralized MRS

In [17], a hybrid approach to controlling a swarm of robots is implemented, in which a part of the swarm acts as a control object, evaluating global information in order to influence the behavior of the remaining agents and increase the performance of tasks. At the same time, adding a control object leads to centralization and requires additional mechanisms to ensure the stability of communication.
As an analogue in this article, we will consider the classical analytical method of task distribution for a group of MRSs. The distribution of tasks is based on the principle of choosing the nearest points of location of tasks. And the process of agreeing a solution between agents consists of comparing the distances between the agent and the task.

Further, a simplified model of the MRS agent, working environment and elementary subtask is introduced.

Let the MRS consist of a set of agents $A = \{A_1, A_2, \dots, A_{n_A}\}$. Every agent $A_i \in A$ is characterized by:

- coordinates on the plane $(x_{A_i}, y_{A_i}) \in \mathbf{R} \times \mathbf{R}$;
- a set of types of performed operations $O_{A_i} \subseteq O$, where O is a set of possible operation types;
- performance P_{A_i} of performing operations of various types, reflecting the amount of work performed by the agent per unit of time:

$$P_{A_i} = \{p_{A_i}(a)\ \forall a \in O\}, \tag{1}$$

where $p(a)$ – performance on operation type a:

$$\forall a \in O_{A_i}\ p_{A_i}(a) \in \mathbf{R}_+,$$
$$\forall a \notin O_{A_i}\ p_{A_i}(a) = 0;$$

- the speed of movement v_{A_i} (within the framework of the simplified model, we take the speed of the agent to be constant in any direction and not dependent on external factors).

The working environment is a two-dimensional plane. Each elementary subtask m_j from the set of subtasks M' is characterized by:

- coordinates in the environment $\left(x_{m_j}, y_{m_j}\right) \in \mathbf{Z} \times \mathbf{Z}$;
- operation type o_{m_j};
- job volume $V(m_j) \in \mathbf{R}_+$;
- status S_{m_i} ($S_{m_i} = 0$ – subtask is free, $S_{m_i} = 1$ – subtask is seized).

Since the subtasks are located at integer points of the work plane, such a model corresponds to dividing the global task by territoriality into square areas of 1×1 size. At the same time, the regions of several elementary subtasks of the same or different types can be located at the same integer point. At a given moment in time, an agent is able to perform only one subtask of one type, however, subtasks located at one point can be simultaneously executed by different agents. Within the accepted model, collisions between agents are not possible and are excluded from consideration.

As an analogue algorithm, we consider a greedy subtask distribution algorithm, which consists in the choice by each agent with the least labor input of a subtask from the free:

1. Initial state:

$$\forall m_j \in M'\ S_{m_j} = 0,\ 0 \le x_{m_j}, y_{m_j} < C_{\max},$$
$$\forall A_i \in A\ \ S_{A_i} = 0,\ 0 \le x_{A_i}, y_{A_i} < C_{\max},$$

where $C_{\max}$ is environment size constraint.

2. For each agent $A_i \in A$ and every subtask $m_j, m_k \in M'$ sets of work intensity metrics are calculated:

2.1. $H_{0,A_i} = \left\{h_{A_i,m_j} = \frac{l(A_i,m_j)}{v_{A_i}} + \frac{V(m_j)}{p_{A_i}(o_{m_j})}\right\}$ is time for agent A_i to reach and complete task m_j from the initial position,

where $l(A_i, m_j) = \sqrt{\left(x_{m_j} - x_{A_i}\right)^2 + \left(y_{m_j} - y_{A_i}\right)^2}$ is distance between two points on a plane;

2.2. $H_{A_i} = \left\{h_{A_i,m_j,m_k} = \frac{l(m_j,m_k)}{v_{A_i}} + \frac{V(m_k)}{p_{A_i}(o_{m_k})}\right\}$ – time of reaching and execution by agent A_i task m_k from the position of subtask m_j.

If the operation of a specific subtask m_j cannot be performed by the agent A_i, then $h_{A_i,m_j} = \infty$.

3. At the initial stage, each agent A_i chooses the task $m_{A_i,\min}$ with the smallest value of the work intensity metric, declares it busy and heads to execute it:

$$m_{A_i,\min} \leftarrow \arg\min_{m_j|S_{m_j}=0} h_{A_i,m_j}, \tag{2}$$
$$S_{m_{A_i,\min}} \leftarrow 1. \tag{3}$$

Agents select subtasks sequentially, thereby ensuring the absence of collisions.

4. When the current subtask $m_{A_i,\min}$ is executed, the agent announces its exclusion from the list, then it selects the following subtask with the minimum metric of work intensity relative to the current position::

$$M' \leftarrow M' \backslash \{m_{A_i,\min}\}, \tag{4}$$
$$m_{A_i,\min} \leftarrow \arg\min_{m_j|S_{m_j}=0} h_{A_i,m_{A_i,\min},m_j}, \tag{5}$$
$$S_{m_{A_i,\min}} \leftarrow 1. \tag{6}$$

5. Step 4 is repeated by different agents while $M' \neq \emptyset$.

This algorithm does not guarantee the optimality of the distribution of elementary subtasks in terms of minimizing the total execution time. Although the efficiency of the greedy algorithm may be sufficient for practical use, its disadvantage is the dependence of the resulting distribution of tasks on the order in which agents select tasks for execution.

The proposed method is an effective and reliable method for finding a feasible solution when distributing tasks between agents. It should be noted that the method is analytical, and when refined, it can search for an optimal solution. Since the system is decentralized, an increase in the number of agents in proportion to the number of tasks does not affect the computational complexity; however, with an increase in the number of tasks in relation to the number of agents, the computational complexity of the problem selection algorithm will grow exponentially.

To reduce the dependence of the computational complexity of the algorithm on the number of tasks, artificial neural networks (ANNs) operating on the equipment of each agent can be used. ANN preparation is carried out in the process of simulating the functioning of the MRS and is based on the methods of machine learning with reinforcement. The advantages of this approach are the absence of a training sample prepared in advance, as well as the absence of the need to explicitly define the control algorithms of agents.

3. The proposed subtasks distribution method

The proposed solution to improve the efficiency of the global task execution by the MRS agents is an algorithm based on collective decision making by the MRS agents.

The metric $r_{A_i}(m_j)$ is used as a metric of the potential reward of the agent $A_i \in A$. It estimates the performance of the subtask m_j execution:

$$r_{A_i}(m_j) = \frac{V(m_j)}{h_{A_i,m_j}}, \tag{7}$$

where n_0 is a number of subtasks from $M' \backslash m_i$.

Since with a large number of subtasks calculating $R_{A_i}(m_j)$ in real time can be difficult, it is proposed to use an ANN to form the expected reward, the input of which is information about the current state of subtasks and the agent, and the output is the value of the expected reward (figure 2).

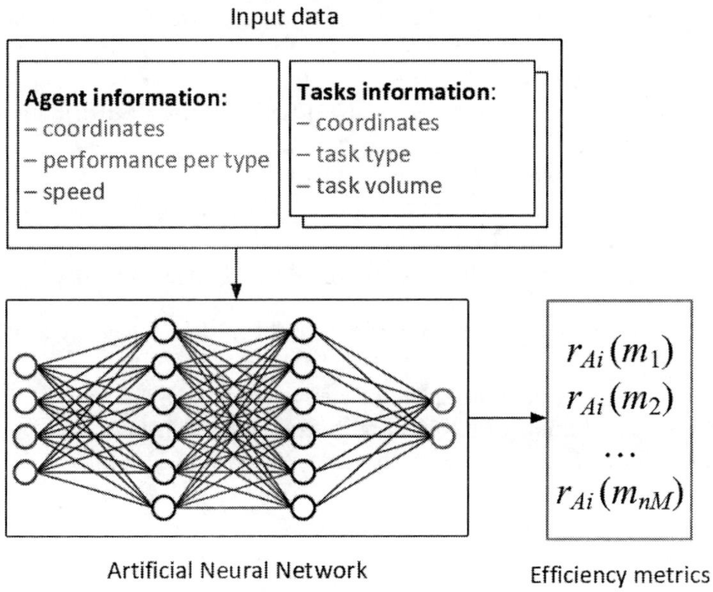

Figure 2. Artificial neural network usage scenario for efficiency metrics generation

Algorithm (2) for the distribution of tasks using collective decision making consists of the following stages:

1. Based on information about the state of the environment, each agent $A_i \in A$ determines the set $\boldsymbol{R}_{A_i}$ of potential reward values $r_{A_i}(m_j)$ for each subtask, sorted in descending order:

$$\boldsymbol{R}_{A_i} \leftarrow \left\{R_{A_i,k} = \left(m_j, r_{A_i}(m_j)\right)_k \; \forall m_j \in M'\right\}, \tag{8}$$

where k is the number of element in $\boldsymbol{R}_{A_i}$.

2. For $n \in \overline{1, n_{M'}}$ the sequence of actions is repeated:

2.1 Agent A_i chooses the subtask $m_{A_i} \in R_{A_i,n}$ as a candidate for execution and informs the rest of the agents about the choice.

2.2 If a response of the form $\left(m_{A_i}, r_{A_x}(m_j)\right)$ comes from at least one agent $A_x \in A$ and $r_{A_x}(m_j) > r_{A_i}(m_j)$, then $n \leftarrow n + 1$ and transition to 2.1.

3. If all subtasks were distributed with the best values of the efficiency metric, i.e. $n = (n_{M'} + 1)$, agent A_i stops until the next run of the task distribution algorithm using CDM. Otherwise, the agent starts executing subtask m_{A_i}. During the execution of the subtask, the agent informs the other MRS agents about the current value $V(m_{A_i})$.

Algorithm (2) is launched with a certain periodicity t, and also every time when agent $A_i \in A$ stops working on the current subtask, both in the case of its completion ($V(m_{A_i}) = 0$), and in the case of displacement by another agent with higher performance metric value.

The use of an ANN of reinforcement learning makes it possible to use this method in the absence of a training sample prepared in advance, and also in the absence of the need to explicitly define the control algorithms of agents.

4. Computational experiment

On the basis of the considered algorithms, the following decision-making methods by MRS agents can be formed:

- selection of a free task with the minimum expected execution time;
- selection of a free task with the maximum expected efficiency;
- selection of a task with the minimum expected execution time using the CDM method;
- selection of a task with the maximum expected efficiency using the CDM method.

Computational experiments were carried out to compare the effectiveness of these methods.

For each experiment, size of the working field is 15x15 cells, number of agents is 8, number of tasks is 100, number of task types is 3.

Using a pseudorandom number generator (PRNG) with a uniform distribution, the following are generated (figure 3):

- initial positions of agents and tasks in range [0; 14];
- types of tasks (1, 2 or 3 – presented in red, green and blue, respectively);
- volumes of tasks in the range [100; 255] units of work;
- the performance of each agent when executing a task of a specific type in range [0; 10] units of work per time tick;
- the speed of movement of agents in range [2; 6] units of distance per tick (the unit of distance is equal to the side of the cell).

To obtain statistics characterizing the speed of task execution when using each method, 1000 tests were carried out with the preliminary PRNG initialization with a constant.

Examples of modeling the four described algorithms for selecting subtasks by MRS agents are shown in figure 4.

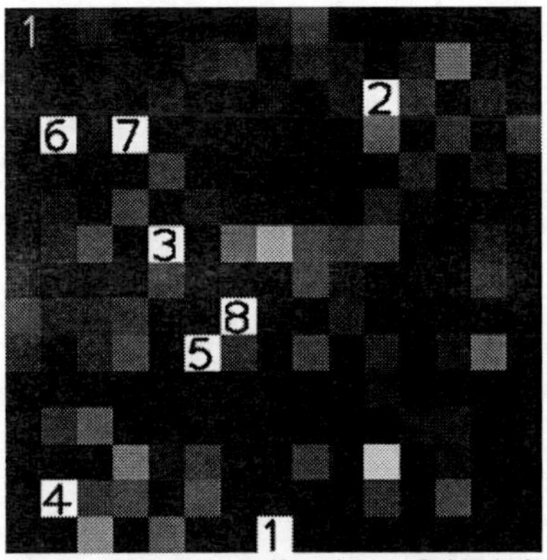

Figure 3. Visualization of the initial state of an experiment

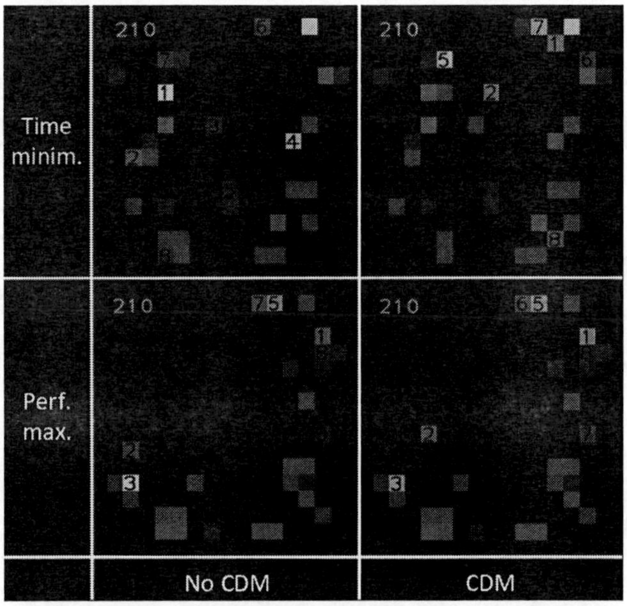

Figure 4. An example of the state of environment at the 210th step of one of the experiments

The greatest differences in decisions made by agents using a method based on maximizing individual efficiency and collective decision-making in comparison with analog methods are observed at the final stages of the experiment, when the areas corresponding to local problems are not densely located.

5. Results

Based on the results of 1000 experiments with identical initial conditions for each method, statistical characteristics were obtained (table 1 and figure 5).

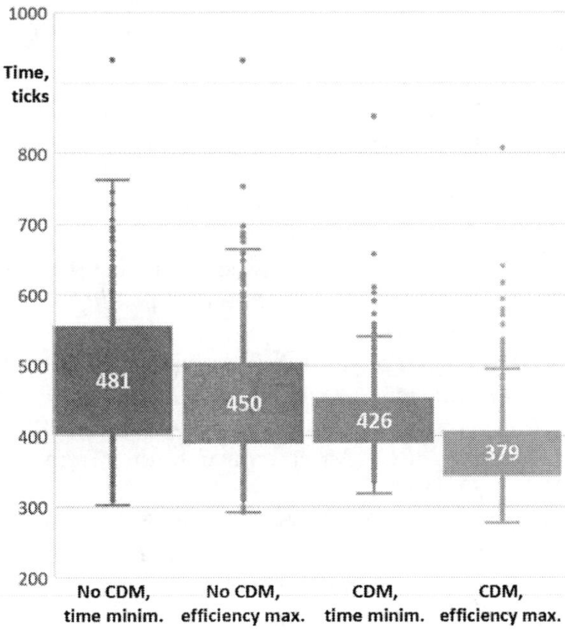

Figure 5. Diagrams of global task completion time distribution for the considered algorithms

Table 1. Statistical performance characteristics of task allocation methods

Method	Task execution time minimization		Efficiency maximization	
	No CDM	CDM	No CDM	CDM
Mean time, ticks	481	426	450	379
RMS deviation	89	50	76	50

The use of the CDM algorithm allows reducing the average execution time of a global task by 11% for the method based on minimizing the individual time of subtask execution, and by 16% for the method based on maximizing the individual efficiency of each agent. In addition, CDM increases the stability of the global task execution time, decreasing the standard deviation of its execution time from the average value. The use of the CDM algorithm and the method of maximizing individual efficiency can reduce the average time to complete a global task by 21% in comparison with the method of minimizing the individual time to complete a subtask without CDM.

6. Conclusion

In this article, a method has been developed for distributing tasks between agents of multi-robotic systems, based on maximizing the individual performance of each agent and using a collective decision-making approach.

The paper considers the basic problems of the development and use of MRS. Emphasis is placed on the decomposition and distribution of tasks in the MRS. The well-known methods for the implementation of these tasks are presented. Analogs of the proposed method are considered.

Based on the results of the computational experiments, it has been established that the implementation of collective decision-making mechanisms can reduce the time for performing a global task by MRS agents by more than 10%; The use of the method of maximizing the individual efficiency of an agent in combination with the CDM algorithm for the considered initial conditions allows reducing the average execution time by more than 20% in comparison with the method of minimizing the individual execution time without CDM.

The prospect for the development of the method lies in the use of ANN not only for the formation of the value of the reward, but also for the distribution of tasks between agents, as well as increasing the depth of the calculation of the next steps; adaptation of the algorithm to more complex models of agents and environments.

References

[1] Zakiev A, Tsoy T and Magid E 2018 Swarm robotics: remarks on terminology and classification *Lecture Notes in Computer Science (including subseries Lecture Notes in Artificial Intelligence and Lecture Notes in Bioinformatics)* **11097** LNAI pp 291–300

[2] Cornejo A and Nagpal R 2015 Distributed range-based relative localization of robot swarms *Springer Tracts in Advanced Robotics* **107** pp 91–107

[3] Krajnik T, Nitsche M, Faigl J, Vanek P, Saska M, Preucil L, Duckett T and Mejail M 2014 A practical multirobot localization system *Journal of Intelligent & Robotic Systems: Theory and Applications* **76** no 3–4 pp 539–62

[4] Iocchi L, Nardi D and Salerno M 2001 Reactivity and deliberation: a survey on multirobot systems *Workshop on Balancing Reactivity and Social Deliberation in Multi-Agent Systems. Lecture Notes in Computer Science* **2103** doi.org/10.1007/3-540-44568-4_2

[5] Navarro I and Matia F 2012 An introduction to swarm robotics *ISRN Robotics* **2013** doi.org/10.5402/2013/608164

[6] Research Areas in Multi-Robot Systems *Multi-Robot Systems* URL: multirobotsystems.org/?q=node/14 (accessed: 13.11.2020)

[7] Zenkevich S L and Galustyan N K 2016 Detsentralizovannoe upravlenie gruppoy kvadrokopterov [Decentralized control of a group of quadcopters] *Mekhatronika, avtomatizatsiya, upravlenie*

[*Mechatronics, automation, control*] **17** no 11 pp 774–82 (in Russian)
[8] Ma H, Honig W, Cohen L, Uras T, Xu H, Kumar T K S, Ayanian N and Koenig S 2017 Overview: a hierarchical framework for plan generation and execution in multirobot systems *IEEE Intelligent Systems* **32** no. 6 pp. 6-12
[9] Sariel-Talay S, Balch T R and Erdogan N 2011 A generic framework for distributed multirobot cooperation *Journal of Intelligent and Robotic Systems* **63** no 2 pp 323–58
[10] Ayanian N, Fitch R, Franchi A and Sabattini L 2017 Multirobot systems [TC Spotlight] *IEEE Robotics & Automation Magazine* **24** no 2 pp 12–6
[11] Sariel S and Balch T 2005 Real time auction based allocation of tasks for multi-robot exploration problem in dynamic environments *Integrating Planning into Scheduling: Papers from the 2005 AAAI Workshop* **WS-05-06** pp 27–33
[12] Erofeeva V A, Ivanskij Y V and Kiyaev V I 2015 Upravlenie roem dinamicheskikh ob'ektov na baze mul'tiagentnogo podkhoda [Control of a swarm of dynamic objects based on a multi-agent approach] *Komp'yuternye instrumenty v obrazovanii* [*Computer tools in education*] no 6 pp 34–42 (in Russian)
[13] Sariel-Talay S, Balch T R and Erdogan N 2011 A generic framework for distributed multirobot cooperation *Journal of Intelligent and Robotic Systems* **63** no 2 pp 323–58
[14] Ayanian N 2011 Coordination of multirobot teams and groups in constrained environments: models, abstractions, and control policies *Publicly Accessible Penn Dissertations* URL: https://repository.upenn.edu/edissertations/373
[15] Beni G and Wang J 1989 Swarm intelligence in cellular robotic systems *Proceed. NATO Advanced Workshop on Robots and Biological Systems* pp. 26–30 doi.org/10.1007/978-3-642-58069-7_38
[16] Magid E A, Matsuno F and Suthakorn J 2019 Disaster management information system in flood and landslide zones using a distributed heterogeneous group of robots *Research grant 19-58-70002 of Russian Foundation of Basic Research*
[17] Trabattoni M, Valentini G and Dorigo M 2018 Hybrid Control of Swarms for Resource Selection *Lecture Notes in Computer Science (including subseries Lecture Notes in Artificial Intelligence and Lecture Notes in Bioinformatics)* **11172** pp 57–70

Development of an algorithm for abnormal human behavior detection in intelligent video surveillance system

D Zh Satybaldina[1], N S Glazyrina[2], K A Kalymova[3], V S Stepanov[4]

[1,2] L.N.Gumilyov Eurasian National University, Nur-Sultan, Pushkin str., 11, 010000, Republic of Kazakhstan

[3] Kazakh National Women's Pedagogical University, Almaty, Ayteke bi str., 99, 05000, Republic of Kazakhstan

[4]Kazdream Technologies LLP, Nur-Sultan, Turkestan str., 10, 010000, Republic of Kazakhstan

E-mail satybaldina_dzh@enu.kz

Abstract. The aim of the work is to develop algorithms for analyzing video data in real time based on computer vision methods, as well as deep learning technologies for artificial neural networks for abnormal human behavior recognition near critical facilities using ATMs as an example. The article provides an overview of the initial research aimed at the choice of data capture devices, neural network architecture, software implementation and selection of experimental conditions (distance and illumination). Static and dynamic hand gestures were used as object's movements. Experimental results show that using the Intel RealSense D435 Depth Camera provides more accurate dynamic gesture recognition under different experimental conditions.

Keywords: Video analytics; Security systems; Abnormal behavior; Pattern recognition; Neural networks

1. Introduction

Video surveillance is an integral component of the security system of modern cities and, in combination with computational analytics, can significantly expand its functionality, including face recognition [1], traffic monitoring [2], detecting suspicious objects and investigating crimes [3].

Key video surveillance applications for public safety, such as monitoring and tracking, need to transmit and process large amounts of video data in real time, which causes significant delays for communication networks [4]. The need for human resources to interpret the data is also growing. This results in the traditional video surveillance systems being designed for offline forensic analysis and used otherwise than as a proactive tool to prevent suspicious activity before damage occurs.

Recently proposed second generation video surveillance systems aim to minimize the role of human operators, instead of which intelligent machine learning algorithms process the collected video frames in the cloud to detect, track and report any unusual circumstances [5]. The introduction of video analytics technologies in recent decades has led to success in various fields, such as emergency detection [6], traffic control and management [7], analysis of the motor activity of people and any objects of interest [8, 9]. The availability, development and price of processors and sensors for depth vision, surveillance cameras with smart functions also led to the widespread use of video analytics tools [5].

In this article, we propose an approach that combines computer vision and pattern recognition methods, as well as modern video capture devices that can provide progress in the development of algorithms for recognizing physical movements of people in real time and alerting security personnel in case of classification of abnormal (suspicious, atypical) behavior near critical facilities using ATMs as an example. In works [10, 11], we have already presented an approach for recognizing static hand gestures based on a deep convolutional neural network using the concept of learning transfer. Experimental studies have shown that the proposed model provides high recognition accuracy (99.4%). The purpose of this work is to develop a software tool for preprocessing data from two types of capture cameras under different experimental conditions in terms of distance to cameras and illumination, implementation of hand gestures classification based on a convolutional neural network.

The rest of this article is structured as follows. Section 2 provides an overview of the related work. Section 3 describes application classes for recognizing static and dynamic hand gestures. Section 4 presents the experimental results. The conclusions made and results obtained are provided in Section 5.

2. Related works

Nowadays, ATMs have become one of the necessities of modern life and are widely used for various banking operations. However, the number of ATM-related crimes has also increased. In general, ATM-related crimes can be divided into three main groups: logical attacks, fraud related to credit cards and currency and physical attacks. In addition, it is important to highlight fraud and illegal actions aimed at people using an ATM. Each type of crime is accompanied by certain movements on the part of the offender and certain suspicious (abnormal) behavior.

In modern video surveillance systems for public safety purposes, smart motion detectors are implemented, capable of classifying only simple types of movements, processes such as "loitering", crowding, falling of a person, holding a weapon in hands and leaving a suspicious object.

Human activity recognition (HAR) is the subject of a significant number of works, which show that each observation environment must satisfy a certain set of requirements [12]. These requirements contributed to the creation of specialized HAR tools, both for equipment (RGB and RGBD sensors, CCTV cameras) and for software (OpenCV libraries, etc.). The most common approaches include face recognition [1, 13], hand gestures [10-11,14], crowd behavior [15, 16], and atypical behavior [17, 18].

Most security approaches focus on the occurrence of crime. The authors of [17] present an algorithm consisting of the stages of capturing motion scenes from a sensor, subtracting the background, tracking pedestrian movements in order to make a decision. This approach divides the frame into eight areas and looks for a speed shift for one of the tracked people, one of whom is attempting a theft. At the same time, the system can only warn when a person has already lost his possessions. The work [18] presents a real approach for detection of 13 patterns of suspicious behavior such as burglary, hand-to-hand combat, gunfire and vandalism. They label patterns into two categories (normal and abnormal) and use a 3D convolutional neural network (CNN) to extract features.

Recent approaches to wireless activity detection provide good results, although they require specialized hardware. The authors of [19] proposed a model for the propagation of WiFi signals in physical space in the presence of human activity. It has been shown that human motor activity interferes with the transmission of wireless signals from the access point to the receiver, while the information about the phase and amplitude in the channel (Channel State Information, CSI) changes considerably. Based on this model, the NotiFi system is proposed, which shows the detection of anomalous human motor activity with an accuracy of 89.2% in the line of sight, 85.6% outside the line of sight and 75.3% in through-the-wall scenarios.

We have previously used the WiFi signal propagation model to study the problem of navigation inside modern buildings with a complex structure [20]. In this work, we have developed our own method for positioning objects and an augmented reality application for indoor navigation.

There are also several previous works related to the definition of abnormal behavior of people near ATMs. The authors of the article [21] proposed a surveillance system for detecting abnormal activity at ATM locations. The system uses Kinect 3D cameras and is capable of classifying a person's aggressive and normal posture. Work [22] presents a system based on image processing and machine learning methods that classify normal and potentially criminal activities associated with theft near ATMs. In [23], a system is proposed that identifies abnormal actions aimed at users of ATMs that occur in rooms with ATMs. The system uses an approach that involves application of Motion history image (MHI) and Hu moments to extract relevant features from the video to identify abnormal human behavior. In [24], the classification of human actions into normal and abnormal is presented based on the histogram of directed gradients (HOG) and the Random forest machine learning algorithm. The authors of [25] developed the ArchCam expert system capable of detecting suspicious behavior in the vicinity of an ATM, such as squatting in front of an ATM, attempting to climb onto an ATM, and detecting a construction belt in a room with an ATM.

Summarizing a brief overview of the state of modern video surveillance systems and research in the field of HAR, it can be assumed that the development of algorithms for analyzing video data in real time, based on computer vision methods, deep learning technologies of artificial neural networks in combination with a WiFi signal propagation model, can be an effective approach for recognizing suspicious human movement near critical facilities.

3. Description of Python applications for recognizing static and dynamic hand gestures

This section presents a description of the main application classes, developed in Python 3.6, for recognizing static and dynamic hand gestures based on video preprocessing and classifying the extracted features using a neural network.

In our previous studies on static gesture recognition, we used an Intel® RealSense ™ D435 depth camera as a data acquisition device, which provides an RGB image as well as depth for each point [10,11]. The application is designed with the use of both depth cameras and RGB cameras as capture sensors. For this, both methods from the RealSense and OpenCV libraries were used, as well as methods implemented by the authors themselves. The application class diagram is shown in Figure 1.

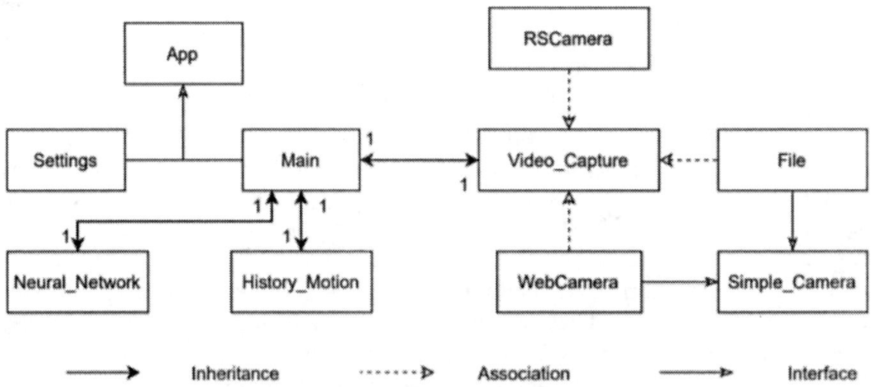

Figure 1. Application class diagram

The methods and functions of the App class are used to build a graphical interface that is implemented using the Tkinter library. The Pillow library is used to draw frames in the graphical interface. This class also contains a keyboard handler.

The Settings class is a graphical application designed for the user to select a video source. Available options are as follows: webcam, Intel camera, and video read. For OS Windows 8 and below, the Intel camera cannot be selected. The GUI inherits from the App class. The Main class is the main program, the link between all classes. Having received a choice of a video source from the Settings class, it initializes the Video_Capture, History_Motion and Neural_Network classes for further work. The Video_Capture, SimpleCamera, WebCamera, File and RSCamera classes are classes responsible for capturing video clips from a depth camera or RGB camera.

The Neural_Network and History_Motion classes are responsible for the preparatory stage and recognition of the gesture in the frame.

The Video_Capture class is the interface between the main class and the class required for the camera to work. The connection to the depth camera is carried out using the RealSense library, which has standard functions for initializing the camera, setting its operation parameters, functions and methods for reading frames from a video stream, calculating the distance from a hand to a depth camera, methods for storing RGB images and depth maps. Connection to the RGB camera is carried out through the OpenCV library. The functions of the class are also intended for capturing the frame, subtracting the background and detecting a hand in the frame. Depending on whether the hand was detected in the fragment or not, either a regular frame or a frame without a background is fed to the main class. This class was developed by us independently, with the main development idea based on the implementation of the possibility of connecting various sensors without changing the code of the main class.

The SimpleCamera class is responsible for connecting to an RGB camera using the OpenCV library. The functions of the class are intended for reading a frame from the video stream, creating an object used to subtract the background from the frames and create a mask, based on which the frame without the background is calculated. The WebCamera and File classes are designed to work with cameras and video files, respectively. All functionality is inherited from the SimpleCamera class.

The RSCamera class is designed to connect to a depth camera using the RealSense library. An RGBD frame is obtained from a video stream and the average brightness of the frame is calculated. If the brightness is below the threshold value, then instead of the RGB frame, the deep one will be fed into the main class. The method for calculating the average brightness was developed by us independently, the idea of the method being to transfer the frame from the RGB system to the HSV (Hue, Saturation, Value) color model where V is the brightness value. HSV is a non-linear RGB conversion. Subtraction of the background is based on comparing the depth pixels with the threshold: if the value is greater than the threshold, then in the resulting frame, the pixel is colored black, otherwise - with the color of the current RGB frame. Hand detection functions are based on calculating the required number of pixels in the frame without the background in the fixed area. Since the entire background is painted black in the frame, the number of pixels that are not equal to zero is calculated. If the required number of pixels is collected, the system fixes the hand in the frame.

The Neural_Network class is responsible for connecting to the neural network using the TENSORFLOW library. The methods of the class are designed to transform the resulting frame with the image of the hand into the format required for feeding it to the input of the neural network. As a result of the work of the connected neural network, an estimate of the convergence with reference sets of movements is issued. This score is passed to the main class to display the result on the screen.

The History_Motion class is intended for displaying frames in dynamics in an informational form. The idea behind the method is to select key frames from a set obtained over a certain period of time. A frame is selected in two cases: if the number of pixels that differ from the frame taken as a basis is more than a certain threshold, or the position of the hand does not change for a certain period of time. Initially, the first frame in the set is taken as the basis for frame comparison. Each subsequently selected frame in this process will be used as a basis until the next frame is selected. After the frame has ceased to be the basis for comparison, it is moved to a new set with a fixed time interval indicating the duration of its use by the basis for comparison. This continues until all the frames in the set have been considered. After that, the arithmetic mean for the time indices is calculated from the selected frames, and those frames in which this indicator is less than the arithmetic one are eliminated from the fixed set. The filtered set is used to map the generated movement. The implementation of the class is based on the methods of the Numpy and OpenCV libraries.

4. Experimental research

The developed application was tested on the problem of static and dynamic hand gestures recognition. We prepared a database that contains images with segmented gestures presented in Figures 2 and 3.

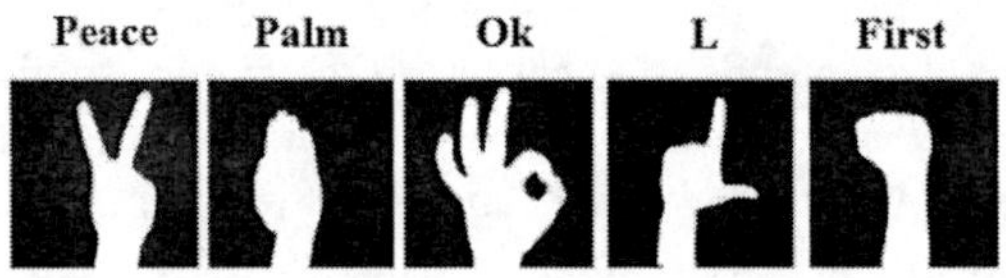

Figure 2. Samples of static gestures

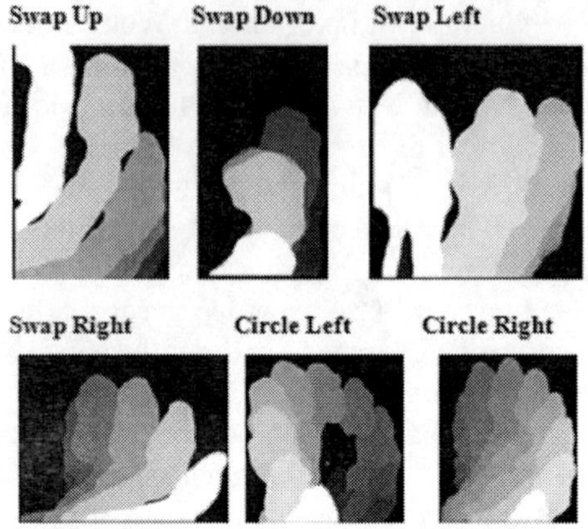

Figure 3. Dynamic gesture samples

A Logitech HD Pro Webcam C920 webcam and an Intel RealSense D435 depth camera were used as a gesture capture device.

Gesture recognition was carried out at a distance of 25 cm, 37.5 cm from the cameras; 43.75 cm; 50 cm, 62.5 cm; 75 cm. Tables 1, 2 show the results of static hand gesture recognition using RGB and RGBD cameras.

Table 1. Recognition of hand gestures based on the Logitech HD Pro Webcam C920

Distance, cm	Fist	L	Okay	Palm	Peace
25	98,899	99,417	99,994	77,766	10,026
37,5	98,284	99,992	99,999	99,461	22,391
43,75	99,99	99,982	100	97,548	63,666
50	98,189	92,826	99,995	90,732	77,887
62,5	99,935	84,518	99,991	41,7559	61,152
75	99,972	39,8749	99,99	0,8199	91,743

In Tables 1, 2, each column represents the percentage of gestures belonging to each class, and along the first diagonal the maximum values of the probability of correct gesture identification are specified, i.e. assignment of the observed object to a certain class of gestures.

Table 2. Accuracy of hand gesture recognition based on the Intel RealSense D435 camera

Distance, cm	Fist	L	Okay	Palm	Peace
25	93,179	89,7115	99,941	49,215	10,584
37,5	97,241	97,662	86,938	99,987	82,866
43,75	99,863	99,99	99,993	99,426	76,338
50	99,685	99,991	99,993	99,981	96,025
62,5	93,025	99,975	99,99	98,253	85,407
75	11,589	99,688	89,909	3,905	4,097

After analyzing the resulting diagrams, we can see that when using the Intel RealSense D435 camera, more accurate gesture recognition is provided.

Recognition of static gestures using the Intel RealSense D435 camera was also carried out under different lighting conditions; the recognition results are shown in Table 3.

Table 3 - Performance of the static gesture recognition system under different lighting conditions

Experimental conditions	Fist	L	Okay	Palm	Peace
late in the evening with no lights on	95,522	98,080	99,402	58,548	75,120
late in the evening with the lights on	63,502	28,072	99,726	57,521	38,320
late in the evening with the lights on and the lamp in front of the camera	95,577	98,308	97,305	43,271	50,324
daytime	98,015	99,713	99,984	98,399	56,74
late evening without the lights on, only the depth pixels used	98,542	99,899	99,729	82,629	97,838

The results of recognition of dynamic hand gestures based on the processing of depth images are presented in Table 4. As can be seen, for all gestures, the value of the elements of the main diagonal is more than 88%.

Considering that the resulting accuracy of the classifier is calculated as the arithmetic mean of its accuracy for all classes, the performance of the proposed approach in terms of the recognition accuracy of dynamic gestures was obtained at the level of 0.9131, or ≈91.31%.

Based on the presented analysis of the results of gesture recognition at various distances from the cameras, under different lighting conditions, as well as studying the technical characteristics of the cameras, Intel RealSense D435 cameras were chosen to detect people in front of the ATM and their suspicious actions.

Table 4. Performance of the dynamic gesture recognition system

	Swap Up	Swap Down	Swap Left	Swap Right	Circle Left	Circle Right
Swap Up	89,112	0,060	0,056	10,353	0,237	0,178
Swap Down	0,154	92,474	0,154	2,002	5,144	0,071
Swap Left	2,177	2,260	92,088	0,939	1,262	1,274
Swap Right	0,021	3,428	3,437	92,906	0,197	0,011
Circle Left	1,0797	3,0598	1,0789	0,3019	88,69	5,772
Circle Right	0,0471	0,7771	0,263	3,4161	2,93	92,561

5. Conclusion

The article provides an overview of the results of application development and its main classes for recognizing static and dynamic hand gestures, experimental studies of the use of the Logitech HD Pro Webcam C920 webcam and the Intel RealSense D435 depth camera as capture sensors under different conditions of distance and illumination are carried out.

It has been shown that the accuracy of gesture recognition depends on the conditions under which hand poses are demonstrated (illumination, distance to the camera).

The use of a convolutional neural network was found to contribute to a sufficiently high classification accuracy even in the case of video capture from an RGB camera. But the obtained higher accuracy of recognition of static and dynamic hand gestures from an RGBD camera under different experimental conditions confirms the potential performance expectations due to the pixels of depth.

Thus, in the course of the research, video data capture sensors for an intelligent video surveillance system were identified in terms of performance.

6. Acknowledgments

This work was supported by the Ministry of Education and Science of the Republic of Kazakhstan within the framework of project No. AP09260786.

References

[1] Ji P, Kim Y, Yang Y and Kim Y S 2016 Face occlusion detection using skin color ratio and LBP features for intelligent video surveillance systems *FedCSIS* (Gdansk) ACSIS (Polish Information Processing Society) V 8 pp.253-259 https://doi.org/10.15439/2016F508
[2] Zhou Y, Liu L, Shao L and Mellor M 2018 Fast automatic vehicle annotation for urban traffic surveillance *IEEE Trans. Intell. Transport. Syst.* (IEEE) vol. 19 no. 6 pp.1973– 1984
[3] Puvvadi U L N, Benedetto K D, Patil A, Kang K-D and Park Y 2015 Cost-effective security support in real-time video surveillance *IEEE Trans Ind. Informat.* (IEEE) vol. 11, no. 6 pp.1457–1465
[4] Prakash P, Suresh R and Pn D. K 2019 Smart city video surveillance using fog computing *Int. J. of Enterprise Network Management* (Springer, Cham) vol. 10, no. 3-4 pp. 389-399
[5] Shidik G. F. et al. 2019 *A* Systematic Review of Intelligence Video Surveillance: Trends, Techniques, Frameworks, and Datasets *IEEE Access* (IEEE) vol. 7 pp.170457-170473
[6] Filonenko A, Hernández D C and Jo K-H 2018 Fast smoke detection for video surveillance using CUDA *IEEE Trans. Ind. Informat.* (IEEE) vol. 14, no. 2 pp. 725–733
[7] Finogeev A, Finogeev A, Fionova L, Lyapin A and Lychagin K A 2019 Intelligent monitoring system for smart road environment *J. Ind. Inf. Integr* (Amsterdam: North-Holland/American Elsevier) vol. 15 pp. 15–20
[8] Murugan A S, Devi K S, Sivaranjani A and Srinivasan P 2018 A study on various methods used for video summarization and moving object detection for video surveillance applications *Multimedia Tools Appl.*(Springer, Singapore) vol. 77, no. 18 pp. 23273–23290
[9] Zhao L, He Z, Cao W and Zhao D 2018 Real-time moving object segmentation and classification from HEVC compressed surveillance video *IEEE Trans. Circuits Syst. Video Technol.* (IEEE) vol. 28, no. 6 pp. 1346–1357
[10] Satybaldina D., Kalymova G. and Glazyrina N. 2020 Application Development for Hand Gestures Recognition with Using the Depth Camera *Communications in Computer and Information Science* (Springer, Cham) V.1243 pp. 55-67
[11] Satybaldina D Zh, Ovechkin G V and Kalymova G A 2020 Static hand gestures recognition system with using depth camera *Vestnik of RSREU* (Ryazan: Ryazan State Radio Engineering University Press) N. 72 pp.93-105. https://doi.org/10.21667/1995-4565-2020-72-93-105
[12] Martínez-Mascorro G A et al. 2005 *Suspicious Behavior Detection on Shoplifting Cases for Crime Prevention by Using 3D Convolutional Neural Networks* (arXiv preprint arXiv:02142v1[cs.CV])
[13] Wang Y, Bao T, Ding C and Zhu M l 2017 Face recognition in real-world surveillance videos with deep learning method *IEEE 2nd International Conference on Image, Vision and Computing (Chengdu)* (IEEE) pp. 239-243
[14] Satybaldina D Zh Kalymova G A Stepanov V S 2020 Application for recognition of static gestures by using deep images *Certificate of state registration of the right to the computer program, № 10319* (National Institute of Intellectual Property)
[15] Direkoglu C, Sah M and O'Connor N E 2017 Abnormal crowd behavior detection using novel optical flow-based features *14th IEEE International Conference on Advanced Video and Signal Based Surveillance(Lecce)* (IEEE) pp. 1-6
[16] Rabiee H et al. 2018 Detection and localization of crowd behavior using a novel tracklet-based model *International Journal of Machine Learning and* Cybernetics (Springer Nature)V. 9, №. 12 pp.1999-2010
[17] Tsushita H and Zin T T 2018 A Study on Detection of Abnormal Behavior by a Surveillance Camera Image *Int. Conference on Big Data Analysis and Deep Learning Applications (Miyazaki)* (Springer, Singapore) pp. 284-291
[18] Sultani W, Chen Ch and Shah M 2018 Real-world anomaly detection in surveillance videos *IEEE/CVF Conference on Computer Vision and Pattern Recognition* (IEEE) pp.6479-6488

IOP Conf. Series: Materials Science and Engineering 1069 (2021) 012046 doi:10.1088/1757-899X/1069/1/012046

[19] Zhu D et al. 2017 A ubiquitous WiFi-based abnormal activity detection system *IEEE International Joint Conference on Neural Networks* pp.1766-1773
[20] Stepanov V S and Glazyrina N S 2018 Development of an application for navigation in the room *Herald of PSU* (Pavlodar: Toraighyrov University Press) №3 pp.171-180
[21] Nar R, Singal A and Kumar P 2016 Abnormal activity detection for bank ATM surveillance *IEEE Int. Conference on Advances in Computing, Communications and Informatics* (IEEE) pp. 2042-2047
[22] Mandal R and Choudhury N 2016 Automatic video surveillance for theft detection in ATM machines: An enhanced approach *3rd International Conference on Computing for Sustainable Global Development, INDIACom*(New Delhi) (IEEE) pp. 2821-2826
[23] Tripathi V, Gangodkar D, Latta V and Mittal A 2015 Robust Abnormal Event Recognition via Motion and Shape Analysis at ATM Installations *Journal of Electrical and Computer Engineering* (Hindawi Publishing Corporation)1(10)
[24] Tripathi V, Mittal A, Gangodkar D and Kanth V 2016 Real time security framework for detecting abnormal events at ATM installations *Journal of Real-Time Image Processing* (Springer Nature)pp. 1-11
[25] Lee W K, Leong C-F, Lai W-K, Leow L K and Yap Th-H 2018 ArchCam: Real Time Expert System for Suspicious Behaviour Detection in ATM Site *Expert Systems with Applications 109* (Elsevier) *pp.* 12-24

IOP Conf. Series: Materials Science and Engineering 1069 (2021) 012047 doi:10.1088/1757-899X/1069/1/012047

A method for circumventing the biometric identification system based on facial recognition

V O Antonov[1], N Y Svistunov[1], A V Garanzha[1] and I P Bychihin[1]

[1]North-Caucasus Federal University, Stavropol, Pushkin str., 1, 355017, Russia

E-mail svart.sky@gmail.com

Abstract. The article deals with the task of identifying wanted people using two-dimensional face recognition. A potential method for circumventing the recognition system is described. "Potential" in this case means the absence of precedents for the use of the described method in governmental systems. An illustration of the algorithm for implementing a potential method for disrupting the functioning of the reference samples database of the facial recognition system is given. The results of testing the effectiveness of the presented method on the Microsoft Azure platform are presented. The results of the analysis of the modified image for obvious signs of montage are presented. As a result, the conclusion is presented that it is impossible to implement effective and operational protection of the reference samples database from the presented potential method of circumventing the facial recognition system.

Keywords: Face recognition; Circumventing; Neural network; Image cloaking

1. Introduction

Facial recognition systems are rapidly becoming more widespread and relevant. They can be used in such tasks as: verification and access control, law enforcement, forensic examination. There are many facial recognition technologies available. All of them are based on the principle of uniqueness of geometric parameters: facial features and the shape of the skull of each person are individual [1]. This area is divided into two directions: 2D (two-dimensional) recognition and 3D (three-dimensional) recognition. For widespread use, for example in a citywide system for recognizing people on the wanted list, 2D recognition methods are used. The advantages of this method are: no need to use expensive scanning equipment; ease of obtaining information for analysis (image from conventional city surveillance cameras); using the most easily accessible biometric unit of a person as a reference for comparison – a face image.

There are also disadvantages: low statistical reliability, high requirements for the quality of lighting, necessarily a frontal image of the face, etc. [1]. However, facial recognition systems based on two-dimensional images are now more popular and widely used, especially in the task of identifying criminals on the streets of the city. Of course, there is also a demand for deception of such recognition systems.

Among the recent works devoted to face recognition, it is worth noting the work [2], in which a probability-theoretic model of a halftone image is constructed and a method of identifying a person from a photo of a face based on the optimal Bayesian rule is applied. In another paper [3], the problem of automatic recognition was solved on the basis of the principle of minimum information mismatch. The method of finding the principal components from the normalized images of the training sample is described in [4]. Work [5] is devoted to the processing of three-dimensional images, but the task of 3D facial recognition has not yet been sufficiently investigated.

2. Facial recognition systems in Russian Federation

In early December 2018, it became known about the intentions of the Moscow city administration to deploy a citywide facial recognition system that will be able to find wanted criminals. The plan includes the creation of a "uniform video space" in which a powerful computer system will receive, store and analyze images from cameras installed in entrances, on streets, at train stations, on transport, at traffic lights and on city highways. According to information given at the beginning of 2019, the Moscow video surveillance network consists of 174 thousand cameras.

The algorithm implemented in the city video surveillance system is based on the use of neural networks and analyzes video recordings from city cameras in real time. The faces on the recordings are scanned so that, if necessary, they can be compared with information in various databases – for example, in the photo databases of law enforcement agencies when it comes to finding an offender.

The video surveillance system operates on a centralized basis [6]. There is a Uniform Data storage and Processing Center (UDSPC), where video streams from all 174 thousand city surveillance cameras flow. The face recognition function is implemented as a separate module of the video surveillance system in the UDSPC. If law enforcement agencies are faced with the task of finding criminals, they use their database of photos, with which, if necessary, they compare the faces that fall into the lenses of cameras. If the images match, the police get a signal about it.

At the same time, the accuracy of recognition depends on many factors: weather conditions, the location of the camera, etc. If the task is to identify a person from a photo from the wanted database and these photos are of good quality, then the system correctly recognizes the person with a probability of 95-97%.

Thus, it can be assumed that there is a possibility of sabotage of this database of reference images by employees who have the right to make changes to the database. Deception of biometric algorithms has two main goals: intruders either try to deceive the system so that it does not recognize them in principle, or to take them for another person. For example, an authorized employee may be asked to upload modified photos of a criminal to the reference database. As a result, the recognition system will no longer be able to react to the criminal on the streets of the city because it does not recognize him because of the distorted reference (standard) in the database, and other employees may not notice the substitution due to the external similarity of the photos.

Thus, it can be assumed that there is a possibility of sabotage of this database of reference images by employees who have the right to make changes to the database. Circumven of biometric algorithms has two main goals: intruders either try to deceive the system so that it does not recognize them in principle, or to take them for another person. For example, an authorized employee may be asked to upload modified photos of a criminal to the reference database. As a result, the recognition system will no longer be able to react to the criminal on the streets of the city because it does not recognize him because of the distorted reference (standard) in the database, and other employees may not notice the substitution due to the external similarity of the photos.

In support of the assumption of sabotage, some statistics can be cited. According to research by the analytical center "Infowatch" (Russia), such types of incident as "access abuse" and "disloyal behavior" are more common in public authorities and law enforcement agencies than in financial organizations and industry [7]. A comparative infographic is shown in figure 1.

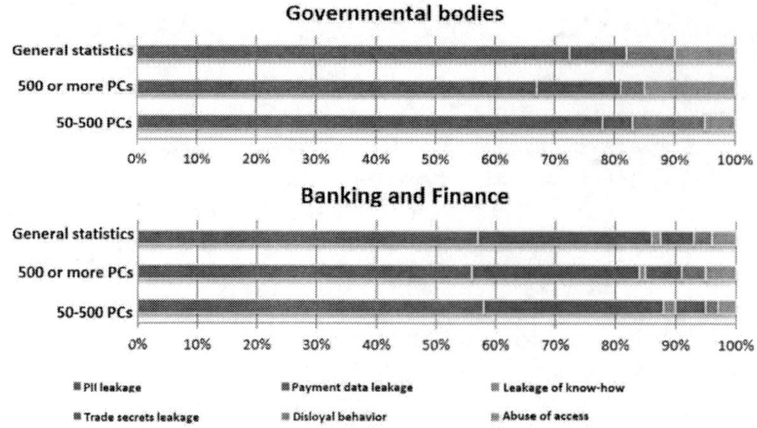

Figure 1. Sectoral map of incidents [7]

Moreover, there are already known cases of abuse of access to the UDSPC. In Moscow, a criminal case was opened against police officers who, according to investigators, illegally distributed information about citizens obtained from the facial recognition system.

3. Method of circumventing the reference database

2D (two-dimensional) facial recognition technology is based on flat two-dimensional images. Face recognition algorithms use anthropometric parameters of the face, graph models of faces or elastic 2D models of faces, as well as images with faces represented by a certain set of physical or mathematical features (Figure 2).

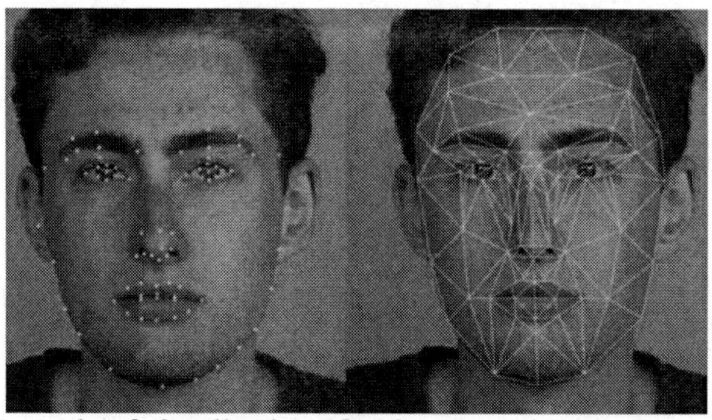

Figure 2. Example of visualization of anthropometric parameters of the face

The considered method consists of distorting the geometry of a person's face in a photo in such a way as to confuse facial recognition systems, while allowing other people to recognize the person. In other literature, this phenomenon is called "cloaking" (Image cloaking, Photo cloaking). Testing has shown that the "cloaking" effect is difficult to recognize when training a neural network and it does not cause errors in training [8]. In other words, the human operators of the facial recognition system will not suspect anything amiss.

As a tool for manually performing the "cloaking" operation, it is proposed to use a graphic editor (for example, Adobe Photoshop [9]), but the final efficiency will depend on the talent of the performer. Automatic execution of "cloaking" can be performed by the De-identification algorithm authored by Facebook AI Research [10], or the Fawkes algorithm authored by the University of Chicago [8]. However, at the moment, the effectiveness of the latter is questionable [11]. It should be noted that the original purpose of these algorithms was not to circumvent governmental information systems by forgery of the reference photo in the database. The primary reason for creating such tools was the need to anonymize people in photos and protect their identity from unauthorized facial recognition systems.

The result is achieved by adding changes to the bitmap image. At its core, "cloaking" does not occur at the pixel level, but at the level of the feature space, that is, pixel changes actually have a deep nature. At the same time, the photo must be modified in a non-obvious way, that is, to preserve the main recognizable at first glance facial features in the image. This is necessary to reduce the probability of detecting a fake photo by a person (human operator). These features include: position, emotional expression, skin tone, hair color, distinguishing features (moles, scars), occlusion (bite), lighting and shadows.

A visual algorithm for a potential method of circumventing the biometric identification system based on facial recognition is presented in figure 3.

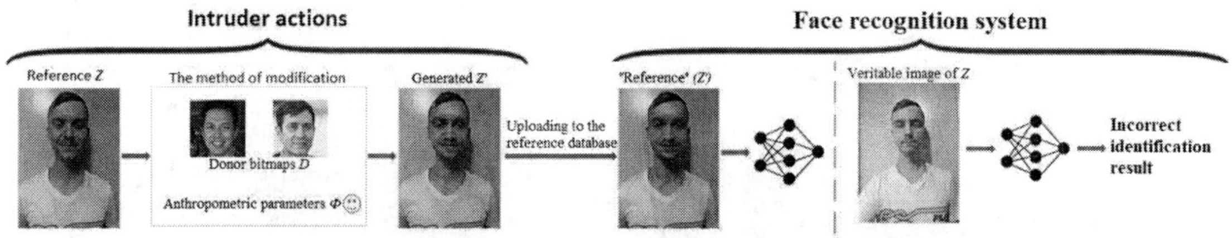

Figure 3. Algorithm for implementing a potential method of circumventing the biometric identification system based on facial recognition

A modification method (represented as a function for extracting anthropometric parameters Φ and donor bitmaps D) is applied to the reference photo of the intruder Z to generate a new image Z'.

The database of reference photos of criminals receives a modified image Z' by uploading this image by an accomplice of forgery (access abuse). Thus, there is a deliberately incorrect training of the intruder Z recognition system. As a result, when the system receives new veritable images of the intruder Z from city surveillance cameras, the recognition system either does not recognize the person at all, or "confuses" him with someone similar in anthropometric parameters from the image Z'. Please note that the presence of image D in the reference database is not a mandatory factor. At the same time, D must have the anthropometric parameters of the face as opposite as possible from Z.

An explanation of why the model trained on distorted photos does not recognize faces on the originals in 2D space, of the four features A, B, U, T, is presented in figure 4 [8]. On the left are the boundaries of decision-making when learning on the originals, on the right are the boundaries of decision-making when learning after "cloaking".

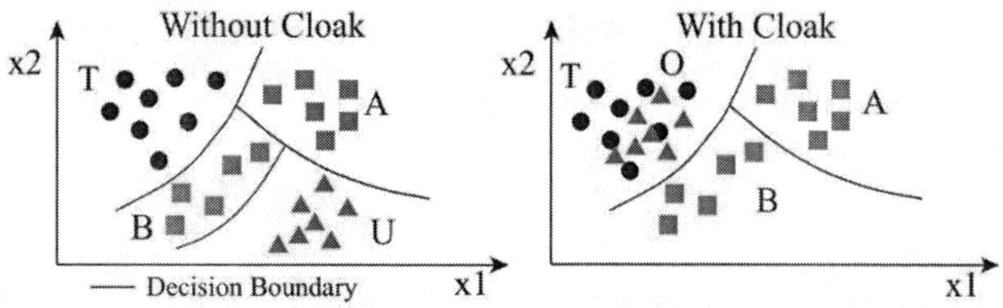

Figure 4. Visualization of decision boundaries by the face recognition system [8]

According to the developers of such algorithms, the efficiency is 100%, and they easily deceive modern face recognition services from Microsoft (Azure), Amazon (Rekognition) and Megvii (Face ++) [8].

At the same time, it is worth noting that the fewer training (reference) images the system uses for each person, the lower its accuracy. The situation when the system uses only one reference image leads to a significant decrease in recognition accuracy. In the scientific community this problem appears under the abbreviations SSPP (Single Sample per Person) [12] or OSS (One Sample Size) [13]. Among the tasks of automatic face recognition, the problem of SSPP occupies a special place.

4. Validating the effectiveness of the presented method

To test the effectiveness of the presented method, the Microsoft Azure [14] platform was used. The source data is the original photo of the co-author of the article Ivan Bychihin, and the same photo modified with the help of Adobe Photoshop software [9]. The tools "Plastic" and "Restoring brush" were used. It is important to note that the distinctive features that are important for human perception are preserved. These features include: position, emotional expression, skin tone, hair color, distinguishing features (moles, scars), occlusion (bite), lighting and shadows.

The result of comparing images (Figure 5) is «The two faces belong to different people». At the same time, the Confidence parameter remained quite large (for identical photos it is equal to 1, for photos of dissimilar people it is close to 0).

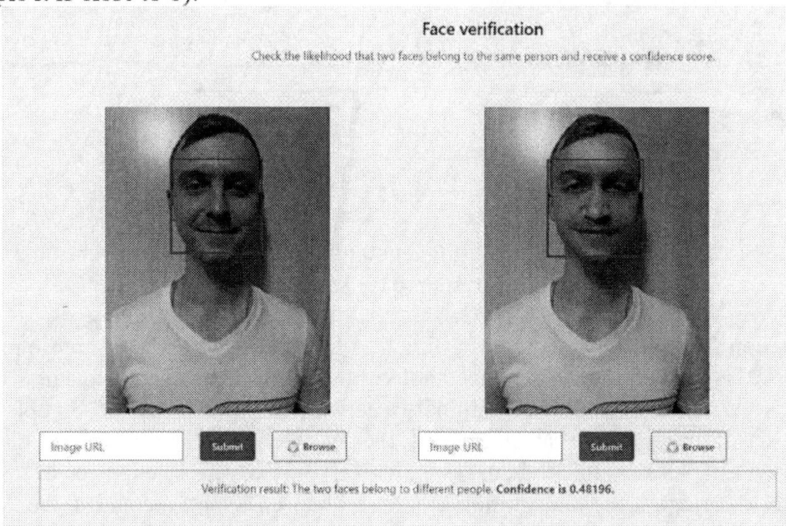

Figure 5. The result of the images comparison on the Microsoft Azure platform

Thus, it is safe to say that the algorithm of a potential method of circumventing the biometric identification system based on facial recognition will work.

5. Forensic analysis of the modified reference (standard)

Based on the result obtained, a question arises about methods for detecting a circumvent attempt. It can be assumed that a large number of changes are received in the reference database, so it is not possible to view each change manually. Then the only way is an automatic forensic analysis of the photo uploaded to the database. But at the moment, such tools do not exist, and to create them, you need a very clear result of using forensic filters on the image.

Based on this assumption, the modified image was analyzed for obvious signs of editing. The web tool Forensically was used for the analysis [15].

As a result of using the Noise Level Analysis, Error Level Analysis and Luminance Gradient tools, a clear sign of editing was detected only when considering the result of processing with the Error Level Analysis tool (Figure 6).

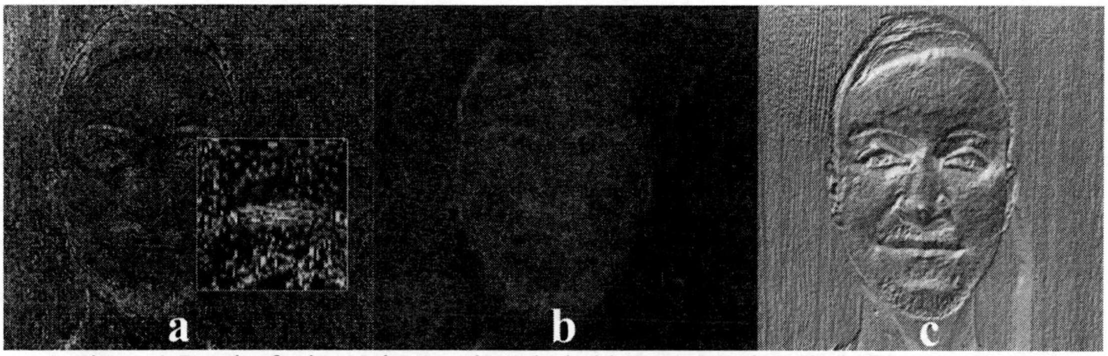

Figure 6. Result of using Noise Level Analysis (a), Error Level Analysis (b) and Luminance Gradient (c)

Thus, we can conclude that it is possible to automate the process of checking photos by the Error Level parameter and there is a prospect of creating such software.

6. Conclusion

During the review of the principle of functioning of Russian facial recognition systems, it was concluded that there is a potential possibility of unauthorized access or abuse of authority by an employee in order to bypass the algorithm for recognizing the faces of people who are wanted. "Potential" in this case means the absence of precedents for using the described method.

An experiment on the Microsoft Azure platform showed the effectiveness of the described method of sabotaging the database of standards of the facial recognition system.

The experiment on the rapid detection of a modified photo using the tools of forensic image analysis gave a result in one test out of three, which means that there is a prospect for automating the analysis process.

Thus, at the moment there is no effective and operational method of protecting the reference database from the presented potential method of circumventing the facial recognition system.

Developers and administrators of such databases are recommended to pay attention to the method presented in this article and tighten control over changes and additions made to the database.

References

[1] Antonchik A and Deryushev A 2009 Review of methods of face recognition in the image *Minsk: Reports of Belarusian State University of Informatics and Radioelectronics*

[2] Viola P 2004 Robust realtime face detection *International Journal of Computer Vision* pp 137-154

[3] Viola P 2001 Rapid object detection using a boosted cascade of simple features *IEEE Conf. on Computer Vision and Pattern Recognition* pp 511-518

[4] Turk M and Pentland A 1991 Face Recognizing Using Eigenfaces *Proc. IEEE* pp 586-591

[5] Baev A, Morozovskiy K and Rozhentsov A 2013 Modificirovannoe obobshhjonnoe preobrazovanie Hafa dlja obrabotki trjohmernyh izobrazhenij s neizvestnymi parametrami vrashhenija i masshtabirovanija [Modified generalized Hough transform for processing three-dimensional images with unknown parameters of rotation and scaling] (*Avtometriya* vol 2) pp 30-41 (in Russian)

[6] Official site of the Moscow Mayor *Conversation with the head of the city video surveillance department of the Moscow Information Technology Department* URL: https://www.mos.ru/dit/documents/view/220179220 (accessed: 10.09.2020) (in Russian)

[7] Information security in corporate information systems *InfoWatch Analytical report* URL: https://www.infowatch.ru/analytics/reports/4609 (accessed: 14.09.2020) (in Russian)

[8] Shawn S, Emily W, Jiayun Z, Huiying Li, Haitao Z and Ben Zhao 2020 Fawkes: Protecting

Privacy against Unauthorized Deep Learning Models *29th USENIX Security Symposium*
[9] The official resource of the software provider Adobe Photoshop URL: https://www.adobe.com/ru/products/photoshop.html (accessed: 30.09.2020)
[10] Oran G, Lior W and Yaniv T 2019 Live Face De-Identification in Video *Facebook Research*
[11] Does the photo protection from face recognition systems works? URL: https://habr.com/ru/post/520194/ (in Russian)
[12] Tan X 2006 Recognition from a Single Sample per Person with Multiple SOM Fusion *Lecture Notes in Computer Science vol 3972/2006* pp 128-133
[13] Štruc V 2009 Using Regression Techniques for Coping with the One-Sample-Size Problem of Face Recognition *Electrotechnical review vol 76* pp 7–12
[14] Microsoft Azure AI service for analyzing faces in images URL: https://azure.microsoft.com/ru-ru/services/cognitive-services/face/
[15] Forensically Web-based tool for analyzing raster images URL: https://29a.ch/photo-forensics

The dependence of the performance of imaging tasks on the forensic imager interface type

F B Tebueva, M G Ogur, M V Zubkov, M G Moisov

North-Caucasus Federal University, Stavropol, Pushkin str., 1, 355017, Russia

E-mail mogur@ncfu.ru

Abstract. Forensic research in the investigation of information security incidents, the production of forensic examinations and many other areas of activity related to computer forensics require the maximum possible preservation of the integrity of the investigated data. To do this, write blockers are used - programs or devices that do not allow writing anything to the drive under investigation. The need to use such means comes both from the requirements of procedural legislation (for example, the Criminal Procedure Code of the Russian Federation), and from various recommendations of a methodological and other nature, as well as from standards (for example, STO BR IBBS-1.3-2016). Some aspects of the functioning of write blockers will be discussed in this article. And on the example of the criminalistic lock the question of influence of the used interface for data transmission in tasks of creation of images of data carriers will be considered.

Keywords: forensic; forensic imager, interface, performance, hardware bridge, software bridge.

1. Introduction

In the modern world the special relevance is acquired by a problem of commission of crimes with use of the computer equipment. As every year information volume grows in the world, criminalists at investigation of such crimes are forced to work with huge data array. As one of the tools used for the analysis of information is the record lock, important aspect in work with this device is the type of a host interface and according to its capacity. In this regard the purpose of this article is to show influence of capacity of various interfaces on the speed of creation of images of data carriers [2,3,4].

2. Basics of forensic imaging

Many experts in computer criminalistics and lawyers share opinion that hardware bridges of record are more reliable, than program bridges of record; this judgment still can be found, directly or indirectly, in various publications [1,2].

2.1. Principles of functioning

Hardware bridges – all hardware bridges of record can be divided into two groups how they process the teams received from a host:

- working at base of the white list;
- working at base of the black list.

The hardware bridge works at base of the white list when it bridges any team to the store if it is not included in the list of the famous safe teams (which are not making changes to the data stored on the store). In this operating mode the lock of record will block all unknown teams, including specific to the producer (for example, for performing low-level diagnostics of the store) and new (the records which are not realized in the built-in program of the lock yet). Such lock of record can block the new standardized safe teams, interpreting them as unknown.

The hardware bridge works at base of the black list when it bridges the teams included in the list of the famous unsafe teams (which are making changes to the data stored on the store or carrying out other dangerous actions), and allows passing to the store of any other teams. In such operating mode the lock of record will allow unknown (specific to the producer or new standardized) unsafe teams to the store.

Besides, all hardware bridges of record can be divided into two other groups depending on details of their realization:

- working as the translator of teams;
- working as the modules providing access to the block device.

The hardware lock of record works as the translator of teams when it just broadcasts the allowed teams received from the interface source by their repetition in the interface recipient. For example, the simple lock of SATA-to-USB record can receive SCSI commands from the USB interface (using a set of the teams "SCSI transparent command set" for the class "mass storage"), and then to make inquiry for each allowed SCSI command to the SATA controller through AHCI, redirecting any answers back to a host under the SCSI protocol.

The hardware bridge works as the module providing access to the block device when it contains the full operating system of general purpose, the connected store is defined as the block device in this operating system, and access to reading from this block device is divided with a host via the special driver. Such lock of record with USB connection to a host will define the connected store as the block device, and then will use the USB gadget for emulation of the USB store with use of the specified block device as data source for the emulated store. In such configuration the hardware lock of record does not carry out direct broadcast of the teams received from a host to the store, it broadcasts the teams received from a host in the internal inquiries used for data reading from the block device source. Thus, the multiple teams of reading received from a host can be connected in one request for reading leading to sending to the connected store of one team of reading. Besides, the built-in program of the lock of record can carry out anticipatory reading (read-ahead) in a cache, it leads to the fact that one team of reading received from a host can not lead to immediate sending the corresponding team of reading to the connected store as the requested data were already read and added to a cache by the built-in program.

Besides, hardware bridges of record can provide special functions for some typical and atypical applications:

- work in the reading record mode;
- permission of commands to write, saving the modified data on other store;
- representation of the store to a host with a mark of write protection (it adds one more level of protection as it is expected that the operating system will not write on the store with a mark of write protection);
- concealment of write errors;
- providing access to hidden by means of HPA or DCO areas of data;
- permission of some unsafe teams used for opening of access to the hidden areas of data (removal of DCO or permanent removal of HPA);
- transparent proofreading of data from the bad (damaged) sectors of the store, transparent work with faulty stores.

Program bridges – details of realization of program bridges of record depend on the used operating system. In the operating systems working in the real mode like DOS, program bridges of record intercept the interruption of BIOS 0x13 used for reading and data recording of a disk, filtering requests for record and causing the initial handler of interruption for requests for reading. Modern operating systems like Windows and GNU/Linux use drivers of direct access for interaction with stores, interruption of BIOS

0x13 are used by the loader for reading a kernel and other data (like drivers of direct access) only at an early stage of loading. Thus, there are set of ways for realization of functionality of blocking of record, for example:

- the driver working in the mode "only reading" for a certain class of stores (PATA, SATA, SCSI, USB, etc.);
- the program (driver) filtering requests for record for their ways to the driver of a certain class of stores of lower level;
- the driver providing the block device in the mode "only reading" for the chosen store or the section with parallel existence in the operating system of the block device in the reading record mode for the same store or the section (it is supposed that use of various programs will happen in relation to the block device in the mode "only reading").

Depending on realization, program bridges of record can analyze or block the following types of inquiries:

- reading, record, dumping of a cache and other inquiries in the unified format specific to the operating system (which is not based on the protocol of interaction with the store at a low level);
- inquiries in the format based on the protocol used for interaction with the store (for example, SCSI) or which directly realizes this protocol.

Besides, the program lock of record can present the store to the operating system with a mark "only reading".

In modern operating systems the following realization of blocking of record was met:

2.2. Windows:

- installation of the filtering driver for packages of I/O-inquiries which are used for transfer of SCSI commands low-level port driver of the store (the operating system is used the SCSI protocol for interaction about port drivers of stores, packages of inquiries are broadcast if it is necessary, in the protocol used by the concrete equipment).

2.3. Linux:

- installation of the device - "loop" in the mode "only reading" for the block device of the store (or undressed), in this situation the driver of the device - "loop" filters the requests for record going to the main block device (the kernel uses own structure for the description of requests to the store on reading, record, dumping of a cache, etc.), use of the device - "loop" for access to data is supposed;
- a kernel patching for the purpose of filtering out of the requests for record going to the block device in the mode "only reading".

During the work with the inquiries created on the basis of the native protocol (like SCSI in Windows), the program lock of record can work with black and white lists as it was shown earlier.

In general, program bridges of record have to intercept inquiries or in one point (for example, before inquiry transfer to one of many drivers of stores), or in all places at once (for example, in all drivers of stores).

It should be noted that the program lock of record cannot block all possible ways of transfer of the unsafe command to the store. For example, the kernel can provide the interface for sending "crude" inquiries to the store (like the SG_IO interface in Linux), or the architecture of drivers of stores can allow any driver to send a request bypassing the filtering driver (as in Windows), or the low-level driver can independently send unsafe teams, or any program can just switch-off the record lock. And though it is possible to take some measures against such shortcomings, the idea is simple - there is always a way bypassing the program lock of record.

2.4. Program quasibridges of record

There is an opportunity to create such operating system which will not send unsafe teams to the connected stores to time and after loading, except situations when the user obviously starts the program sending unsafe requests. In that case there is no component blocking record, but there are also no teams which are subject to blocking (and such operating systems often carry to containing the record blocking mechanism therefore in this section and designation "quasi" is used)

Unfortunately, some products are described how the records containing the program lock, but in fact in them are not present such lock, and depending on various circumstances the similar product can send to the connected store unsafe teams.

2.5. Comparison of hardware bridges of record with program

The following important differences between hardware and program bridges of record can be noted:

- Hardware bridges of record do a gap between a host and the store, as a result unsafe commands from a host are subject to blocking regardless of their origin. Program bridges of record can be bypassed by the malicious application as it was shown earlier.
- Program bridges of record for drivers of direct access are inactive during an early stage of loading: blocking of record is absent when the loader uses interruption BIOS 0x13 or EFI services for loading of a kernel and other components of the modern operating system. At the same time hardware bridges of record before completion of their initialization do not process any teams.

There is an opinion that program bridges of record cannot be reliable because depend on the fragile program environment: for example, updating of the operating system, either updating of the driver, or a mistake (in hardware or the software) can interfere with the record blocking mechanism; hardware bridges of record, on the other hand, are considered as reliable because contain the stable, well tested hardware and the built-in software[4].

3. Formation of tasks of an influence research a host interfaces on final performance

The following actions will be made for achievement of an objective:

- Copying of the test hard drive SATA III connected via the record lock via USB 3.0 interface.
- Copying of the test hard drive SATA II connected directly via the USB 3.0 interface.
- Copying of the test hard drive SATA III of the disk connected directly via the SATA interface.
- Copying of the test hard drive SATA II of the disk connected directly via the SATA interface.
- Comparison with SATA SSD

As the criminalistic lock of record Tableau T3iu which connection is carried out via the USB interface is used. The used software for copying is the X-Ways forensics complex [5,6].

3.1. Practical research

In the first case the copying of the classical hard drive on 1tb, connected via the record lock is considered.

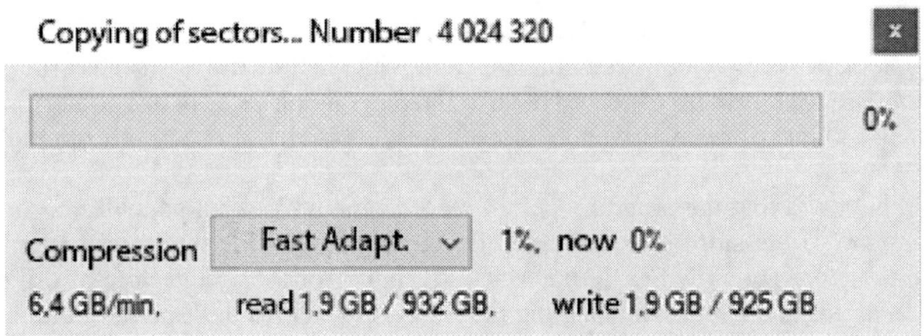

Figure 1. Copying of SATA III of the hard drive via the USB 3.0 interface.

Copying is made on other hard drive connected directly via the SATA III interface. In this situation speed of performance of copying rests against the maximum capacity of the USB 3.0 interface that when copying enough volume data carriers can slow down implementation of the analysis of collected data. At the exit we receive speed in 6.4 GB/min. that will allow to make an image of this disk for 2.4 h.

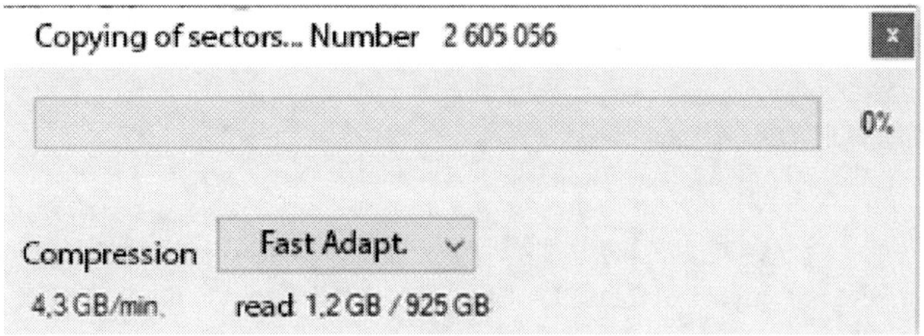

Figure 2. Copying of SATA II of the hard drive via the USB 3.0 interface.

As well as in the first case the record is made on the same disk, but as copied the disk which does not support the SATA III mode was used. At the exit we receive speed in 4.3 GB/min. that for 33% more slowly than a disk from the first experience. Therefore, time spent for copying will be 3.6 h.

In the third case the hard drive by volume on 1tb, but connected directly via the SATA II interface owing to the fact that it does not support the SATA III mode was used [9].

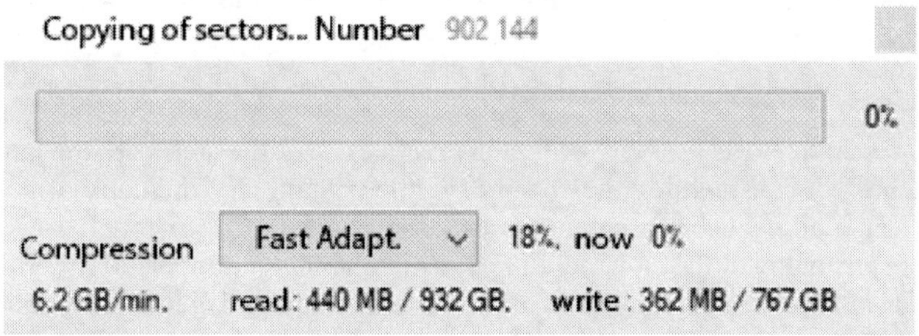

Figure 3. Copying of the hard drive connected directly via the SATA II interface.

At this investigation phase it is possible to tell that direct connection to the SATA II interface does not concede on copying speed to a disk with support of SATA III which was connected via the lock to the USB 3.0 interface. Concerning the first experience the speed is 0.2 GB/min. less that at the level of an error of the taken measurements. The image of a disk will be created in 2.5 hours.

Figure 4. Copying of the hard drive connected directly via the SATA III interface.

At connection of a disk to the SATA III interface the natural growth of speed of copying as capacity grew from 3 Gbps to 6 Gbps is observed. But despite the increased capacity of high speeds it is impossible to achieve owing to the structure of the hard drive and existence in it of mobile parts. Speed is 7.1 GB/min. that is 10% faster than a disk from the first experience. The image will be created in 2.2 hours.

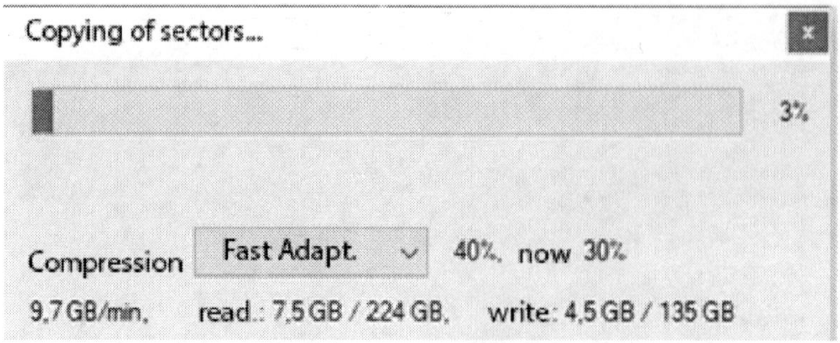

Figure 5. Copying of the SSD disk connected directly via the SATA III interface.

In the last experience the copying of a SSD disk with the SATA III interface is considered. Speed of copying is much higher than classical stores with mobile parts that allows to achieve such result. Speed of creation of an image is 9.7 GB/min. that quicker:

1. About 34% faster than the first disk;
2. 51.5% faster than the second disk;
3. 36% faster than the third disk;
4. 26.8% faster than the fourth disk.

It is not possible to compare time of creation of an image as the volume of the considered SSD disk is about 4.2 times less [6,7,8].

4. Conclusion

During this research it was established that in practice use of the lock with the interface of connection of USB 3.0 can affect considerably the speed of implementation of the copy of a disk that in turn can increase time expenditure by implementation of the analysis of collected data [10,11]. Thus, it is expedient to use an analog of the considered lock which will have support of the SATA interface thanks to what decrease in time for the analysis of data as creation of an image for further work with it will take less time owing to the increased speeds will become possible [12].

Also, during the research, the following was revealed:

- the Disk SATA III connected via the lock to the USB 3.0 interface more slowly similar, but connected directly to SATA III for 10%
- the Disk SATA II connected via the lock to the USB 3.0 interface more slowly similar, but connected directly to SATA II for 36%

References

[1] Gary C. Kessler, Gregory H. Carlton 2014 *A Study of Forensic Imaging in the Absence of Write-Blockers* (Journal of Digital Forensics, Security and Law: Vol. 9 : No. 3 , Article 4.)

[2] Carrier, B. 2019 *Open Source Digital Forensics Tools,*

[3] James, J.I.; Gladyshev, P. 2015 *Automated inference of past action instances in digital investigations* Int. J. Inf. Secur. 2015, 14) pp 249-261. [CrossRef]

[4] Sitompul, O.S.; Handoko, A.; Rahmat, R.F. 2018 *File Reconstruction in Digital Forensic* (TELKOMNIKA Indones. J. Electr. Eng. 2018, 16) pp 776-794. [CrossRef]

[5] Kalber, S.; Dewald, A.; Freiling, F.C. 2013 *Forensic application-fingerprinting based on file system metadata* (Proceedings of the Seventh International Conference on IT Security Incident Management and IT Forensics, Nuremberg, Germany, 12-14 March 2013) pp 98-112.

[6] Bang, J.; Yoo, B.; Kim, J.; Lee, S. 2009 *Analysis of time information for digital investigation* (Proceedings of the Fifth International Joint Conference on INC, IMS and IDC, Seoul, Korea, 25-27 August 2009) pp 1858-1864.

[7] Chabot, Y.; Bertaux, A.; Nicolle, C.; Kechadi, M.T. 2014 *A complete formalized knowledge representation model for advanced digital forensics timeline analysis* (Digit. Investig. 2014, 11)

pp 95-105. [CrossRef]

[8] Hargreaves, C.; Patterson, J. 2012 *An automated timeline reconstruction approach for digital forensic investigations* (Digit. Investig. 2012, 9) pp 69-79. [CrossRef]

[9] Brady, O.; Overill, R. 2015 *DESO: Addressing volume and variety in large scale criminal cases* (Digit. Investig. 2015, 88) pp 72-825. [CrossRef]

[10] Brady, O.; Overill, R.; Keppens, J. 2014 *Addressing the increasing volume and variety of digital evidence using an ontology* (Proceedings of the 2014 IEEE Joint Intelligence and Security Informatics Conference, Hague, The Netherlands, 24-26 September 2014) pp 176-183.

[11] Debinski, M.; Breitinger, F.; Mohan, P. 2019 *Timeline2GUI: A Log2timeline CSV parser and training scenarios* (Digit. Investig. 2019, 28) pp 34-43. [CrossRef]

[12] Soltani, S.; Seno, S.A.H.; Yazdi, H.S. 2019 *Event reconstruction using temporal pattern of file system modification* (IET Inf. Secur. 2019, 13) pp 201-212. [CrossRef]

AUTHOR INDEX